APPLIED CATEGORICAL DATA ANALYSIS AND TRANSLATIONAL RESEARCH

APPLIED CATEGORICAL DATA ANALYSIS AND TRANSLATIONAL RESEARCH

Second Edition

CHAP T. LE
School of Public Health
University of Minnesota, Twin Cities
Minneapolis, Minnesota

A JOHN WILEY & SONS, INC., PUBLICATION

Published by John Wiley & Sons, Inc., Hoboken, New Jersey
Published simultaneously in Canada

Library of Congress Cataloging-in-Publication Data:

Applied categorical data analysis and translational research: second edition / Chap T. Le
p. cm.
Includes bibliographical references and index.
ISBN 978-0-470-37130-5 (cloth)

Printed in the United States of America

10 9 8 7 6 5 4 3 2 1

To my wife, Minh-Hà, and my daughters, Mina and Jenna,
with deepest love and appreciation

CONTENTS

PREFACE

The first edition of this book was well received, and we believe that it was well received because of its "applied" features, one of only a few books written for the training of graduate students in epidemiology, environmental health, and other fields in public health, as well as serving as a reference book for practicing biomedical research workers. But it has been a decade and it's time to improve a good product. After thorough discussion with the publisher, we decided to get even further into the applied direction where it would be truly needed. Except for necessary improvements where needed, we decided to keep the original seven chapters almost intact. The lone exception is the addition of three sections: one on "Modeling a Probability" as a way to introduce logistic regression (Chapter 4), one on "Quantal Bioassays" (Chapter 4), and one on "Competing Risks" (Chapter 9).

The main feature of the revision leading to this second edition consists of two new chapters: Chapter 7 on "Categorical Data and Translational Research" and Chapter 8 on "Categorical Data and Diagnostic Medicine." We hope that we turn in the right direction and students and users support us as enthusiastically as they did with the first edition.

Note: Data sets are available on the related Web site: `www.biostat.umn.edu/~chap`.

CHAP T. LE

Edina, Minnesota
2009

PREFACE TO THE FIRST EDITION

This book is intended to meet the need of practitioners and students in applied fields for a single, fairly thin volume covering major, updated methods in the analysis of categorical data. It is written for the training of graduate students in epidemiology, environmental health, and other fields in public health, as well as for biomedical research workers. It is designed to offer sufficient details to provide a better understanding of the various procedures as well as the relationships among different methods. In addition, the level of mathematics has been kept to a minimum level. As a book for students in applied fields and as a reference book for practicing biomedical research workers, *Applied Categorical Data Analysis* is application oriented. It introduces applied research areas, with a large number of real-life examples and questions, most of which are completely solved; samples of computer programs are included.

I would like to express my thanks to colleagues for their extensive comments and suggestions; the comments from many of my graduate students at the University of Minnesota also have been most helpful. At the University of Minnesota, we offer a three-quarter course sequence for students in Epidemiology, and Environmental and Occupational Health, and a few other graduate programs in the Academic Health Center. The last quarter of this sequence, devoted to the study of techniques used in analyzing categorical data, and the contents herein, represent my many-year effort in teaching that course.

CHAP T. LE

1

INTRODUCTION

The purpose of most research is to assess relationships among a set of variables, and choosing an appropriate statistical technique depends on the type of variables under investigation. Suppose we have a set of numerical values for a variable:

1. If each element of this set may lie only at a few isolated points, we have a discrete or categorical data set. In other words, a categorical variable is one for which the measurement scale consists of a set of categories; examples are race, sex, counts of events, or some sort of artificial grading.
2. If each element of this set may theoretically lie anywhere on the numerical scale, we have a continuous data set. Examples are blood pressure, cholesterol level, or time to a certain event such as death.

This text focuses on the analysis of categorical data and multivariate problems when at least three variables are involved. The first section of this chapter shows a simple example of real-life problems to which some of the methods described in this book can be applied. This example shows a potential complexity when the data involve more than two variables with a phenomenon known as effect modification. We will return to this example later when illustrating some methods of analysis for categorical data in Chapters 2, 3 and 4. The second section briefly reviews some likelihood-based statistical methods to be used in subsequent chapters with various

Applied Categorical Data Analysis and Translational Research, Second Edition, By Chap T. Le

regression models. The last section summarizes special features of this text, its objectives, and for whom it is intended; we also briefly points out special features of this second edition.

1.1 A PROTOTYPE EXAMPLE

Many research outcomes can be classified as belonging to one of two possible categories: for example, Presence and Absence, White and Nonwhite, Male and Female, Improved and Not Improved. Of course, one of these two categories is usually identified as of primary interest to the researcher; for example, Presence in the "Presence and Absence" classification, Nonwhite in the "White and Nonwhite" classification. We can, in general, relabel the two outcome categories as Positive (or $+$) and Negative (or $-$). An outcome is positive if the primary category is observed and is negative if the other category is observed. Health decisions are frequently based on the "proportion" of positive outcomes defined by

$$p = x/n$$

where x in the above equation is the number of positive outcomes from the n observations made on n individuals: $0 \leq p \leq 1$ because $1 \leq x \leq n$. Proportion is a number used to describe a group of individuals according to a dichotomous (or binary) characteristic under investigation and the following example provides an illustration of its use in the health sciences.

Comparative studies are intended to show possible differences between two or more groups. Data for comparative studies may come from different sources, with the two fundamental designs being *retrospective* and *prospective*. Retrospective studies gather past data from selected cases and controls to determine differences, if any, in the exposure to a suspected risk factor. They are commonly referred to as *case–control studies*; a "case" is a person with the disease under investigation and a "control" is a person without that disease. In a case–control study, cases of a specific disease are ascertained as they arise from population-based registers or lists of hospital admissions and controls are sampled either as disease-free individuals from the population at risk, or as hospitalized patients having a diagnosis other than the one under study. The advantages of a retrospective study or case–control study are that it is economical and it is possible to obtain answers to research questions relatively quickly because the cases are already available. Major limitations are due to the inaccuracy of the exposure histories and uncertainty about the appropriateness of the control sample; these problems sometimes hinder retrospective studies and make them less preferred than prospective studies. The following example introduces a retrospective study concerning occupational health.

■ **Example 1.1** A case–control study was undertaken to identify reasons for the exceptionally high rate of lung cancer among male residents of coastal Georgia (Blot et al., 1978). Cases (of lung cancer) were identified from these sources:

(i) diagnoses since 1970 at the single large hospital in Brunswick, (ii) diagnoses during 1975 and 1976 at three major hospitals in Savannah, and (iii) death certificates for the period 1970–1974 in the area.

Controls (or control subjects) were selected from admissions to the four hospitals and from death certificates in the same period for diagnoses other than lung cancer, bladder cancer, or chronic lung cancer. Data are tabulated separately for smokers and nonsmokers as follows:

Smoking	Shipbuilding	Cases	Controls
No	Yes	11	35
	No	50	203
Yes	Yes	84	45
	No	313	270

The exposure under investigation, "Shipbuilding," refers to employment in shipyards during World War II. By a separate tabulation, with the first half of the table for nonsmokers and the second half for smokers, we treat *smoking* as a potential *confounder*. A confounder is a factor that may be an exposure by itself, not under investigation but related to the disease (in this case, lung cancer) and the exposure (shipbuilding); previous studies have linked smoking to lung cancer and construction workers are more likely to be smokers. The term *exposure* is used here to emphasize that employment in shipyards is a *risk factor*; however, the term would also be used in studies where the factor under investigation has beneficial effects.

In an examination of the smokers in the above data set, the numbers of people employed in shipyards, 84 and 45, tell us little because the sizes of the two groups, cases and controls, are different. Adjusting these absolute numbers for the group sizes, we have the following:

(1a) For the controls,

$$\begin{aligned}\text{Proportion of exposure} &= 45/315 \\ &= 0.143 \text{ or } 14.3\%\end{aligned}$$

(2a) For the cases,

$$\begin{aligned}\text{Proportion of exposure} &= 84/397 \\ &= 0.212 \text{ or } 21.2\%\end{aligned}$$

The results reveal different exposure histories: the proportion of exposure among cases was higher than that among controls. It is not in any way yet a conclusive proof, but it is a good clue indicating a possible relationship between the disease (lung cancer) and the exposure (employment in shipbuilding industry—a possible occupational hazard).

Similar examination of the data for nonsmokers shows that, by taking into consideration the numbers of cases and of controls, we have the following figures for employment:

(1b) For the controls:

$$\begin{aligned}\text{Proportion of exposure} &= 35/238\\ &= 0.147 \text{ or } 14.7\%\end{aligned}$$

(2b) For the cases:

$$\begin{aligned}\text{Proportion of exposure} &= 11/61\\ &= 0.180 \text{ or } 18.0\%\end{aligned}$$

Again, the results also reveal different exposure histories: the proportion of exposure among cases was higher than that among controls.

The above analyses also show that the difference (cases versus controls) between proportions of exposure among smokers, that is,

$$21.2\% - 14.3\% = 6.9\%$$

is different from the difference (cases versus controls) between proportions of exposure among nonsmokers, which is

$$18.0\% - 14.7\% = 3.3\%$$

The differences, 6.9% and 3.3%, are measures of the *strength of the relationship* between the disease and the exposure, one for each of the two strata—the two groups of smokers and nonsmokers, respectively. The above calculation shows that the possible effects of employment in shipyards (as a suspected risk factor) are different for smokers and nonsmokers. This difference of the two case–control differences (6.9% versus 3.3%), if confirmed, is called an "interaction" or an *effect modification*, where smoking alters the effect of employment in shipyards as a risk factor for lung cancer. In that case, smoking is not only a confounder, it is an *effect modifier,* which modifies the effects of shipbuilding (on the possibility of having lung cancer).

In some extreme examples, a pair of variables may even have their marginal association in a different direction from their partial association (the association between them as seen at each and every level of a confounder or effect modifier). This interesting phenomenon is called *Simpson's paradox*, which further emphasizes the analysis complexity when we have data involving more than two variables.

1.2 A REVIEW OF LIKELIHOOD-BASED METHODS

Problems in biological and health sciences are formulated mathematically by considering the data that are to be used for making a decision as the observed values of a certain random variable X. The distribution of X is assumed to belong to a certain family of distributions specified by one or several parameters; a *parameter* can be defined as an (unknown) numerical characteristic of a population. The problem for decision makers is to decide on the basis of the data which members of the family could represent the distribution of X; that is, to predict or estimate the value of a certain parameter θ (or several parameters). The magnitude of a parameter often represents the effect of a risk or environmental factor, and knowing its value, even approximately, would shed some light on the impact of such a factor. The *likelihood function* $L(x; \theta)$ for a random sample (x's) of size n from the probability density function (or pdf) $f(x; \theta)$ is

$$L(x;\theta) = \prod_{i=1}^{n} f(x_i;\theta)$$

The *maximum likelihood estimator* (MLE) of θ is the value $\hat{\theta}$ for which $L(x; \theta)$ is maximized. Calculus suggests setting the derivative of $L(x; \theta)$ with respect to θ equal to zero and solving the resulting equation.

Since

$$\frac{dL}{d\theta} = (L)\frac{d(\ln L)}{d\theta}$$

$dL/d\theta = 0$ if and only if $d(\ln L)/d\theta = 0$ because L is never zero. Thus we can find the possible maximum of L by maximizing $\ln L$; it is often easier to deal mathematically with a sum than with a product.

$$\ln L = \sum_{i=1}^{n} \ln f(x_i;\theta)$$

The MLE has a number of good properties, which we will state without proofs; readers can skip this entire section without having any discontinuity.

1. MLE is consistent.
2. If an efficient estimator exists, it is the MLE.
3. The MLE is asymptotically distributed as normal. The variance of this asymptotic distribution is given by the following formula:

$$\mathrm{Var}(\hat{\theta}) = \frac{1}{E\left\{-\dfrac{d^2 \ln L}{d\theta^2}\right\}}$$

in which $E\{.\}$ denotes the *expected value* and the value of the denominator, called Fisher's information matrix (in this case, it is a number), often needs to be estimated using the MLE of θ.

This is an asymptotic distribution; that is, results are good for large samples only. If a closed-form solution does not exist, the iterative solution may be obtained by first solving for an additive correction,

$$\Delta\hat{\theta} = -\left(\frac{d \ln L}{d\theta}\right) \Big/ \left(\frac{d^2 \ln L}{d\theta^2}\right)$$

using numerical values of the derivatives. The iterative solution by this Newton–Raphson method would proceed as follows:

Step 1: Provide an initial value of $\hat{\theta}$, denoted by $\hat{\theta}^{(0)}$.
Step 2: Determine the value of $\Delta\hat{\theta}$ by evaluating the derivatives at $\hat{\theta}^{(0)}$.
Step 3: Add $\Delta\hat{\theta}$ to the initial value to obtain a new value for $\hat{\theta}$, that is, $\theta^{(1)} = \theta^{(0)} + \Delta\theta$.
Step 4: Repeat Steps 2 and 3 using $\hat{\theta}^{(1)}$ to obtain $\hat{\theta}^{(2)}$ and stop when results from successive steps are very close (below certain previously set threshold).

After a final solution has been obtained, an estimate of its variance is then given by

$$\widehat{\mathrm{Var}}(\hat{\theta}) = \frac{1}{-\dfrac{d^2 \ln L}{d\theta^2}}$$

where the second derivative is evaluated using the value of the MLE of θ.

For example, we have for a binomial distribution with unknown probability π

$$L(x;\pi) = \binom{n}{x}\pi^x(1-\pi)^{n-x}$$

$$\ln L(x;\pi) = \ln\binom{n}{x} + x\ln\pi + (n-x)\ln(1-\pi)$$

$$\frac{d}{dx}\ln L(x;\pi) = \frac{x}{\pi} - \frac{n-x}{1-\rho}$$

$$-\frac{d^2}{dp^2}\ln L(x;p) = \frac{x}{\pi^2} + \frac{n-x}{(1-\pi)^2}$$

$$E\left\{-\frac{d^2}{dp^2}\ln L(x;\pi)\right\} = \frac{np}{\pi^2} + \frac{n-np}{(1-\pi)^2}$$
$$= \frac{n}{\pi(1-\pi)}$$

The results are

$$\hat{\pi} = p$$
$$= \frac{x}{n} \quad \text{(sample proportion)}$$
$$\text{Var}(p) = \frac{1}{E\left\{-\frac{d^2}{dp^2}\ln L(x;\pi)\right\}}$$
$$= \frac{\pi(1-\pi)}{n}$$

Consider the case of a two-parameter model with probability density function $f(.)$. Let $L(x;\ \theta_1,\ \theta_2)$ be the likelihood function defined from a random sample $\{x_i\}$ by

$$L(x;\theta_1,\theta_2) = \prod_{i=1}^{n} f(x_i;\ \theta_1,\theta_2)$$

The MLEs θ_1 and θ_2 are the values $\hat{\theta}_1$ and $\hat{\theta}_2$ of θ_1 and θ_2 for which $L(x;\ \theta_1,\ \theta_2)$ is maximized. These estimators are obtained by solving the following equations:

$$\frac{\delta}{\delta\theta_1}\ln L = 0 \quad \text{and} \quad \frac{\delta}{\delta\theta_2}\ln L = 0$$

If closed-form solutions do not exist, the iterative solutions to these equations may be obtained by the Newton–Raphson method, which is similar to that for the above one-parameter model.

The variance–covariance matrix of the estimators $\hat{\theta}_1$ and $\hat{\theta}_2$ can be obtained from the Fisher's information matrix, which is defined as

$$I = \begin{bmatrix} E\left(-\frac{\delta^2}{\delta\theta_1^2}\right)\ln L & E\left(-\frac{\delta^2}{\delta\theta_1\delta\theta_2}\ln L\right) \\ E\left(-\frac{\delta^2}{\delta\theta_1\delta\theta_2}\ln L\right) & E\left(-\frac{\delta^2}{\delta\theta_2^2}\right)\ln L \end{bmatrix}$$

In obtaining numerical variance–covariance estimates, all expected values of the partial derivatives are replaced by numerical evaluations of those partial derivatives using MLE values for the parameters or values from the last iteration if iterative solutions are required.

$$\begin{bmatrix} \widehat{\text{Var}}(\hat{\theta}_1) & \widehat{\text{Cov}}(\hat{\theta}_1, \hat{\theta}_2) \\ \widehat{\text{Cov}}(\hat{\theta}_1, \hat{\theta}_2) & \widehat{\text{Var}}(\hat{\theta}_2) \end{bmatrix} = \begin{bmatrix} -\frac{\delta^2}{\delta\theta_1^2}\ln L & \frac{\delta^2}{\delta\theta_1\delta\theta_2}\ln L \\ -\frac{\delta^2}{\delta\theta_1\delta\theta_2}\ln L & -\frac{\delta^2}{\delta\theta_2^2}\ln L \end{bmatrix}^{-1}$$

Of course, the maximum likelihood procedure, as explained for the two-parameter model, can be easily generalized to models with more than two parameters.

As an example of two-parameter models, let us consider a random sample of size n, $\{x_i\}$, from the normal distribution with mean μ and variance $\theta = \sigma^2$. We have

$$\begin{aligned} \theta &= \sigma^2 \\ L(x; \mu, \theta) &= \prod_{i=1}^{n} \frac{1}{\theta^{1/2}\sqrt{2\pi}} \exp\left[-\frac{(x_i-\mu)^2}{2\theta}\right] \\ \ln L(x; \mu, \theta) &= -\frac{n}{2}\ln\,\theta - \frac{n}{2}\ln(2\pi) - \frac{1}{2\theta}\sum_{i=1}^{n}(x_i-\mu)^2 \\ \frac{\delta}{\delta\mu}\ln L &= \frac{1}{\theta}(x_i-\mu)^2 \\ \frac{\delta}{\delta\theta}\ln L &= \frac{1}{2\theta}\left\{-n + \frac{1}{\theta}\sum(x_i-\mu)^2\right\} \end{aligned}$$

The results are

$$\begin{aligned} \hat{\mu} &= \bar{x} \\ \hat{\sigma}^2 &= \hat{\theta} \\ &= \frac{1}{n}(x_i-\mu)^2 \\ &= \frac{n-1}{n}s^2 \end{aligned}$$

From these derivatives, we find

$$\begin{aligned} \frac{\delta^2}{\delta\mu^2}\ln L &= -n/\theta \\ \frac{\delta^2}{\delta\mu\,\delta\theta}\ln L &= -\frac{1}{\theta^2}\left\{\sum x_i - n\mu\right\} \\ \frac{\delta^2}{\delta\theta^2}\ln L &= \frac{n}{2\theta^2} - \frac{1}{\theta^3}\sum_{i=1}^{n}(x_i-\mu)^2 \end{aligned}$$

And from their expected values, we can easily derive the variances and covariance:

$$\begin{aligned}
\text{Var}(\bar{x}) &= \frac{\sigma^2}{n} \\
\text{Var}(\hat{\sigma}^2) &= \frac{2\sigma^4}{n} \\
\text{Cov}(\bar{x}, \hat{\sigma}^2) &= 0
\end{aligned}$$

A multiple regression model involves many parameters, the unknown regression coefficients, β. Once we have fit such a multiple regression model and obtained estimates for the various parameters of interest using the above method, we want to answer questions about the contributions of various factors to the prediction of the response variable. There are three types of questions:

1. *An Overall Test.* Taken collectively, does the entire set of explanatory or independent variables contribute significantly to the prediction of the response? The null hypothesis for this test may be stated as: "all k independent variables considered together do not explain the variation in the responses." In other words, the null hypothesis is

$$H_0\colon\ \beta_1 = \beta_2 = \cdots = \beta_k = 0$$

Two likelihood-based statistics can be used to test this *global* null hypothesis; each has an asymptotic chi-squared distribution with k degrees of freedom under the null hypothesis H_0:

(a) *Likelihood Ratio Test.*

$$X^2_{\text{LR}} = 2[\ln L(\hat{\boldsymbol{\beta}}) - \ln L(\mathbf{0})]$$

(b) *Score Test.*

$$X^2_{\text{S}} = \left[\frac{\delta}{\delta\beta}\ln L(\mathbf{0})\right]\left[-\frac{\delta^2}{\delta\beta^2}\ln L(\mathbf{0})\right]^{-1}\left[\frac{\delta}{\delta\beta}\ln L(\mathbf{0})\right]$$

Both statistics are provided by most standard computer programs such as SAS and they are asymptotically equivalent yielding identical statistical decisions most of the time.

2. *Test for the Value of a Single Factor.* Does the addition of one particular variable of interest add significantly to the prediction of response over and above that achieved by other independent variables? Let us assume that we now wish to test whether the addition of one particular independent variable of interest adds sig-

nificantly to the prediction of the response over and above that achieved by other factors already present in the model. The null hypothesis for this test may be stated as: "factor X_i does not have any value added to the prediction of the response given that other factors are already included in the model." In other words,

$$H_0\colon \beta_i = 0$$

To test such a null hypothesis, one can perform a likelihood ratio chi-squared test, with one degree of freedom, similar to that for the above global hypothesis:

$$X_{\mathrm{LR}}^2 = 2[\ln L(\hat{\boldsymbol{\beta}}; \text{all } X\text{'s}) - \ln L(\hat{\boldsymbol{\beta}}; \text{all other } X\text{'s with } X_i \text{ deleted})]$$

A much easier alternative method is to use.

$$z_i = \frac{\hat{\beta}_i}{SE(\hat{\beta}_i)}$$

where $\hat{\beta}_i$ is the corresponding estimated regression coefficient and $SE(\hat{\beta}_i)$ is the estimate of the standard error of $\hat{\beta}_i$, both of which are printed by standard packaged computer programs. In performing this test, we refer the value of the z statistic to percentiles of the standard normal distribution.

3. *Test for Contribution of a Group of Variables.* Does the addition of a group of variables add significantly to the prediction of response over and above that achieved by other independent variables? This testing procedure addresses the more general problem of assessing the additional contribution of two or more factors to the prediction of the response over and above that made by other variables already in the regression model. In other words, the null hypothesis is of the form

$$H_0\colon \beta_1 = \beta_2 = \cdots = \beta_m = 0$$

To test such a null hypothesis, one can perform a likelihood ratio chi-squared test, with m degrees of freedom:

$$X_{\mathrm{LR}}^2 = 2[\ln L(\hat{\boldsymbol{\beta}}; \text{all } X\text{'s}) - \ln L(\hat{\boldsymbol{\beta}}; \text{all other } X\text{'s with } X\text{'s under investigation deleted})]$$

This "multiple contribution" procedure is very useful for assessing the importance of potential explanatory variables. In particular, it is often used to test whether a similar group of variables, such as "demographic characteristics," is important for the prediction of the response; these variables have some trait in common. Another application would be a collection of powers and/or product terms (referred to as interaction variables). It is often of interest to assess the interaction effects collectively before trying to consider individual interaction terms in a model as previously suggested. In fact, such use reduces the total number of tests to be performed and this, in turn, helps to provide better control of overall Type I error rates, which may be inflated due to multiple testing.

In many applications, we wish to identify from many available factors a small subset of factors that relate significantly to the outcome, for example, the disease under investigation. In that identification process, of course, we wish to avoid a large Type I (or false positive) error. In a regression analysis, a Type I error corresponds to including a predictor that has no real relationship to the outcome; such an inclusion can greatly confuse the interpretation of the regression results. In a standard multiple regression analysis, this goal can be achieved by using a strategy that adds into or removes from a regression model one factor at a time according to a certain order of relative importance. Therefore the two important steps are:

1. Specifying a criterion or criteria for selecting a model. The selection is often based on the likelihood ratio chi-squared statistic.
2. Specifying a strategy for applying the chosen criterion or criteria. Such a strategy is concerned with whether a particular variable should be added to a model or whether any variable should be deleted from a model at a particular stage of the process (stepwise regression). As computers became more accessible and more powerful, these practices became more popular.

1.3 INTERVAL ESTIMATION FOR A PROPORTION

Recall the following results on the maximum likelihood estimation of an unknown probability or proportion π:

$$\begin{aligned}\hat{\pi} &= p \\ &= \frac{x}{n} \quad \text{(sample proportion)} \\ \mathrm{Var}(p) &= \frac{1}{E\left\{-\dfrac{d^2}{dp^2}\ln L(x;\pi)\right\}} \\ &= \theta \\ &= \frac{\pi(1-\pi)}{n}\end{aligned}$$

Consider the usual estimate of variance of p:

$$\mathrm{var}(p) = \frac{p(1-p)}{n}$$

We can see that

$$
\begin{aligned}
E\{\text{var}(p)\} &= \frac{1}{n}\{E(p) - E(p^2)\} \\
p &= \frac{\sum x_i}{n};\ x_i = 0/1 \\
p^2 &= \frac{\{\sum x_i^2 + 2\sum x_i x_j\}}{n^2} \\
E(p) &= \pi \\
E(p^2) &= \frac{\{\pi + (n-1)\pi^2\}}{n} \\
E\{\text{var}(p)\} &= \left\{\frac{n-1}{n}\right\}\frac{\pi(1-\pi)}{n} \\
&= \left\{1 - \frac{1}{n}\right\}\pi
\end{aligned}
$$

The results have the following meaning:

1. $\text{var}(p)$ is a biased estimate of $\text{Var}(p)$; an unbiased estimate of $\text{Var}(p)$ is $p(1-p)/(n-1)$, with denominator $(n-1)$ similar to the sample variance of a continuous sample.
2. However, $\text{var}(p)$, with denominator n, is asymptotically unbiased and its use is popular.

In summary, we have an approximate 95% confidence interval for a population proportion π:

$$p \pm 1.96\, SE(p)$$

where the standard error of the sample proportion, $SE(p)$, is calculated as

$$SE(p) = \sqrt{\frac{p(1-p)}{n}}$$

■ **Example 1.2** Consider the problem of estimating the prevalence of malignant melanoma in 45–54-year-old women in the United States. Suppose a random sample of $(n = 5000)$ women is selected from this age group and $(x = 28)$ are found to have the disease. Our point estimate for the prevalence of this disease is 0.0056 (=28/5000); its standard error is

$$
\begin{aligned}
SE(p) &= \sqrt{\frac{(0.0056)(1-0.0056)}{5000}} \\
&= 0.0011
\end{aligned}
$$

Therefore a 95% confidence interval for the prevalence π of malignant melanoma in 45–54-year-old women in the United States is given by

$$0.0056 \pm (1.96)(0.0011) = (0.0034, 0.0078) \quad \text{or} \quad (0.34\%, 0.78\%)$$

1.4 ABOUT THIS BOOK

This book is intended to meet the needs of practitioners and students in applied fields by covering major, updated methods in the analysis of categorical data. It is also intended to meet the needs of clinicians and students in biomedical sciences with some basic introduction to a reemerging field called "translational research." It is written for beginning graduate students in biostatistics, epidemiology, and environmental health, as well as for biomedical research workers. As a book for biostatistics and statistics students, it is designed to offer some details for a better understanding of the various procedures as well as the relationships among different methods. However, the mathematics have been kept to an absolute minimum. As a book for students in applied fields and as a reference book for practicing biomedical research workers, this book is very application oriented. It introduces applied research areas and a large number of real-life examples, most of which are completely solved with samples of computer programs.

The book is divided into nine chapters including this introductory chapter.

Chapter 2 covers basic methods and applications of two-way contingency tables including etiologic fraction, the evaluation of ordinal risks, and the Mantel–Haenszel method. Compared to the first edition, the first section (on screening tests) is moved and expanded to form a new chapter, Chapter 8.

Chapter 3 is devoted to loglinear models; topics covered include the selection of the best model for three-way tables, and selection of a model for higher-dimensional tables, with or without the identification of a dependent variable.

Chapter 4 is focused on logistic regression models, both binary and ordinal responses. Topics covered include the stepwise procedure, measures of goodness-of-fit, and the use of logistic models for different designs. Compared to the first edition, the new edition represents a major overhaul of Chapter 4: (i) we added a new introductory Section 4.1, "Modeling a Probability," to include other models such as probit; (ii) we moved old Section 4.2.5, "ROC Curve," to the new Chapter 8; and (iii) we added a new Section 4.5, "Quantal Bioassays," an important topic in translational research.

Chapter 5 covers similar topics as those in Chapters 2–4, but for matched designs, singly or multiply, including the conditional logistic regression model.

Chapter 6 covers analytical methods for count data including the Poisson regression model. Topics covered in this chapter include overdispersion and how to fit overdispersed models.

Chapter 7, "Categorical Data and Translational Research," is a new chapter. Topics covered represent the core material of translational research—early phase clinical trials. These topics include, among others, the standard design, the sequential monitoring of toxicity, and one-stage and two-stage designs for Phase II clinical trials.

Chapter 8, "Categorical Data and Diagnostic Medicine", is another new addition. Topics covered include examples and description of the disease screening process, some basic issues, the ROC curve and the corresponding optimization problem, and the roles of covariates.

Chapter 9 presents a brief introduction to survival analysis and Cox's regression model. This inclusion is partly to show the difference between categorical data and survival data, and partly to serve as a brief introduction to the field of survival analysis, which is an important part of translational research.

In each of the nine chapters, numerous examples are provided for illustration.

2

CONTINGENCY TABLES

This chapter presents basic inferential methods for categorical data, the analysis of two-way contingency tables, and the combination of several two-way contingency tables.

Let X_1 and X_2 denote two categorical variables, X_1 having I levels (or categories) and X_2 having J levels (categories); there are IJ combinations of classifications, each combination corresponds to a category of X_1 and a category of X_2. We display the data in a rectangular table having I rows for the I categories of X_1 and J columns for the J categories of X_2. The IJ cells in this 2×2 table represent the IJ combinations of categories or levels. Their probabilities are π_{ij} values, where π_{ij} denotes the probability that the values X_1 and X_2 fall in the cell in row i and column j; in other words, the characteristic or variable X_1 is observed at level or category i and the characteristic or variable X_2 is observed at level or category j. When the cells contain frequencies of outcomes, the table is called a contingency table or cross-classified table, also referred to as an I-by-J or $I \times J$ table. Let start with the simplest case, the 2-by-2 tables—each such table has two rows and two columns, and revisit the same example presented in Chapter 1. The data in this example form 2-by-2 tables, one at each level of a (potential) confounder or confounding variable.

Applied Categorical Data Analysis and Translational Research, Second Edition, By Chap T. Le

Example 2.1 A case–control study was undertaken to identify reasons for the exceptionally high rate of lung cancer among male residents of coastal Georgia. The exposure under investigation, "shipbuilding," refers to employment in shipyards during World War II. Data are tabulated separately for smokers and nonsmokers as follows:

Smoking	Shipbuilding	Cases	Controls
No	Yes	11	35
	No	50	203
Yes	Yes	84	45
	No	313	270

The following example presents another 2-by-2 table, the result of a screening test or diagnostic procedure. Following these diagnostic procedures, individuals are classified as healthy or as falling into one of a number of disease categories. Almost all such tests are imperfect, in the sense that healthy individuals will occasionally be classified wrongly as being ill, while some individuals who are really ill may fail to be detected. A 2-by-2 table can be used to cross-classify the two sets of outcomes: the "test result" versus the true "disease status." More details of screening tests and similar applications, statistical methods in diagnostic medicine, can be found in Chapter 8.

Example 2.2 A cytological test was undertaken to screen women for cervical cancer (May 1974). Consider a group of 24,103 women consisting of 379 women whose cervices are abnormal (to an extent sufficient to justify concern with respect to possible cancer) and 23,724 women whose cervices are acceptably healthy. A test was applied and results are tabulated as follows:

	Test Result	
Disease Status	Negative	Positive
Negative	23,362	362
Positive	225	154

2.1 SOME SAMPLING MODELS FOR CATEGORICAL DATA

In this section we describe briefly several discrete distributions that are often chosen as models for categorical data. Since this section is to be used only as a brief reference, we assume that its readers have some familiarity with basic statistical terminology.

2.1.1 The Binomial and Multinomial Distributions

The binomial model applies when the result of each trial of an experiment belongs to two possible categories. Many outcomes can be classified as belonging to one of two possible categories: Presence and Absence, White and Nonwhite, Male and Female, Improved and Not Improved. The resulting data are called *binary data* or *dichotomous data*. Of course, one of these two categories is usually identified as of primary interest; for example, Presence in the Presence-and-Absence classification, Nonwhite in the White-and-Nonwhite classification. We can always, in general, relabel the two outcome categories as Positive (+) and Negative (−). An outcome is positive if the primary category is observed and is negative if the other category is observed (the two possible outcomes are also referred to as "Failure" and "Success"; one has a success when the primary outcome is observed).

Let the probabilities of failure and success be $(1-\pi)$ and π, respectively, and we "code" these two outcomes as 0 (zero successes) and 1 (one success). The experiment consists of n repeated trials (n is a finite integer) satisfying the following assumptions:

(i) The n trials are all independent.
(ii) The parameter π is the same for each trial (i.e., π is a constant).

The model is concerned with the total number of successes in n trials as a random variable, denoted by X. To put it in a simple way, observations are made on a group of individuals—say, patients—and the result consists of x positive outcomes or successes. It can be seen that

$$\begin{aligned}\Pr(X=x) &= \binom{n}{x}\pi^x(1-\pi)^{n-x} \\ &= b(x;n,\pi)\end{aligned}$$

where $\binom{n}{x}$ is the number of combinations of x objects or subjects selected from a set of n objects or subjects,

$$\binom{n}{x} = \frac{n!}{x!(n-x)!}$$

The above formula for $\Pr(X=x)$ is intuitively obvious because there is one factor π for each of the x successes, and one factor $(1-\pi)$ for each of the $(n-x)$ failures; these factors are all multiplied together by virtue of the assumption that the n trials are independent. In addition, the products of n factors applies to any sequence of n trials in which there are x successes and $(n-x)$ failures in any order; the number of such sequences is $\binom{n}{x}$. We refer to $b(x;n,\pi)$ as the *binomial density function* and to the corresponding distribution, the distribution of the random variable X, as the *binomial distribution* $B(x;n,\pi)$. This distribution is specified by two parameters n and π, the number of trials, and the probability of success for each trial.

The mean and variance of the binomial distribution are

$$\begin{aligned} E(X) &= \mu \\ &= n\pi \\ \mathrm{Var}(X) &= \sigma^2 \\ &= n\pi(1-\pi) \end{aligned}$$

(In some real-life problems, the true variance may be larger than $n\pi(1-\pi)$, a phenomenon known as *overdispersion*. We will deal with this within the context of logistic regression analysis in Chapter 4.) If X is a binomial random variable with parameter n and π, then the maximum likelihood estimate of π is the sample proportion p. The proportion p and its standard error are given by

$$p = \frac{x}{n}$$

$$SE(p) = \sqrt{\frac{\pi(1-\pi)}{n}}$$

from which we can estimate the standard error $SE(p)$, when replacing π by p, and form confidence intervals for π.

An obvious generalization of the binomial is the *multinomial distribution*, which arises when each trial has more than two possible outcomes. Let k be the number of mutually exclusive outcomes or categories of the outcome whose respective probabilities are $\pi_1, \pi_2, \ldots, \pi_k$ with $\sum \pi_i = 1$. Let $X_1, X_2, \ldots, X_k$ be the k random variables defined so that X_i is the number of times (out of a total of n trials) that the ith outcome occurs; for example, in the binomial distribution case, $k = 2$, X_1 is the number of successes, and X_2 is the number of failures. For the general case, with any k, the joint probability function of the X's is given by

$$\Pr(X_1 = x_1, X_2 = x_2, \ldots, X_k = x_k) = \frac{n!}{x_1! x_2! \cdots x_k!} \pi_1^{x_1} \pi_2^{x_2} \cdots \pi_k^{x_k}$$

$$x_i = 1, 2, \ldots, n$$

$$\sum_{i=1}^{k} x_i = n$$

Using this multinomial probability density function, it can be shown that the means, variances, and covariances of the multinomial distribution are given by

$$
\begin{aligned}
E(X_i) &= \mu_i \\
&= n\pi_i \\
\mathrm{Var}(X_i) &= \sigma_i^2 \\
&= n\pi_i(1-\pi_i) \\
\mathrm{Cov}(X_i, X_j) &= \sigma_{ij} \\
&= -n\pi_i\pi_j
\end{aligned}
$$

That is, for $i \neq j$, X_i and X_j are not independent.

■ **Example 2.3** The unknown number X of cells in a solution is estimated using the following experiment:

Step 1: A number Y of "beads" was added to the solution (Y is known).

Step 2: n Units from the solution were sampled, which was found to contain x cells and $(n - x)$ beads.

If the solution was well mixed, we have

$$\frac{X}{Y} = \frac{x}{n-x}$$

from which we can solve to obtain an estimate of X, the unknown number of cells:

$$\hat{X} = (Y)\left(\frac{x}{n-x}\right)$$

where p is the sample proportion $p = x/n$. It can be shown that the cells counting experiment is "optimal", that is, when the coefficient of variation of the estimate of X is minimized, if $p = \frac{1}{2}$.

2.1.2 The Hypergeometric Distributions

Sampling (selecting objects) may be done with or without replacement. To develop a formula or model analogous to that of the binomial probability, which applies to sampling without replacement, in which case successive trials are not independent, let us consider a set or population having $(a + b)$ elements of which a are labeled "success" and b are labeled "failure." Suppose a sample of n objects are taken from this population and let X denote the number of successes in this sample.

Then it can be shown that

$$\begin{aligned}\Pr(X = x) &= \frac{\binom{a}{x}\binom{b}{n-x}}{\binom{a+b}{n}}; \quad x = 1, 2, \ldots, n \\ &= h(x; n, a, b)\end{aligned}$$

subject to the restrictions that x cannot exceed a and $(n - x)$ cannot exceed b. We refer to $h(x;\, n,\, a,\, b)$ as the *hypergeometric density function* and the corresponding distribution, the distribution of the random variable X, as the *hypergeometric distribution* $H(x;\, n,\, a,\, b)$. This distribution is specified by three parameters (n, a, and b), the number of trials, and the sizes of a and b of the two finite subpopulations.

The mean and variance of the hypergeometric distribution are

$$\begin{aligned}E(X) &= \mu \\ &= (n)\left(\frac{a}{a+b}\right) \\ \text{Var}(X) &= \sigma^2 \\ &= \frac{nab(a+b-n)}{(a+b)^2(a+b-1)} \\ &= (n)\left(\frac{a}{a+b}\right)\left(1-\frac{a}{a+b}\right)\left(\frac{a+b-n}{a+b-1}\right)\end{aligned}$$

In the case of a large population and n small as compared to a and b (n is very small as compared to $(a + b)$), these results are similar to those of the binomial with $\pi = a/(a + b)$. In the general case, Wise (1954) gave the following approximation for cumulative hypergeometric probabilities:

$$\sum_{i=1}^{k} h(x; n, a, b) = \sum_{i=1}^{k} b(x; n, \pi)$$

in which the probability π is given by

$$\pi = \frac{a-k/2}{a+b-n/2+0.5}$$

For example, let $a = 6, b = 12, n = 5$ (in this example, n is not very small as compared to $(a + b)$), and $k = 1$, we have 0.439 for the left-hand side (the hypergeometric cumulative probability) and 0.440 for the right-hand side (the binomial cumulative probability).

The binomial distribution was generalized to the multinomial distribution. An obvious similar generalization of the hypergeometric is the *multivariate hypergeometric distribution,* which arises when the population has more than two types of individuals. Consider a population of N subjects, of which N_i are of type i, $1 \leq i \leq k$; in the case of the hypergeometric distribution, $k=2$, $N_1=a$, and $N_2=b$. If a sample of size n is taken without replacement and we let X_1, X_2, ..., X_k be the k random variables defined so that X_i is the number of subjects (out of a total of n subjects in the sample) that belong to type i, then the joint probability function of the Xs is given by

$$\Pr(X_1 = x_1, X_2 = x_2, \ldots, X_k = x_k) = \frac{\prod_{i=1}^{k} \binom{N_i}{x_i}}{\binom{N}{n}}$$

$$x_i = 1, 2, \ldots, N_i$$

$$\sum_{i=1}^{k} x_i = n$$

Using this probability density function, it can be shown that the means, variances, and covariances of the multivariate hypergeometric distribution are given by

$$\begin{aligned} E(X_i) &= \mu_i \\ &= (n)\left(\frac{N_i}{N}\right) \\ \mathrm{Var}(X_i) &= \sigma_i^2 \\ &= (n)\left(\frac{N_i}{N}\right)\left(1-\frac{N_i}{N}\right)\left(\frac{N-n}{N-1}\right) \\ \mathrm{Cov}(X_i, X_j) &= -(n)\left(\frac{N_i}{N}\right)\left(\frac{N_j}{N}\right)\left(\frac{N-n}{N-1}\right) \end{aligned}$$

That is, for $i \neq j$, X_i and X_j are not independent, similar to the case of the multinomial distribution.

For the hypergeometric distribution and its multivariate generalization, the multivariate hypergeometric distribution, the parameter estimation procedures are less often needed; these models are often used when conditional arguments are being used or conditional tests are being considered (see Section 2.2.6).

2.2 INFERENCES FOR 2-BY-2 CONTINGENCY TABLES

Data for 2-by-2 tables may be obtained by a different way. In the following two examples, the first is the result of a case–control study (two independent binomial samples) and the second comes from a cross-sectional survey (one multinomial sample with four categories).

■ **Example 2.4** The role of smoking in the etiology of pancreatitis has been recognized for many years. In order to provide estimates of the quantitative significance of these factors, a hospital-based study was carried out in eastern Massachusetts and Rhode Island between 1975 and 1979 (Yen et al., 1982). Ninety-eight patients who had a hospital discharge diagnosis of pancreatitis were included in this unmatched case–control study. The control group consisted of 451 patients admitted for diseases other than those of the pancreas and biliary tract. Risk factor information was obtained from a standardized interview with each subject, conducted by a trained interviewer.

The following are some data for the males:

Use of Cigarettes	Cases	Controls
Current smokers	38	81
Never or ex-smokers	15	136
Total	53	217

■ **Example 2.5** In 1979 the U.S. Veterans Administration conducted a health survey of 11,230 veterans (True et al., 1988). The advantages of this survey are that it includes a large random sample with a high interview response rate and it was done before the recent public controversy surrounding the issue of the health effects of possible exposure to Agent Orange. The following are data relating Vietnam service to having sleep problems among the 1787 veterans who entered the military service between 1965 and 1975.

	Service in Vietnam		
Sleep Problems	Yes	No	Total
Yes	173	160	333
No	599	851	1450
Total	772	1011	1783

2.2.1 Comparison of Two Proportions

Perhaps the most common problem involving categorical data is the comparison of two proportions as in Example 2.4. In this type of problem, we have two independent samples of binary data (n_1, x_1) and (n_2, x_2), where the n's are adequately large sample sizes that may or may not be equal, the x's are the numbers of "positive" outcomes in the two samples, and we consider the null hypothesis:

$$H_0: \pi_1 = \pi_2$$

expressing the equality of the two population proportions.

To perform a test of significance for H_0, we proceed with the following steps:

Step 1: Decide whether a one-sided test, say,

$$H_A: \pi_1 > \pi_2$$

or a two-sided test,

$$H_A: \pi_1 \neq \pi_2$$

is appropriate.

Step 2: Choose a significance level alpha (α), a common choice being $\alpha = 0.05$.

Step 3: Calculate the z-score,

$$z = \frac{p_2 - p_1}{\sqrt{p(1-p)(1/n_1 + 1/n_2)}}$$

where p (without a subscript) is the *pooled proportion*, a proportion obtain by pooling the two samples together—an estimate of the common proportion under H_0; p is defined by

$$p = \frac{x_1 + x_2}{n_1 + n_2}$$

Step 4: Refer to any table for standard normal distribution for selecting a cut point. For example, if the choice of α is 0.05, then the rejection region is determined by:

(i) For the one-sided alternative $H_A: \pi_1 > \pi_2$, $z > +1.65$.
(ii) For the one-sided alternative $H_A: \pi_1 < \pi_2$, $z < -1.65$.
(iii) For the two-sided alternative $H_A: \pi_1 \neq \pi_2$, $z \leq -1.96$ or $z \geq 1.96$.

What we are doing here follows a process that can be informally described as follows:

1. The basic term of $(p_2 - p_1)$ measures the difference between the two samples.
2. Its expected hypothesized value of this difference (i.e., under H_0) is zero.
3. The denominator of z is the standard error of $(p_2 - p_1)$, a measure of how good $(p_2 - p_1)$ is as an estimate of $(\pi_2 - \pi_1)$; z measures how many standard errors $(p_2 - p_1)$ is from zero, its hypothesized value.
4. The z-score measures the number of standard errors that $(p_2 - p_1)$, the evidence, is away from its hypothesized value.

■ **Example 2.6** A study was conducted to see whether an important public health intervention would significantly reduce the smoking rate among men. Of $n = 100$ males sampled in 1965 at the time of the release of the Surgeon General's report on the health consequences of smoking, $x = 51$ were found to be smokers. In 1980 a second random sample of $n = 100$ males, similarly gathered, indicated that $x = 43$ were smokers. An application of the above method yields:

$$p = \frac{51 + 43}{100 + 100}$$
$$= 0.47$$
$$z = \frac{0.51 - 0.43}{\sqrt{(0.47)(0.53)\left(\frac{1}{100} + \frac{1}{100}\right)}}$$
$$= 1.13$$

It can be seen that the observed rate was reduced from 51% to 43%, but the reduction is not statistically significant at the 0.05 level ($z = 1.13 < 1.65$; two-sided p-value $= 0.2585$).

In the two-sided form, the square of the z-score, denoted X^2 is more often used. The test is referred to as the *chi-squared test* and the null hypothesis is rejected at the 0.05 level when $X^2 > 3.84$ (since $3.84 = (1.96)^2$, the decision remains the same). With data arranged in the form of a 2×2 table, the test statistic can also be obtained using the shortcut formula with its denominator being the product of the four marginal totals.

Exposure	Sample #1	Sample #2	Total
Yes	a	c	$a + c$
No	b	d	$b + d$
Total	$a + b$	$c + d$	n

The chi-square test statistic is

$$X^2 = \frac{n(ad-bc)^2}{(a+b)(c+d)(a+c)(b+d)}$$

The chi-squared test and the two-sided "z-test" give an identical decision (the chi-squared value is equal to the square of the z-score); the z-score had the advantage that it can be used against a one-sided alternative.

■ **Example 2.7** An investigation was made into fatal poisonings of children by two drugs, which were among the leading causes of such deaths. In each case, an inquiry was made as to how the child had received the fatal overdose and responsibility for the accident was assessed. Results were as follows:

	Drug A	Drug B
Child responsible	8	12
Child not responsible	31	19

We have the proportions of cases for which the child is responsible,

$$\begin{aligned} p_{\mathrm{A}} &= \frac{8}{8+31} \\ &= 0.205 \\ p_{\mathrm{B}} &= \frac{12}{12+19} \\ &= 0.387 \end{aligned}$$

suggesting that they are not the same and that a child seems more prone to taking drug B (38.7%) than drug A (20.5%). However, the chi-squared statistic,

$$\begin{aligned} X^2 &= \frac{(39+31)[(8)(19)-(31)(12)]}{(39)(31)(20)(50)} \\ &= 2.80 \end{aligned}$$

shows that the difference is not statistically significant at the 0.05 level (p-value $= 0.0943$).

■ **Example 2.8** In Example 2.1, a case–control study was conducted to identify reasons for the exceptionally high rate of lung cancer among male residents of coastal Georgia. The primary risk factor under investigation was employment in shipyards during World War II, and the following table provides data for nonsmokers:

Shipbuilding	Cases	Controls
Yes	11	35
No	50	203
Total	51	238

With current smoking being the exposure, we have for the cases

$$p_2 = \frac{11}{58} = 0.180$$

and for the controls

$$p_1 = \frac{35}{238} = 0.147$$

An application of the procedure yields a pooled proportion of

$$p = \frac{11+35}{61+238} = 0.154$$

leading to

$$z = \frac{0.180-0.147}{\sqrt{(0.154)(0.846)\left(\frac{1}{61}+\frac{1}{238}\right)}} = 0.64$$

It can be seen that the rate of employment for the cases (18.0%) was higher than that for the controls (14.7%) but the difference is not statistically significant at the 0.05 level ($z = 0.64 < 0.65$; p-value $= 0.52218$).

■ **Example 2.9** The role of smoking in the etiology of pancreatitis has been recognized for many years. In order to provide estimates of the quantitative significance of these factors, a hospital-based study was carried out in eastern

Massachusetts and Rhode Island between 1975 and 1979. Ninety-eight patients who had a hospital discharge diagnosis of pancreatitis were included in this unmatched case–control study. The control group consisted of 451 patients admitted for diseases other than those of the pancreas and biliary tract. Risk factor information was obtained from a standardized interview with each subject, conducted by a trained interviewer.

The following are some data for the males:

Use of Cigarettes	Cases	Controls
Never	2	56
Ex-smokers	13	80
Current smokers	38	81
Total	53	217

With current smoking being the exposure, we have for the cases and for the controls

$$\begin{aligned} p_2 &= \frac{38}{53} \\ &= 0.717 \\ p_{1\cdot} &= \frac{81}{217} \\ &= 0.373 \end{aligned}$$

An application of the procedure yields a pooled proportion of

$$\begin{aligned} p &= \frac{38+81}{53+217} \\ &= 0.441 \end{aligned}$$

leading to

$$\begin{aligned} z &= \frac{0.717-0.373}{\sqrt{(0.441)(0.559)\left(\frac{1}{53}+\frac{1}{217}\right)}} \\ &= 4.52 \end{aligned}$$

It can be seen that the proportion of smokers among the cases (71.7%) was higher than that for the controls (37.7%) and the difference is highly statistically significant (p-value < 0.001).

2.2.2 Tests for Independence

When we examine the Veterans Administration data in Example 2.5, the method of Section 2.2.1 does not seem to be applicable because we do not have two independent binomial samples. What we have in Example 2.5 are not two independent samples but a multinomial sample with probabilities π_{ij}. For contingency tables, the maximum likelihood (ML) estimates of cell probabilities are the sample cell proportion, and the ML estimates of marginal probabilities are sample marginal proportions. When two categorical variables forming the two-way table are independent, we have

$$\pi_{ij} = \pi_{i+}\pi_{+j}$$

where the two terms forming the product on the right-hand side are the two marginal probabilities.

The ML estimate of π_{ij} under this condition is

$$\begin{aligned}\hat{\pi}_{ij} &= \hat{\pi}_{i+}\hat{\pi}_{+j} \\ &= p_{ij}p_{ij} \\ &= \frac{x_{ij}x_{ij}}{N}\end{aligned}$$

For a multinomial sample of size n over $2 \times 2 = 4$ cells, an individual cell count x_{ij} has the binomial distribution with size n and probability π_{ij}. The mean (or *expected value* or *expected frequency*) of this binomial is $m_{ij} = n\pi_{ij}$, which has ML estimate, under the assumption of independence

$$\begin{aligned}e_{ij} &= n\hat{\pi}_{ij} \\ &= \frac{x_{ij}x_{ij}}{N} \\ &= \frac{(\text{row total})(\text{column total})}{\text{total sample size}}\end{aligned}$$

That formula gives the estimated expected frequencies, the frequencies we expect to have under the null hypothesis of independence. They have the same marginal totals as do the observed data.

We use the estimated expected frequencies in tests of independence through Pearsons chi-squared statistic or the likelihood ratio chi-squared statistic; these are two different ways, in which to compare the observed cell frequencies versus the expected cell frequencies. Pearson's chi-squared statistic is given by

$$X^2 = \sum_{i,j} \frac{(x_{ij} - e_{ij})^2}{e_{ij}}$$

and the likelihood ratio chi-squared statistic is given by

$$G^2 = 2\sum_{i,j} x_{ij} \ln \frac{x_{ij}}{e_{ij}}$$

For large samples, both X^2 and G^2 have approximately a chi-squared distribution with 1 degree of freedom under the null hypothesis of independence; greater values lead to a rejection of H_0. Pearson's chi-squared statistic is more often used; as applied to the studies with two binomial samples, it is identical to the z-test of Section 2.2.1. That means, with data arranged in the form of a 2×2 table,

	Category 1	Category 2	Total
Category 1	a	c	$a + c$
Category 2	b	d	$b + d$
Total	$a + b$	$c + d$	n

that Pearson's chi-squared statistic is given by the same formula introduced in the previous subsection:

$$X^2 = \frac{n(ad-bc)^2}{(a+b)(c+d)(a+c)(b+d)}$$

■ **Example 2.10** Refer to the Veterans Administration data of Example 2.5:

	Service in Vietnam		
Sleep Problems	Yes	No	Total
Yes	173	160	333
No	599	851	1450
Total	772	1011	1783

$$e_{11} = \frac{(333)(772)}{1783} = 144.18$$

$$e_{12} = 333-144.12 = 188.82$$

$$e_{21} = 772-144.18 = 627.82$$

$$e_{22} = 1011-188.82 = 882.18$$

leading to

$$X^2 = \frac{(173-144.18)^2}{144.18} + \frac{(160-188.82)^2}{188.82} + \frac{(599-627.82)^2}{627.82} + \frac{(851-822.18)^2}{822.18}$$

$$G^2 = 2\left\{173 \ln\frac{173}{144.18} + 160 \ln\frac{160}{188.82} + 599 \ln\frac{599}{627.82} + 851 \ln\frac{851}{822.18}\right\}$$

We have $X^2 = 12.49$ and $G^2 = 12.40$; these statistics, both with 1 degree of freedom, indicate a significant correlation ($p < 0.001$) relating Vietnam service to having sleep problems among veterans.

Statistical decisions based on Pearson's chi-squared statistic (and the likelihood ratio chi-squared statistic for the same purpose) make use of the percentiles of the chi-squared distribution. Since chi-squared is a continuous distribution and categorical data are discrete, some statisticians use a version of Pearson's statistic with a continuity correction, called the *Yates corrected chi-squared test*, which can be expressed as

$$X_c^2 = \sum_{i,j} \frac{\left(|x_{ij}-e_{ij}|-0.5\right)^2}{e_{ij}}$$

Statisticians still disagree about whether or not a continuity correction is needed (Conover, 1974). Generally, the corrected version is more conservative and more widely used in applied literature.

2.2.3 Fisher's Exact Test

Even with a continuity correction, the goodness-of-fit test statistics such as Pearson's X^2 and the likelihood ratio G^2 are not suitable when the sample is small. Generally, statisticians suggest using them only if no expected frequency in the table is less than 5. For studies with small samples, we will introduce a method known as Fisher's exact test. For tables in which the use of the chi-square test X^2 is appropriate, the two tests give very similar results.

Our purpose is to find the exact significance level associated with an observed table. The central idea is to enumerate all possible outcomes consistent with a given set of marginal totals and add up the probabilities of those tables more extreme than the one observed. Conditional on the margins, a 2×2 table is a one-dimensional random variable having a hypergeometric distribution so the exact test is relatively easy to implement. The probability of observing a table with cells a, b, c, and d (with total n) is

$$\Pr(a, b, c, d) = \frac{(a+b)!(c+d)!(a+c)!(b+d)!}{n!a!b!c!d!}$$

The process for doing hand calculations would be as follows:

Step 1: Rearrange the rows and columns of the observed table so the smaller total is in the first row and the smaller column total is in the first column.

Step 2: Start with the table having 0 in the (1,1) cell (top left cell). The other cells in this table are determined automatically from the fixed row margins and column margins.

Step 3: Construct the next table by increasing the (1,1) cell from 0 to 1 and decreasing all other cells accordingly.

Step 4: Continue to increase the (1,1) cell by 1 until one of the other cells become 0. At that point we have enumerated all possible tables.

Step 5: Calculate and add up the probabilities of those tables with cell (1,1) having values from 0 to the observed frequency (left tail for a one-tailed test); double the smaller tail for a two-tailed test.

In practice, the calculations are often tedious and should be left to a computer program to implement; a sample of the SAS program instruction is given at the end of the following example.

■ **Example 2.11** A study on deaths of men aged over 50 yields the following data (numbers in parentheses are expected frequencies):

	Type of Diet		
Cause of Death	High Salt	Low Salt	Total
Non-CVD	2(2.92)	23(22.08)	25
CVD	5(4.08)	30(30.92)	35
Total	7	53	60

An application of Fisher's exact test yields a one-tailed p-value of 0.375 or a two-tailed p-value of 0.688; we cannot say, on the basis of this limited amount of data, that there is a significant association between salt intake and cause of death even though the proportions of CVD deaths are different (71.4% vs. 56.6%). For implementing hand calculations, we would focus on the tables where cell (1,1) equals 0, 1, and 2 (observed value); the probabilities for these tables are 0.017, 0.105, and 0.252, respectively.

Note: A SAS program would include these instructions

```
DATA;
DO CVD=1 TO 2;
DO DIET=1 TO 2;
INPUT COUNT @@; OUTPUT;
```

```
END;
END;
CARDS;
2 23
5 30;
PROC FREQ;
WEIGHT COUNT;
TABLES CVD*DIET;
```

The output also includes Pearson's test ($X_2 = 0.559; p = 0.455$) and likelihood ratio test ($G_2 = 0.581$; $p = 0.446$) as well.

2.2.4 Relative Risk and Odds Ratio

A ratio is a computation of the form where a and b are *similar quantities* measured from *different groups* or under *different circumstances*,

$$\text{Ratio} = \frac{a}{b}$$

An example is the male-to-female ratio of smoking rates; each ratio is positive but, unlike proportions, it may well exceed 1.0.

One of the most often used ratios in epidemiological studies is the *relative risk*, a concept for the comparison of two groups or populations with respect to a certain unwanted event (disease or death). The traditional method of expressing it in prospective studies is simply the ratio of the incidence rates:

$$\text{Relative risk} = \frac{\text{Disease incidence in group 1}}{\text{Disease incidence in group 2}}$$

However, ratio of disease prevalences as well as follow-up death rates can also be formed. Usually, group 2 is under standard conditions—such as nonexposure to a certain risk factor—against which group 1 (exposed) is measured. A relative risk that is greater than 1.0 indicates harmful effects, whereas a relative risk that is less than 1.0 indicates beneficial effects. For example, if group 1 consists of smokers and group 2 nonsmokers, then we have a *relative risk due to smoking*.

The *relative risk*, also called *risk ratio*, is an important index in epidemiological studies because in such studies it is often useful to measure the *increased* risk (if any) of incurring a particular disease if a certain factor is present. In cohort studies such an index is readily obtained by observing the experience of groups of subjects with and without the factor as shown above. In a case–control study the data do not present an immediate answer to this type of question, and we now consider how to obtain a useful shortcut solution.

Suppose that each subject in a large study, at a particular time, is classified as positive or negative according to some risk factor, and as having or not having a certain disease under investigation. For any such categorization the population may be enumerated in a 2×2 table as follows:

Factor	Disease		Total
	Yes (+)	No (−)	
Exposed (+)	A	B	$A + B$
Unexposed (−)	C	D	$C + D$
Total	$A + C$	$B + D$	$N = A + B + C + D$

The entries A, B, C, and D in the table are sizes of the four combinations of disease presence-and-absence and factor presence-and-absence and the number N at the lower right corner of the table is the total population size. The relative risk (RR) is

$$RR = \frac{A}{A+B} \div \frac{C}{C+D}$$

$$= \frac{A(C+D)}{C(A+B)}$$

In many situations, the number of subjects classified as disease positive is very small as compared to the number classified as disease negative; that is,

$$C + D \cong D$$

$$A + B \cong B$$

($\cong$ means "almost equal to") and therefore the relative risk can be approximated as follows:

$$RR \cong \frac{AD}{BC}$$

$$= \frac{A/B}{C/D} = \frac{A/C}{B/D}$$

where the slash denotes division. The resulting ratio, AD/BC, is an approximate relative risk, but it is often referred to as *odds ratio* because

(i) A/B and C/D are the *odds* in favor of having disease from groups with or without the factor.

(ii) A/C and B/D are the odds in favor of having exposure to the factors from groups with or without the disease.

The two odds in (ii) can easily be estimated using case–control data, by using sample frequencies. For example, the odds A/C can be estimated by a/c where a is the number of exposed cases and c is the number of exposed controls in the sample of cases in a case–control design.

For the many diseases that are rare (most are), the terms relative risk (RR) and odds ratio (OR) are used interchangeably because of the above-mentioned approximation. The relative risk is an important epidemiological index used to measure seriousness, or the magnitude of the harmful effect of suspected risk factors. For example, if we have

$$RR = 3.0$$

then we can say that the exposed individuals have a risk of contracting the disease which is approximately three times the risk of unexposed individuals. A perfect 1.0 indicates no effect and beneficial factors result in relative risk values that are smaller than 1.0. From data obtained by a case–control or retrospective study, it is impossible to calculate the relative risk that we want, but if it is reasonable to assume that the disease is rare (prevalence is less than 0.05, say), then we can calculate the *odds ratio* as a "stepping stone" and use it as an approximate *relative risk* (we use the notation "$\cong$", meaning *almost equal to*, for this purpose). In these cases, we interpret the calculated odds ratio just as we would do the relative risk.

■ **Example 2.12** The role of smoking in the etiology of pancreatitis has been recognized for many years. In order to provide estimates of the quantitative significance of these factors, a hospital-based study was carried out in eastern Massachusetts and Rhode Island between 1975 and 1979. Ninety-eight patients who had a hospital discharge diagnosis of pancreatitis were included in this unmatched case–control study. The control group consisted of 451 patients admitted for diseases other than those of the pancreas and biliary tract. Risk factor information was obtained from a standardized interview with each subject, conducted by a trained interviewer.

The following are some data for the males:

Use of Cigarettes	Cases	Controls
Never	2	56
Ex-smokers	13	80
Current smokers	38	81
Total	53	217

For the above data for this example, the approximate relative risks or odds ratios are calculated as follows:

(i) For ex-smokers,

$$OR_e = \frac{13/2}{80/56} = \frac{(13)(56)}{(80)(2)} = 4.55$$

(The subscript "e" in the notation OR_e indicates that we are calculating the odds ratio for ex-smokers, but will interpret this as relative risk.)

(ii) For current smokers,

$$OR_c = \frac{38/2}{81/56} = \frac{(38)(56)}{(81)(2)} = 13.14$$

(The subscript "c" in the notation OR_c indicates that we are calculating the odds ratio for current smokers.)

In these calculations, the nonsmokers (who never smoked) are used as references. These values indicate that the risk of having pancreatitis for current smokers is approximately 13.14 times the same risk for people who never smoked. The effect for ex-smokers is smaller (4.55 times) but it is still very high (as compared to 1.0—the no-effect baseline for relative risks and odds ratios). In other words, if the smokers quit smoking they would reduce their own risk (from 13.14 times to 4.55 times), but *not* to the normal level for people who never smoked.

In general, data from a case–control study, for example, may be summarized in a 2×2 table as follows:

Exposure	Cases	Controls
Exposed	a	b
Unexposed	c	d

We have the following:

(i) The odds that a case was exposed is

$$\text{Odds for cases} = \frac{a}{b}$$

(ii) The odds that a control was exposed is

$$\text{Odds for controls} = \frac{c}{d}$$

Therefore the (observed) odds ratio from the samples is

$$OR = \frac{a/b}{c/d} = \frac{ad}{bc}$$

Confidence intervals are derived from the normal approximation to the sampling distribution of the odds ratio on the log scale, $\ln(OR)$, with variance:

$$\text{Variance}[\ln(OR)] = \frac{1}{a} + \frac{1}{b} + \frac{1}{c} + \frac{1}{d}$$

Consequently, an approximate 95% confidence interval, on the log scale, for odds ratio is given by

$$\ln\left(\frac{ad}{bc}\right) \pm (1.96)\sqrt{\frac{1}{a} + \frac{1}{b} + \frac{1}{c} + \frac{1}{d}}$$

(again, ln is logarithm to base e, also called the "natural" logarithm). A 95% confidence interval for the odds ratio under investigation is obtained by "exponentiating" (the reverse log operation or antilog) the two endpoints, one with the minus sign and one with the plus sign.

■ **Example 2.13** The role of smoking in the etiology of pancreatitis has been recognized for many years. In order to provide estimates of the quantitative significance of these factors, a hospital-based study was carried out in eastern Massachusetts and Rhode Island between 1975 and 1979. Ninety-eight patients who had a hospital discharge diagnosis of pancreatitis were included in this unmatched case–control study. The control group consisted of 451 patients admitted for diseases other than those of the pancreas and biliary tract. Risk factor information was obtained from a standardized interview with each subject, conducted by a trained interviewer.

The following are some data for the males:

Use of Cigarettes	Cases	Controls
Never	2	56
Ex-smokers	13	80
Current smokers	38	81
Total	53	217

We have the following from the data in the above table:

(i) For ex-smokers, compared to those who never smoked,

$$\begin{aligned} OR &= \frac{(13)(56)}{(80)(2)} \\ &= 4.55 \end{aligned}$$

A 95% confidence interval for the population odds ratio on the log scale is from (−0.01),

$$\ln(4.55) - 1.96\sqrt{\frac{1}{13} + \frac{1}{56} + \frac{1}{80} + \frac{1}{2}} = -0.01$$

to (3.04),

$$\ln(4.55) + 1.96\sqrt{\frac{1}{13} + \frac{1}{56} + \frac{1}{80} + \frac{1}{2}} = 3.04$$

Therefore, the corresponding 95% confidence interval for the population odds ratio is (0.99, 20.96).

(ii) For current smokers, compared to those who never smoked,

$$\begin{aligned} OR &= \frac{(38)(56)}{(81)(2)} \\ &= 13.14 \end{aligned}$$

A 95% confidence interval for the population odds ratio on the log scale is from (1.11),

$$\ln(13.14) - 1.96\sqrt{\frac{1}{38} + \frac{1}{56} + \frac{1}{81} + \frac{1}{2}} = 1.11$$

to (4.04),

$$\ln(4.55) + 1.96\sqrt{\frac{1}{38} + \frac{1}{56} + \frac{1}{81} + \frac{1}{2}} = 4.04$$

and hence the corresponding 95% confidence interval for the population odds ratio is (3.04, 56.70).

The result for ex-smokers is "inconclusive," with the confidence interval (0.99, 20.96) showing a slim possibility that the odds ratio could be "1.0," due to a small cell frequency: only 2 cases who have never smoked. Unlike data on a continuous scale, we need to have all four frequencies larger in order to have a narrow confidence interval; just a large total sample size is not enough.

■ **Example 2.14** Toxic shock syndrome (TSS) is a disease first recognized in the 1980s, characterized by sudden onset of high fever (over 102°F), vomiting, diarrhea, and rapid progression to hypotension and, in most cases, shock. Because of the striking association with menses, several studies have been undertaken to look at various practices associated with the menstrual cycle. In a study by the Centers for Disease Control, 30 of 40 TSS cases and 30 of 114 controls used a single brand of tampons, the Rely brand. Data are presented in a 2×2 table as follows:

Brand	Cases	Controls
Rely	30	30
Others	10	84
Total	40	114

We have

$$OR = \frac{(30)(84)}{(10)(30)}$$

$$= 8.4$$

A 95% confidence interval for the population odds ratio on the log scale is from (1.30),

$$\ln(8.4) - 1.96\sqrt{\frac{1}{30} + \frac{1}{10} + \frac{1}{30} + \frac{1}{84}} = 1.30$$

to (2.96),

$$\ln(8.4)+1.96\sqrt{\frac{1}{30}+\frac{1}{10}+\frac{1}{30}+\frac{1}{84}}=2.96$$

Therefore the corresponding 95% confidence interval for the population odds ratio is (3.67, 19.30), indicating a very high risk elevation for Rely users.

The above method for confidence interval of odds ratio follows a three-step process that is normally used for the interval estimation of ratios (of statistics):

Step 1: Take the log of the ratio; that would turn a ratio into a difference.
Step 2: Calculate and estimate the variance on the log scale.
Step 3: Form confidence intervals on the log scale, then exponentiate the endpoints.

In the above process, the second step is achieved by using the error propagation method (also called the delta method). For example,

$$\begin{aligned}\mathrm{Var}(\ln p) &\cong \frac{1}{p^2}\frac{p(1-p)}{n} \\ &= \left(\frac{1}{n}\right)\left(\frac{1-p}{p}\right)\end{aligned}$$

In the third step, we treat logarithms of the numerator and the denominator as normally distributed statistics. In the specific example of an odds ratio, we treat the "log of the odds" as a normally distributed statistic, and that is asymptotically correct. When data are available, this method for interval estimation of odds ratio does not apply to the interval estimation of relative risk. We simply cannot treat the log of the proportion as a normally distributed statistic; the proportion itself is distributed asymptotically as normal according to the Central Limit Theorem. By wrongly replacing a normal distribution (for a proportion, the numerator and denominator of a relative risk) we would end up with a confidence interval that is inefficiently too long (Lui, 2007).

Instead of a log transformation, we could use the *Fieller Theorem* to determine confidence interval of a relative risk with the following process.

If $r=p_2/p_1$ is an estimate of a relative risk R—a ratio with normally distributed numerator and denominator—we consider the statistic $C=p_2-Rp_1$ that is distributed as normal because both p_2 and p_1 are normally distributed and R, a parameter, is a constant.

We derive the mean and variance of that statistic, which leads to the confidence limits for R as follows:

$$R = \frac{\pi_2}{\pi_1}$$

$$r = \frac{p_2}{p_1}$$

$$C = p_2 - Rp_1$$

$$\mathrm{Var}(C) = \frac{\pi_2(1-\pi_2)}{n_2} + R^2 \frac{\pi_1(1-\pi_1)}{n_1}$$

$$\widehat{\mathrm{Var}}(C) = \frac{p_2(1-p_2)}{n_2} + r^2 \frac{p_1(1-p_1)}{n_1}$$

$$\frac{\{p_2-(R)p_1\}^2}{\widehat{\mathrm{Var}}(C)} = z^2_{1-\alpha/2}$$

Then by solving the last quadratic equation, the two roots of this equation would form the $(1-\alpha)100\%$ confidence interval of the relative risk R. The result is

$$r \pm z_{1-\alpha}\left(\frac{1}{p_1}\right)\left(\frac{p_2(1-p_2)}{n_2} + r^2 \frac{p_1(1-p_1)}{n_1}\right)$$

which is often much shorter (i.e., more efficient) than the result based on the log transformation.

■ **Example 2.15** In a trial of diabetic therapy, patients were either treated with Phenformin or a placebo. The numbers of patients and deaths from cardiovascular causes were as follows:

Result	Phenformin	Placebo	Total
Cardiovascular death	26	2	28
Not a death	178	62	240
Total	204	64	268

Suppose we want to investigate the difference in cardiovascular mortality between the Phenformin and placebo groups and to express it in the form of a 95% confidence interval for the relative risk.

Applying the above method, we have

$$p_1 = \frac{2}{64} = .03$$

$$p_2 = \frac{26}{204} = .127$$

$$\text{Point estimate } r = \frac{0.127}{0.03} = 4.230$$

$$\begin{aligned}
95\%\ \text{Confidence interval} &= 4.23 \pm (1.96)\left(\frac{1}{0.03}\right) \\
&\quad \times \left(\frac{(0.127)(1-0.127)}{204} + (4.23)^2 \frac{(0.03)(1-0.03)}{64}\right) \\
&= (3.663, 4.797)
\end{aligned}$$

2.2.5 Etiologic Fraction

The *Etiologic fraction* or *population attributable risk* λ is a measure of the impact of an exposure "on the population" (whereas the relative risk measures the impact of the exposure "on the exposed subpopulation"). The etiologic fraction is defined as the proportion of disease cases attributable to the risk factor. For example, we are interested in the proportion of lung cancer cases attributable to smoking. At a given time, let N be the population size and N_1 be the number of cases. If p_1 and p_0 are the disease rates of the exposed and nonexposed subpopulations, respectively, then

$$\lambda = \frac{(N_1 - Np_0)}{N_1}$$

because if the exposure has no effect then the expected number of cases would be Np_0. Since

$$N_1 = N\{p_e p_1 + (1-p_e)p_0\}$$

where p_e is the population exposure rate (the percentage of smokers in the target population), we have:

$$\lambda = \frac{p_e(RR-1)}{1 + p_e(RR-1)}$$

with RR being the relative risk associated with the exposure.

This result shows that the attributable risk, or etiologic fraction, depends on the effect of the exposure, through RR, as well as the exposure rate, p_e:

(i) it is an increasing function of RR, $\lambda = 0$ if $RR = 1$, that is no excess risk

(ii) it is also an increasing function of p_e, which makes it an important parameter for public health intervention policy; $\lambda = 0$ if $p_e = 0$

However, the above formula is often not useful in practice because the population exposure rate, p_e, is unknown and not estimable accurately from case–control studies (we can estimate it using the controls, but it is only an approximation). Using the Bayes theorem, we can express the etiologic fraction in a different but equivalent form:

$$\lambda = p_{1e}\{1-1/RR\}$$

where p_{1e} is the exposure rate of the subpopulation of cases which can be estimated using the sample of cases.

When data from a case–control study, for example, may be summarized in a 2×2 table as follows:

Exposure	Cases	Controls
Exposed	a	b
Unexposed	c	d
Sample size	n_1	n_0

then we have

$$\begin{aligned}\hat{\lambda} &= \frac{a}{n_1}\left\{1-\frac{bc}{ad}\right\}\\ &= 1-\frac{cn_0}{dn_1}\end{aligned}$$

and a 95% confidence interval for λ is given by

$$1-(1-\hat{\lambda})\exp\left\{\pm 1.96\sqrt{\frac{a}{cn_1}+\frac{b}{dn_0}}\right\}$$

■ **Example 2.16** A case–control study was conducted in Auckland, New Zealand, to investigate the effects of alcohol consumption on both nonfatal myocardial infarction and coronary death in the 24 hours after drinking, among regular drinkers (Jackson et al., 1992). The following table shows coronary death data for men.

Drink in the Last 24 hours	Cases	Controls
Yes	69	159
No	103	135
Sample size	172	294

Here we have $n_1 = 172$ ($a = 69$, $c = 103$) and $n_0 = 294$ ($b = 159$, $d = 135$), leading to a point estimate of $\lambda = -0.304$ and a 95% confidence interval of $(-0.552, -0.096)$. The results indicate a protective effect; that is, drinking (in the last 24 hours) reduces coronary deaths in the population of male drinkers in New Zealand between 9.6% and 55.2%.

2.2.6 Crossover Designs

The two-period crossover design is often used in clinical trials to improve the sensitivity of the trial by eliminating the individual patient effects. For two experimental treatments A and B, a set of N patients is randomly subdivided into n_A patients assigned to receive treatment sequence (A, B) (treatment A in the first period and treatment B in the second period; there may be a "washout" period in between) and n_B assigned to (B, A) (treatment B in the first period and treatment A in the second period; the reverse of the other sequence): $N = n_A + n_B$ (usually, $n_A = n_B = N/2$). For the case of a binary response, coded $0/1$ for negative/positive responses, Gart (1969) proposed the following so-called logistic response probabilities:

1. For the sequence (A, B),

   (i) Treatment A:

$$\Pr(\text{Response} = 1) = \frac{\exp(\beta_i + \lambda + \tau)}{1 + \exp(\beta_i + \lambda + \tau)}; \quad i = 1, 2, \ldots, n_A$$

   (ii) Treatment B:

$$\Pr(\text{Response} = 1) = \frac{\exp(\beta_i - \lambda - \tau)}{1 + \exp(\beta_i - \lambda - \tau)}; \quad i = 1, 2, \ldots, n_A$$

2. For sequence (A, B),

   (i) Treatment A:

$$\Pr(\text{Response} = 1) = \frac{\exp(\beta_i - \lambda + \tau)}{1 + \exp(\beta_i - \lambda + \tau)}; \quad i = 1, 2, \ldots, n_A$$

   (ii) Treatment B:

$$\Pr(\text{Response} = 1) = \frac{\exp(\beta_i + \lambda - \tau)}{1 + \exp(\beta_i + \lambda - \tau)}; \quad i = 1, 2, \ldots, n_A$$

   in which the β's are the individual patient effects and λ and τ are the order and treatment effects, respectively.

Under this response model, Gart showed that optimum inferences about λ (for order effects) and τ (for treatment effects), regarding the β's as nuisance parameters, are based on those patients with unlike responses in consecutive periods (those patients with outcomes (1, 0) or (0, 1), a conditional approach). Among these patients with unlike outcomes, let

(i) For sequence (A, B)

$$\begin{aligned} y_A &= \text{number of patients with outcome } (1,0) \\ y_B &= \text{number of patients with outcome } (0,1) \\ n &= y_A + y_B = n_A \end{aligned}$$

(ii) For sequence (B, A)

$$\begin{aligned} y_A &= \text{number of patients with outcome } (0,1) \\ y_B &= \text{number of patients with outcome } (1,0) \\ n' &= y'_A + y'_B = n_B \end{aligned}$$

Then it can be shown that, with n and n' being fixed, y_A and y'_A have the binomial distributions $B(n, \pi)$ and $B(n', \pi')$, respectively, where the probability parameters π and π' are given by

$$\pi = \frac{1}{1+\exp\{-2(\lambda+\tau)\}}$$

$$\pi' = \frac{1}{1+\exp\{2(\lambda-\tau)\}}$$

The result leads to simple ways to test for the null hypothesis of no treatment effects and the null hypothesis of no order effects:

(i) With data presented as in the following 2 × 2 table,

Positive Responses Coming from	Treatment Sequence (A, B)	(B, A)	Total
First period	y_A	y'_B	$y_A + y'_B$
Second period	y_B	y'_A	$y_B + y'_A$
Total	$n = y_A + y_B$	$n' = y'_B + y'_A$	$n + n'$

it can be seen that y_A has the hypergeometric distribution $H(y_A + y'_B, n, n')$ under the null hypothesis:

$$H_0: \ \tau = 0$$

of no treatment effects and conditional on the marginal totals. This can be tested, in the data format of the above 2 × 2 table, using Pearson's chi-squared test or Fisher's exact test. This null hypothesis is often primary.

(ii) With data presented as in the following 2×2 table,

Positive Responses Coming from	Treatment Sequence (A, B)	(B, A)	Total
Treatment A	y_A	y'_A	$y_A + y'_A$
Treatment B	y_B	y'_B	$y_B + y'_B$
Total	$n = y_A + y_B$	$n' = y'_A + y'_B$	$n + n'$

it can be seen that y_A has the hypergeometric distribution $H(y_A + y'_A, n, n')$ under the null hypothesis:

$$H_0: \lambda = 0$$

of no order effects and conditional on the marginal totals. This can be tested, in the data format of the above 2×2 table, using Pearson's chi-squared test or Fisher's exact test.

2.3 THE MANTEL–HAENSZEL METHOD

We are often interested only in investigating the relationship between two binary variables, for example a disease and an exposure; however, we have to control for confounders. A confounding variable is a variable that may be associated with either the disease or exposure or both. For example, in Example 1.1, a case–control study was undertaken to investigate the relationship between lung cancer and employment in shipyards during World War II among male residents of coastal Georgia. In this case, smoking is a confounder; it has been found to be associated with lung cancer and it may be associated with employment because construction workers are likely to be smokers. Specifically, we want to know

(i) among smokers, whether or not shipbuilding and lung cancer are related, and
(ii) among nonsmokers, whether or not shipbuilding and lung cancer are related.

The underlying question is the question concerning conditional independence between lung cancer and shipbuilding; however, we do not want to reach separate conclusions, one at each level of smoking. Assuming that the confounder, smoking, is not an effect modifier (i.e., smoking does not alter the relationship between lung cancer and shipbuilding), we want to pool data for a combined decision. When both the disease and the exposure are binary, a popular method to achieve this task is the *Mantel–Haenszel method.* The process can be summarized as follows:

(i) We form 2×2 tables, one at each level of the confounder.
(ii) At a level of the confounder, we have the following:

Exposure	Disease Classification		Total
	Positive	Negative	
Yes	a	b	r_1
No	c	d	r_2
Total	c_1	c_2	n

Under the null hypothesis and fixed marginal totals, frequency of the upper left cell a is distributed with the following mean and variance:

$$E_0(a) = \frac{r_1 c_1}{n}$$

$$\text{Var}_0(a) = \frac{r_1 r_2 c_1 c_2}{n^2(n-1)}$$

and the Mantel–Haenszel test is based on the z-statistic:

$$z = \frac{\sum a - \sum \frac{r_1 c_1}{n}}{\sqrt{\sum \frac{r_1 r_2 c_1 c_2}{n^2(n-1)}}}$$

where the summation ($\sum$) is across levels of the confounder. Of course, one can use the square of the z-score, a chi-squared test at 1 degree of freedom, for two-sided alternatives. When the above test is statistically significant, the association between the disease and the exposure is *real*. Since we assume that the confounder is not an effect modifier, the odds ratio is constant across its levels. The odds ratio at each level is estimated by ad/bc; the Mantel–Haenszel procedure pools data across levels of the confounder to obtain a combined estimate:

$$OR_{\text{MH}} = \frac{\sum \frac{ad}{n}}{\sum \frac{bc}{n}}$$

■ **Example 2.17** A case–control study was conducted to identify reasons for the exceptionally high rate of lung cancer among male residents of coastal Georgia as first presented in Example 1.1. The primary risk factor under investigation was employment in shipyards during World War II, and data are tabulated separately for three levels of smoking as follows:

Smoking	Shipbuilding	Cases	Controls
No	Yes	11	35
	No	50	203
Moderate	Yes	70	42
	No	217	220
Heavy	Yes	14	3
	No	96	50

1. We begin with the 2 × 2 table for *nonsmokers*:

Smoking	Shipbuilding	Cases	Controls	Total
No	Yes	11 (*a*)	35 (*b*)	46
	No	50 (*c*)	203 (*d*)	253
	Total	51	238	299 (*n*)

We have, for the nonsmokers,

$$a = 11$$

$$\frac{r_1 c_1}{n} = \frac{(46)(61)}{299} = 9.38$$

$$\frac{r_1 r_2 c_1 c_2}{n^2(n-1)} = \frac{(46)(253)(61)(238)}{(299)^2(298)} = 6.34$$

$$\frac{ad}{n} = \frac{(11)(203)}{299} = 7.47$$

$$\frac{bc}{n} = \frac{(35)(50)}{299} = 5.85$$

2. For moderate smokers,

Smoking	Shipbuilding	Cases	Controls	Total
Moderate	Yes	70 (*a*)	42 (*b*)	112
	No	217 (*c*)	220 (*d*)	437
	Total	287	262	549 (*n*)

$$a = 70$$

$$\frac{r_1 c_1}{n} = \frac{(112)(287)}{549} = 58.55$$

$$\frac{r_1 r_2 c_1 c_2}{n^2(n-1)} = \frac{(112)(437)(287)(262)}{(549)^2(548)} = 22.28$$

$$\frac{ad}{n} = \frac{(70)(220)}{549} = 28.05$$

$$\frac{bc}{n} = \frac{(42)(217)}{549} = 16.60$$

3. For heavy smokers,

Smoking	Shipbuilding	Cases	Controls	Total
Heavy	Yes	14 (*a*)	3 (*b*)	17
	No	96 (*c*)	50 (*d*)	146
	Total	110	53	163 (*n*)

$$a = 14$$

$$\frac{r_1 c_1}{n} = \frac{(17)(110)}{163} = 14.47$$

$$\frac{r_1 r_2 c_1 c_2}{n^2(n-1)} = \frac{(17)(146)(110)(53)}{(163)^2(162)} = 3.36$$

$$\frac{ad}{n} = \frac{(14)(50)}{163} = 4.29$$

$$\frac{bc}{n} = \frac{(3)(96)}{163} = 1.77$$

The above results from the three levels of the confounder (smoking) are combined to obtain the z-score:

$$
\begin{aligned}
z &= \frac{(11-9.38)+(70-58.55)+(14-11.47)}{\sqrt{6.34+22.28+3.36}} \\
&= 2.76
\end{aligned}
$$

and a z-score of 2.76 yields a one-sided p-value of 0.0029, which is beyond the 1% level. This result is stronger than those for tests at each level because it is based on more information where all data at all three smoking levels are used. The combined odds ratio estimate is

$$
\begin{aligned}
OR_{\text{MH}} &= \frac{7.47+28.05+4.28}{5.85+16.60+1.77} \\
&= 1.64
\end{aligned}
$$

This combined estimate of the odds ratio, 1.64, represents an approximate increase of 64% in lung cancer risk for those employed in the shipbuilding industry.

Note: A SAS program would include these instructions:

```
DATA;
INPUT SMOKE SHIP CANCER COUNT;
CARDS;
1 1 1 11
1 1 2 35
1 2 1 50
1 2 2 203
2 1 1 70
2 1 2 42
2 2 1 217
2 2 2 220
3 1 1 14
3 1 2 3
3 2 1 96
3 2 2 50;
PROC FREQ;
WEIGHT COUNT;
TABLES SMOKE*SHIP*CANCER/CMH;
```

The result is given in a chi-squared form ($X^2 = 7.601, p = 0.006$); CMH stands for Cochran-Mantel-Haenszel statistic.

The following is another similar example aiming at the possible effects of oral contraceptive use on myocardial infarction.

■ **Example 2.18** A case–control study was conducted to investigate the relationship between myocardial infarction (MI) and oral contraceptive (OC) use.

The data, stratified by cigarette smoking, were as follows:

Smoking	OC Use	Cases	Controls
No	Yes	4	52
	No	34	754
Yes	Yes	25	83
	No	171	853

1. We begin with the 2 × 2 table for *nonsmokers*:

Smoking	OC Use	Cases	Controls	Total
No	Yes	4 (*a*)	52 (*b*)	56
	No	34 (*c*)	754 (*d*)	788
	Total	38	806	844 (*n*)

$$\begin{aligned} a &= 4 \\ \frac{r_1 c_1}{n} &= \frac{(56)(38)}{844} \\ &= 2.52 \\ \frac{r_1 r_2 c_1 c_2}{n^2(n-1)} &= \frac{(56)(788)(38)(806)}{(844)^2(843)} \\ &= 2.25 \end{aligned}$$

2. For smokers,

Smoking	OC Use	Cases	Controls	Total
Yes	Yes	25 (*a*)	83 (*b*)	108
	No	171 (*c*)	853 (*d*)	1024
	Total	196	936	1132 (*n*)

$$\begin{aligned} a &= 25 \\ \frac{r_1 c_1}{n} &= \frac{(108)(196)}{1132} \\ &= 18.70 \\ \frac{r_1 r_2 c_1 c_2}{n^2(n-1)} &= \frac{(108)(1024)(196)(936)}{(1132)^2(1131)} \\ &= 14.00 \end{aligned}$$

The above results are combined to obtain the z-score:

$$z = \frac{(4-2.52)+(25-18.70)}{\sqrt{2.25+14.00}} = 1.93$$

and a z-score of 1.93 yields a one-sided p-value of 0.0268, which is beyond the 5% level. This result is stronger than those for tests at each level because it is based on more information where all data at all three smoking levels are used.

We have for nonsmokers

$$\frac{ad}{n} = \frac{(4)(754)}{844} = 3.57$$

$$\frac{bc}{n} = \frac{(52)(34)}{845} = 2.09$$

and for smokers,

$$\frac{ad}{n} = \frac{(25)(853)}{1132} = 18.84$$

$$\frac{bc}{n} = \frac{(83)(171)}{1132} = 12.54$$

The above results from the two levels of the confounder (smoking) are combined to estimate the common odds ratio:

$$OR_{\text{MH}} = \frac{3.57+18.84}{2.09+12.54} = 1.53$$

This combined estimate of the odds ratio, 1.53, represents an approximate increase of 53% in myocardial infarction risk for oral contraceptive users.

2.4 INFERENCES FOR GENERAL TWO-WAY TABLES

Perhaps the most frequent use of the chi-square distribution is to test the null hypothesis that two categorical variables are independent. Two variables are independent if the distribution of one is the same no matter what the level of the other. For example, if socioeconomic status and area of residence are independent, we would expect to find the same proportion of families in the low, medium, and high socioeconomic groups in all areas of the city.

2.4.1 Comparison of Several Proportions

This is a straight extension of the method for the comparison of two proportions, an analog of the one-way analysis of variance (ANOVA) procedure for binary data. The relationship between the following method and that in Section 2.3.1 is similar to the relationship between the one-way ANOVA F-test and the the two-sample t-test.

In this type of problem, we have k ($k = 2$) independent samples of binary data $\{(n_1, x_1), (n_2, x_2), \ldots, (n_k, x_k)\}$, where the n's are the k sample sizes and the x's are the numbers of positive outcomes in the k samples. With these k independent binomial samples, we consider the null hypothesis:

$$H_0\colon \ \pi_1 = \pi_2 = \cdots = \pi_k$$

expressing the equality of the k population proportions, the π's.

Let

$$p_i = \frac{x_i}{n_i}$$

be the sample proportion of group i ($1 = i = k$) and p be the "pooled proportion," an estimate of the common proportion under the null hypothesis, defined by

$$p = \frac{\sum x_i}{\sum n}$$

The test statistic is given by the following formula:

$$X^2 = \frac{\sum n_i(p_i - p)^2}{p(1-p)}$$

where the summation is over the k groups and the decision is made referring to the chi-square distribution with $(k - 1)$ degrees of freedom. If we apply this procedure to the comparison of two proportions, we would have the same result as if we were to use the method of Section 2.3.1; it can be verified that this chi-squared statistic is equal to the square of the z-score.

■ **Example 2.19** A study was undertaken to investigate the roles of blood-borne environmental exposures on ovarian cancer from assessment of consumption of coffee, tobacco, and alcohol (Whittemore et al., 1988). Study subjects consisted of 188 women in the San Francisco Bay area with epithelial ovarian cancers diagnosed in 1983–1985, and 539 control women. Of the 539 controls, 280 were hospitalized women without overt cancer, and 259 were chosen from the general population by random telephone dialing. Data for coffee consumption are summarized as follows:

Coffee Drinkers	Cases	Hospital Controls	Population Controls	Total
Yes	177	249	233	659
No	11	31	26	68
Total	188	280	259	727

In this example, we have:

$$k = 3$$
$$n_1 = 189, \quad x_1 = 177$$
$$p_1 = \frac{177}{189} = 0.937$$
$$n_2 = 280, \quad x_2 = 249$$
$$p_2 = \frac{249}{280} = 0.865$$
$$n_3 = 259, \quad x_3 = 233$$
$$p_3 = \frac{233}{259} = 0.900$$
$$p = \frac{659}{727} = 0.906$$

leading to

$$X^2 = \frac{189(0.937-0.906)^2 + 280(0.865-0.906)^2 + 259(0.900-0.906)^2}{(0.906)(1-0.906)}$$
$$= 0.662$$

The difference between the three groups is not significant; coffee consumption does not seem to be associated with epithelial ovarian cancer.

2.4.2 Testing for Independence in Two-Way Tables

Data in a $2 \times k$ table may come from different sampling models, including the case of independent binomial samples (called product binomial in previous section).

For example, we may have two multinomial samples (a case–control study with a multinomial risk factor) or one multinomial sample with $2k$ categories (a cross-sectional survey). If data come from a sampling model other than the product binomial, or if the contingency table has more than two rows and more than two columns, then the method of the previous section does not seem to apply because we simply do not have k proportions to compare. In these general cases of an $I \times J$ table, however, we can use the e_{ij} in tests of independence through Pearson's chi-squared statistic:

$$X^2 = \sum_{i,j} \frac{(x_{ij}-e_{ij})^2}{e_{ij}}$$

or the likelihood ratio chi-squared statistic:

$$G^2 = 2\sum_{i,j} x_{ij} \ln \frac{x_{ij}}{e_{ij}}$$

For large samples, both X^2 and G^2 have approximately a chi-squared distribution with the degrees of freedom under the null hypothesis of independence given by

$$\text{df} = (I-1)(J-1)$$

Greater values of the test statistics lead to a rejection of the null hypothesis. This is applicable regardless of the sampling model; in fact, as applied to the data coming from a product binomial, Pearson's chi-squared X^2 yields the same result as that of the previous section.

■ **Example 2.20** The following table shows the results of a survey. Each subject of a sample of 300 adults was asked to indicate which of three policies they favored with respect to smoking in public places (data were taken from Daniel, 1987). The numbers in parentheses are expected frequencies.

	Smoking Policy Favored				
Highest Education Level	No Restriction on Smoking	Smoking Allowed in Designated Areas Only	No Smoking at All	No Opinion	Total
College	5(8.75)	44(46)	23(13.25)	3(4.5)	75
High school	15(17.5)	100(92)	30(26.5)	5(9)	150
Grade school	15(8.75)	40(46)	10(13.25)	10(4.5)	75
Total	35	184	53	18	300

An application of Pearson's chi-squared test, at 6 degrees of freedom, yields

$$X^2 = \frac{(5-8.75)^2}{8.75} + \frac{(44-46)^2}{46} + \frac{(23-13.25)^2}{13.25} + \cdots + \frac{(10-4.5)^2}{4.5}$$
$$= 25.50$$

The result indicates a high correlation between education levels and preferences about smoking in public places ($p = 0.001$). Similar result can be obtained using the likelihood ratio chi-squared test ($G^2 = 20.60$, $p = 0.002$).

Note: A SAS program would include these instructions:

```
DATA;
INPUT EDUCAT POLICY COUNT;
CARDS;
1 1 5
1 2 44
1 3 23
1 4 3
2 1 15
2 2 100
2 3 30
2 4 5
3 1 15
3 2 40
3 3 10
3 4 10;
PROC FREQ;
WEIGHT COUNT;
TABLES EDUCAT*POLICY/CHISQ;
```

2.4.3 Ordered 2-by-*k* Contingency Tables

This section presents an efficient method for use with ordered $2 \times k$ contingency tables, tables with 2 rows and with k columns having a certain natural ordering. Let us first consider an example concerning the use of seat belts in automobiles. Each accident in this example is classified according to whether a seat belt was used and the severity of injuries received: none, minor, major, or death. The data were as follows:

	Extent of Injury Received			
Seat Belt	None	Minor	Major	Death
Yes	75	160	100	15
No	65	175	135	25
Total	130	335	235	40

To compare the extent of injury from those who used seat belts and those who did not, we can perform a chi-squared test as presented in the previous section. Such an application of the chi-squared test yields

$$X^2 = 9.26$$

with 3 degrees of freedom ($p = 0.026$). Therefore the difference between the two groups is significant at the 5% level but not at the 1% level. However, the usual chi-squared calculation takes no account of the fact that the extent of injury has a natural ordering: (none $<$ minor $<$ major $<$ death). In addition, the percentage of seat belt users in each injury group decreases from level "none" to level "death":

$$\text{None:}\quad \frac{75}{130} = 58\%$$

$$\text{Minor:}\quad \frac{160}{235} = 48\%$$

$$\text{Major:}\quad \frac{100}{235} = 43\%$$

$$\text{Death:}\quad \frac{15}{40} = 38\%$$

We now present a special procedure specifically designed to detect such a "trend" and will use the same example to show that it attains a higher degree of significance. In general, consider an ordered $2 \times k$ table with frequencies as follows:

	Column Level				
Row	1	2	...	k	Total
1	a_1	a_2	...	a_k	A
2	b_1	b_2	...	b_k	B
Total	n_1	n_2	...	n_k	N

The number of *concordances* (or concordant pairs) is calculated by

$$C = a_1 \sum_{i=2}^{k} b_i + a_2 \sum_{i=3}^{k} b_i + \cdots + a_{k-1} b_k$$

(The term "concordant" pair as used in the above example corresponds to a less severe injury for the seat belt user.) The number of *discordances* (or discordant pairs) is

$$B = b_1 \sum_{i=2}^{k} a_i + b_2 \sum_{i=3}^{k} a_i + \cdots + b_{k-1} a_k$$

In order to perform the test, we calculate the statistic:

$$S = C - D$$

then standardize it to obtain

$$z = \frac{S - E_0(S)}{\sqrt{\text{Var}(S)}}$$

where the mean and the variance of S under the null hypothesis are given by

$$\begin{aligned} E_0(S) &= 0 \\ \text{Var}_0(S) &= \left\{ \frac{AB}{3N(N-1)} \left[N^3 - n_1^3 - n_2^3 - \cdots - n_k^3 \right] \right\} \end{aligned}$$

The standardized z-score is distributed as standard normal if the null hypothesis is true; the null hypothesis is rejected at the 5% level if $z < -1.96$ or $z > 1.96$.

■ **Example 2.21** For the above study on the use of seat belts in automobiles, we have the following:

	Extent of Injury Received			
Seat Belt	None	Minor	Major	Death
Yes	75	160	100	15
No	65	175	135	25
Total	130	335	235	40

$$\begin{aligned} C &= 75(175 + 135 + 25) + 160(135 + 25) + (100)(25) \\ &= 53,225 \\ D &= 65(160 + 100 + 15) + 175(100 + 15) + (135)(15) \\ &= 40,025 \end{aligned}$$

Putting these values into the equations of the test statistic, we have

$$\begin{aligned} S &= 53,225 - 40,025 \\ &= 13,200 \end{aligned}$$

leading to

$$E_0(S) = 0$$

$$\text{Var}_0(S) = \left\{\frac{(350)(390)}{3(740)(739)}\left[740^3 - 130^3 - 335^3 - 235^3 - 40^3\right]\right\}$$

$$= (5414.76)^2$$

$$z = \frac{13,200}{5414.76}$$

$$= 2.44$$

which shows a higher degree of significance (one-sided $p = 0.0073$) than that of the chi-squared test with 3 degrees of freedom, which does not take into account ordering levels of injury.

The method seems ideal for the evaluation of ordinal risk factors in case–control studies. In this case, the statistic

$$\theta = C/D$$

serves as a generalized odds ratio measuring the effect of the exposure.

Example 2.22 Prematurity, which ranks as the major cause of neonatal morbidity and mortality, has traditionally been defined on the basis of a birth weight under 2500 g. But this definition encompasses two distinct types of infants: infants who are small because they are born early, and infants who are born at or near term but are small because their growth was retarded. "Prematurity" has now been replaced by (i) *low birth weight* to describe the second type and (ii) *preterm* to characterize the first type (babies born before 37 weeks of gestation).

A case–control study of the epidemiology of preterm delivery was undertaken at Yale–New Haven Hospital in Connecticut during 1977 (Berkowitz, 1981). The study population consisted of 175 mothers of singleton preterm infants and 303 mothers of singleton full-term infants. The following table gives the distribution of age of the mother.

Age	Cases	Controls	Total
14–17	15	16	31
18–19	22	25	47
20–24	47	62	109
25–29	56	122	178
30 and older	35	78	113
Total	175	303	478

We have

$$
\begin{aligned}
C &= 15(25+62+122+78)+22(62+122+78)+47(122+78)+(56)(78) \\
&= 20,911 \\
D &= 16(22+47+56+35)+25(47+56+35)+(62)(56+35)+(122)(35) \\
&= 15,922 \\
S &= 20,911-15,922 \\
&= 4989 \\
E_0(S) &= 0 \\
\mathrm{Var}_0(S) &= \left\{ \frac{(175)(303)}{3(478)(477)} \left[478^3-31^3-47^3-109^3-178^3-113^3\right] \right\} \\
&= (2794.02)^2 \\
z &= \frac{4,989}{2794.02} \\
&= 1.79
\end{aligned}
$$

which shows a significant association between the mother's age and preterm delivery (one-tailed $p = 0.0367$): the younger the mother, the more likely the preterm delivery.

Suppose we want to evaluate an ordinal risk factor in the presence of a confounder assuming that the confounder is not an effect modifier. The data can be presented in a series of ordered $2 \times k$ tables, one at each level of the confounder. For example, the data on the mother's age in Example 2.22 may be tabulated separately for the races, one ordered 2×5 table for whites and one for nonwhites. At a level of the confounder, we can apply the above method to obtain the statistic S and its standard error under the null hypothesis. The combined decision is then based on

$$
z = \frac{\sum \{S - E_0(S)\}}{\sum \mathrm{Var}_0(S)}
$$

where the summation is across levels of the confounder. The statistic z is distributed as standard normal if the null hypothesis is true; so that the null hypothesis is rejected at the 5% level if $z < -1.96$ or $z > 1.96$. This can be considered as a generalized Mantel–Haenszel procedure; it reduces to the Mantel–Haenszel test if the risk factor has two levels. The statistic

$$
\theta = \frac{\sum C/N}{\sum D/N}
$$

serves as a generalized odds ratio that reduces to OR_{MH} if the risk factor has two levels.

2.5 SAMPLE SIZE DETERMINATION

The determination of the size of a sample is a crucial element in the design of a survey or a clinical trial. In designing any study, one of the first questions that must be answered is: "How large must the sample be to accomplish the goals of the study?" Depending on the study goals, the planning of sample size can be approached in two different ways: either in terms of controlling the width of a desired confidence interval for the parameter of interest, or in terms of controlling the risk of making Type II errors.

Suppose the goal of another study is to estimate an unknown population proportion π; say, the smoking rate of a certain well-defined population. For the confidence interval to be useful, it must be short enough to pinpoint the value of the parameter reasonably well with a high degree of confidence. If a study is unplanned or poorly planned, there is a real possibility that the resulting confidence interval will be too long to be of any use to the researcher. In this case, we may determine to have a "margin of error" of the estimate not exceeding d. The 95% confidence interval for the population proportion π is given by

$$p \pm 1.96\sqrt{\frac{p(1-p)}{n}}$$

where p is the sample proportion. Therefore our goal is expressed as

$$1.96\sqrt{\frac{p(1-p)}{n}} \leq d$$

leading to the required minimum sample size:

$$n = \frac{(1.96)^2 p(1-p)}{d^2}$$

(rounded up to the next integer). This required sample size is also affected by three factors:

(i) The degree of confidence, that is, 95% which yields the coefficient, 1.96.
(ii) The maximum tolerated error, d, determined by the investigator(s).
(iii) The proportion p itself.

The third factor is unsettling! In order to find n so as to obtain an accurate value of the proportion, we need the proportion itself. There is no perfect, exact solution for this. Usually, we can use information from similar studies, past studies, or studies on similar populations. If no good prior knowledge about the proportion is available, we can replace $p(1-p)$ by 0.25 and use a "conservative" sample size:

$$n_{\max} = \frac{(1.96)^2(0.25)}{d^2}$$

because $n_{\max}$ is greater or equal to n regardless of the value of π.

Let us now consider the problem where we want to design a study to compare two proportions. For example, a new vaccine will be tested in which subjects are to be randomized into two groups of equal size: a control (e.g., unimmunized) group (group 1), and an experimental (e.g., immunized) group (group 2). Subjects, in both control and experimental groups, will be challenged by a certain type of bacteria and we wish to compare the infection rates. The null hypothesis to be tested is

$$H_0: \ \pi_1 = \pi_2$$

versus

$$H_A: \ \pi_1 > \pi_2$$

How large a total sample should be used to conduct this vaccine study?

Suppose that it is important to detect a reduction of infection rate

$$d = \pi_1 - \pi_2$$

If we decide to preset the size of the study at $\alpha = 0.05$ and the power $(1 - \beta)$ to detect the difference d, then the required sample size is given by the following formula:

$$N = \frac{4\lfloor 2z_{1-\alpha} + z_{1-\beta}\{\pi_1(1-\pi_1) + \pi_2(1-\pi_2)\}\rfloor}{(\pi_1 - \pi_2)^2}$$

In this formula the quantities $z_{1-\alpha}$ and $z_{1-\beta}$ are defined as in the previous section; π is the common value of the proportions under the null hypothesis. It is obvious that the problem of planning sample size is more difficult and a good solution requires a deeper knowledge of the scientific problem: some good idea of the magnitude of the proportions π_1, π_2 themselves.

■ **Example 2.23** A new vaccine will be tested in which subjects are to be randomized into two groups of equal size: a control (unimmunized) group and an experimental (immunized) group. Based on prior knowledge about the vaccine through small pilot studies, the following assumptions are made:

(i) The infection of the control group (when challenged by a certain type of bacteria) is expected to be about 50%, that is, $\pi_1 = 0.50$.

(ii) About 80% of the experimental group is expected to develop adequate antibodies (i.e., at least a twofold increase). If antibodies are inadequate, then the infection rate is about the same as for a control subject. But if an experimental subject has adequate antibodies, then the vaccine is expected to be about 85% effective (that corresponds to a 15% infection rate against the challenged bacteria). Putting these assumptions together, we obtain an expected value of the infection rate of the treated or immunized group:

$$\pi_2 = (0.80(0.15) + (0.20)(0.50)$$
$$= 0.22$$

Suppose also that we decide to preset $\alpha = 0.05$ and want the power $(1 - \beta)$ to be about 90% (i.e., $\beta = 0.10$). In other words, we use $z_{1-\alpha} = 1.96$ and $z_{1-\beta} = 1.65$. From this information, the required total sample size is

$$N = \frac{4[2(1.96)(0.50(0.50) + (1.65)\{(0.50(0.50) + (0.22)(0.78)\}]}{(0.50\text{–}0.22)^2}$$
$$\cong 144$$

so that each group will have 72 subjects. In this solution we use $\pi = 0.50$ because, under the null hypothesis, the vaccine is not effective (or is only as effective as the control) so that the common value of the infection rate is that of the control group, 50%.

EXERCISES

2.1 Self-reported injuries among left-handed and right-handed people were compared in a survey of 1896 college students in British Columbia, Canada. Ninety-three of the 180 left-handed students reported at least one injury and 619 of the 1716 right-handed students reported at least one injury in the same period. Test to compare the proportions of students with injury, left-handed versus right-handed; state clearly your null and alternative hypotheses, and your choice of the test size.

2.2 A study was conducted in order to evaluate the hypothesis that tea consumption and premenstrual syndrome are associated. One hundred and eighty-eight nursing students and 64 tea factory workers were given questionnaires. The prevalence of premenstrual syndrome was 39% among the nursing students and 77% among the tea factory workers. Test to compare the prevalences of premenstrual syndrome, tea factory workers versus nursing students; state clearly your null and alternative hypotheses, and your choice of the test size.

2.3 A study was conducted to investigate drinking problems among college students. In 1983, a group of students were asked whether they had ever driven an automobile while drinking. In 1987, after the legal drinking age was raised, a different group of college students were asked the same question. The results are as follows:

Drove While Drinking	Year of Survey		Total
	1983	1987	
Yes	1250	991	2241
No	1387	1666	3053
Total	2637	2657	5294

Test to compare the proportions of students with a drinking problem, 1987 versus 1983; state clearly your null and alternative hypotheses, and your choice of the test size.

2.4 In August 1976 tuberculosis was diagnosed in a high school student (Index case) in Corinth, Mississippi. Subsequently, laboratory studies revealed that the students disease was caused by drug-resistant tubercule bacilli. An epidemiologic investigation was conducted at the high school. The following table gives the rate of positive tuberculin reactions, determined for various groups of students according to degree of exposure to the index case:

Exposure Level	Number Tested	Number Positive
High	129	63
Low	325	36

Test to compare the proportions of students with infection, high exposure versus low exposure; state clearly your null and alternative hypotheses, and your choice of the test size.

2.5 Epidemic keratoconjunctivitis (EKC) or "shipyard eye" is an acute infectious disease of the eye. A case of EKC is defined as an illness:

(i) consisting of redness, tearing, and pain in one or both eyes for more than 3 days duration, and
(ii) having been diagnosed as EKC by an ophthalmologist.

In late October 1977, one of the two ophthalmologists (Physician A) providing the majority of specialized eye care to the residents of a central Georgia county (population 45,000) saw a 27-year-old nurse who had returned from a vacation in Korea with severe EKC. She received symptomatic therapy and was warned that her eye infection could spread to others; nevertheless, numerous cases of an illness similar to hers soon occurred in the patients and staff of the nursing home (Nursing Home A) where she worked (these individuals came to Physician A for diagnosis and treatment). The following table provides exposure history of 22 persons with EKC between October 27, 1977 and January 13, 1978 (when the outbreak stopped after proper control techniques were initiated). Nursing Home B, included in this table, is the only other area chronic-care facility.

Exposed Cohort	Number Exposed	Number Positive
Nursing Home A	64	16
Nursing Home B	238	6

Using an appropriate test, compare the proportions of cases from the two nursing homes. State clearly your null and alternative hypotheses, and your choice of the test size.

2.6 Consider the data taken from a study that attempts to determine whether the use of electronic fetal monitoring (EFM) during labor affects the frequency of cesarean section deliveries. Of the 5824 infants included in the study, 2850 were electronically monitored and 2974 were not. The outcomes are as follows:

	EFM Exposure		
Caesarean Delivery	Yes	No	Total
Yes	358	229	587
No	2492	2745	5237
Total	2850	2974	5824

Test to compare the rates of cesarean section delivery, EFM-exposed versus nonexposed; state clearly your null and alternative hypotheses, and your choice of the test size. Find a 95% confidence interval for the odds ratio.

2.7 A study was conducted to investigate the effectiveness of bicycle safety helmets in preventing head injury. The data consist of a random sample of 793 individuals who were involved in bicycle accidents during a 1-year period.

	Wearing Helmet		
Head Injury	Yes	No	Total
Yes	17	218	237
No	130	428	558
Total	147	646	793

Test to compare the proportions with head injury, those with helmets versus those without; state clearly your null and alternative hypotheses. Try again as a test of independence. Find a 95% confidence interval for the odds ratio.

2.8 A case–control study was conducted relating to the epidemiology of breast cancer and the possible involvement of dietary fats, along with other vitamins and nutrients. It included 2024 breast cancer cases who were admitted to Roswell Park Memorial Institute, Erie County, New York, from 1958 to 1965. A control group of 1463 was chosen from the patients having no neoplasms and no pathology of gastrointestinal or reproductive systems. The primary factors being investigated were vitamins A and E (measured in international units per month). The following are data for 1500 women over 54 years of age:

Vitamin A (IU/mo)	Cases	Controls
$< 150{,}500$	893	392
$> 150{,}500$	132	83
Total	1025	475

Test to compare the proportions of subjects who consumed less vitamin A ($\leq$ 150,500 IU/mo), cases versus controls; state clearly your null and alternative hypotheses, and your choice of the test size. Find a 95% confidence interval for the odds ratio.

2.9 Risk factors of gallstone disease were investigated in male self-defense officials who received, between October 1986 and December 1990, a retirement health examination at the Self-Defense Forces Fukuoka Hospital, Fukuoka, Japan. The following table provides parts of the data relating to three variables: smoking, drinking, and body mass index (BMI, which is dichotomized into below and above 22.5):

		Number of Men Surveyed	
Factor	Level	Total	Number with Gallstones
Smoking	Never	621	11
	Yes	2108	50
Alcohol	Never	447	11
	Yes	2292	50
BMI (kg/m^2)	Below 22.5	719	13
	Over 22.5	2020	48

For each of the three 2×2 tables, test to investigate the relationship between that criterion and gallstone disease. State clearly your null and alternative hypotheses, and your choice of the test size.

2.10 In 1979 the U.S. Veterans Administration conducted a health survey of 11,230 veterans. The advantages of this survey are that it includes a large random sample with a high interview response rate and it was done before the recent public controversy surrounding the issue of the health effects of possible exposure to Agent Orange. The following are data relating Vietnam service to eight post-traumatic stress disorder symptoms among the 1787 veterans who entered the military service between 1965 and 1975:

		Service in Vietnam	
Symptom	Level	Yes	No
Nightmares	Yes	197	85
	No	577	925
Sleep problems	Yes	173	160
	No	599	851
Troubled memories	Yes	220	105
	No	549	906
Depression	Yes	306	315
	No	465	699
Temper control problem	Yes	176	144
	No	595	868
Life goal association	Yes	231	225
	No	539	786
Omit feelings	Yes	188	191
	No	583	821
Confusion	Yes	163	148
	No	607	864

For each of the symptoms, compare the veterans who served in Vietnam and those who did not. State clearly your null and alternative hypotheses.

2.11 In a seroepidemiologic survey of health workers representing a spectrum of exposure to blood and patients with hepatitis B virus (HBV), it was found that infection increased as a function of contact. The following table provides data for hospital workers with uniform socioeconomic status at an urban teaching hospital in Boston, Massachusetts.

Personnel	Exposure	Not Tested	HBV Positive
Physicians	Frequent	81	17
	Infrequent	89	7
Nurses	Frequent	104	22
	Infrequent	126	11

Test to compare the proportions of HBV infection between the two levels of exposure (frequent exposure, infrequent exposure); perform separate comparisons for groups of physicians and nurses. State clearly your null and alternative hypotheses, and your choice of the test size. In each case, find a 95% confidence interval for the relative risk.

2.12 Postneonatal mortality due to respiratory illnesses is known to be inversely related to maternal age, but the role of young motherhood as a risk factor for respiratory morbidity in infants has not been thoroughly explored. A study was conducted in Tucson, Arizona, aimed at the incidence of lower respiratory tract illnesses during the first year of life. In this study, over 1200 infants were enrolled at birth between 1980 and 1984 and the following data are concerned with wheezing lower respiratory tract illnesses (wheezing LRI: No/Yes):

Maternal Age	Boys		Girls	
(years)	No	Yes	No	Yes
Below 30	277	113	296	85
Over 30	110	20	116	25

For each of the two groups, boys and girls, test to investigate the relationship between maternal age and respiratory illness. State clearly your null and alternative hypotheses, and your choice of the test size.

2.13 Data were collected from 2197 white ovarian cancer patients and 8893 white controls in 12 different U.S. case–control studies conducted by various investigators in the period 1956–1986. These were used to evaluate the relationship of invasive epithelial ovarian cancer to reproductive and menstrual characteristics, exogenous estrogen use, and prior pelvic surgeries. The following are parts of the data related to unprotected intercourse and to history of infertility:

Duration of Unprotected Intercourse	Cases	Controls
Less than 2 years	237	477
2 years and over	346	619

History of Infertility	Cases	Controls
No	526	966
Yes	96	135

For each of the two criteria, duration of unprotected intercourse and history of infertility, test to investigate the relationship between that criterion and ovarian cancer. State clearly your null and alternative hypotheses, and your choice of the test size.

2.14 A case–control study was conducted in Auckland, New Zealand, to investigate the effects of alcohol consumption on both nonfatal myocardial infarction and coronary death in the 24 hours after drinking, among regular drinkers. Data were tabulated separately for men and women.

For each group, men and women, and for each type of event, myocardial infarction and coronary death, test to compare cases versus controls. State, in each analysis, your null and alternative hypotheses, and your choice of the test size.

Data for Men

Drink in the	Myocardial Infarction		Coronary Death	
Last 24 hours	Controls	Cases	Controls	Cases
No	197	142	135	103
Yes	201	136	159	69

Data for Women

Drink in the	Myocardial Infarction		Coronary Death	
Last 24 hours	Controls	Cases	Controls	Cases
No	144	41	89	12
Yes	122	19	76	4

2.15 Adult male residents of 13 counties of western Washington State, in whom testicular cancer had been diagnosed during 1977–1983, were interviewed over the telephone regarding their history of genital tract conditions, including vasectomy. For comparison, the same interview was given to a sample of men selected from the population of these counties by dialing telephone numbers at random. The following data are tabulated by religious background:

Religion	Vasectomy	Cases	Controls
Protestant	Yes	24	56
	No	205	239
Catholic	Yes	10	6
	No	32	90
Others	Yes	18	39
	No	56	96

Compare the cases versus controls vasectomy for each religious group; state your null and alternative hypotheses. Is there any evidence of an effect modification?

2.16 Refer to the data in Exercise 2.15 but assume that gender (male/female) may be a confounder but not an effect modifier. For each type of event, myocardial infarction and coronary death, use the Mantel–Haenszel method to investigate the effects of alcohol consumption. State, in each analysis, your null and alternative hypotheses, and your choice of the test size.

2.17 Since incidence rates of most cancers rise with age, this must always be considered a confounder. The following are stratified data for an unmatched case–control study; the disease was esophageal cancer among men and the risk factor was alcohol consumption.

		Daily Alcohol Consumption	
Age	Sample	80+ g	0–79 g
25–44	Cases	5	5
	Controls	35	270
45–64	Cases	67	55
	Controls	56	277
65 and over	Cases	24	44
	Controls	18	129

Use the Mantel–Haenszel procedure to compare the cases versus the controls. State your null hypothesis and your choice of the test size.

2.18 Postmenopausal women who develop endometrial cancer are, on average, heavier than women who do not develop the disease. One possible explanation is that heavy women are more exposed to endogenous estrogens, which are produced in postmenopausal women by conversion of steroid precursors to active estrogens in peripheral fat. In the face of varying levels of endogenous estrogen production, one might ask whether the carcinogenic potential of exogenous estrogens would be the same in all women. A case–control study has been conducted to examine the relation between weight, replacement estrogen therapy, and endometrial cancer. Results are as follows:

		Estrogen Replacement	
Weight (kg)	Sample	Yes	No
<57	Cases	20	12
	Controls	61	183
57–75	Cases	37	45
	Controls	113	378
>75	Cases	9	42
	Controls	23	140

(a) Compare the cases versus the controls separately for the three weight groups. State your null hypothesis and your choice of the test size.

(b) Use the Mantel–Haenszel procedure to compare the cases versus the controls. State your null hypothesis and your choice of the test size.

2.19 It has been hypothesized that dietary fiber decreases the risk of colon cancer, while meats and fats are thought to increase this risk. A large study was undertaken to confirm these hypotheses. Fiber and fat consumptions are classified as "low" or "high" and data are tabulated separately for males and females as follows ("low" means below median):

	Males		Females	
Diet	Cases	Controls	Cases	Controls
Low fat, high fiber	27	38	23	39
Low fat, low fiber	64	78	82	81
High fat, high fiber	78	61	83	76
High fat, low fiber	36	28	35	27

For each group (males and females), investigate the relationship between the diets and the disease using data from the two 2×4 tables. State clearly your null and alternative hypotheses, and your choice of the test size.

2.20 Prematurity, which ranks as the major cause of neonatal morbidity and mortality, has traditionally been defined on the basis of a birth weight under 2500 g. But this definition encompasses two distinct types of infants: infants who are small because they are born early, and infants who are born at or near term but are small because their growth was retarded. "Prematurity" has now been replaced by (i) low birth weight to describe the second type and (ii) preterm to characterize the first type (babies born before 37 weeks of gestation).

A case–control study of the epidemiology of preterm delivery was undertaken at Yale–New Haven Hospital in Connecticut during 1977. The study population consisted of 175 mothers of singleton preterm infants and 303 mothers of singleton full-term infants. The following table gives the distribution of socioeconomic status:

Socioeconomic Level	Cases	Controls
Upper	11	40
Upper-middle	14	45
Middle	33	64
Lower-middle	59	91
Lower	53	58

(a) Using data in this 2×5 table, investigate the relationship between socioeconomic status and the disease. State your null hypothesis and your choice of the test size.

(b) Is it true, in general, that the poorer the mother the higher the risk? Resolve this problem by taking into account the ordering of the five socioeconomic levels.

3

LOGLINEAR MODELS

Topics in Chapter 2 focused mainly on the relationship between two categorical factors—a response and explanatory variables. However, incorrect conclusions may result from investigating variables two at a time. The Mantel–Haenszel method reaches a little further, allowing us to adjust for a confounder, and it has been a very popular methodology. It is a simple procedure to pool the results from a number of 2×2 tables; however, because of this simplicity, the scope of the Mantel–Haenszel method is rather narrow. Both main factors are binary, only the confounder can have several levels. It also assumes that the odds ratio between the two main factors remains constant across levels of the confounder; that is one assumes that there is no effect modification. The confounder does not modify the effect, say, of the explanatory factor on the response, an assumption that cannot be tested by that method alone (or any other methods in Chapter 2). In addition, the purpose of most research is to assess relationships among a set of several categorical variables, some of which may have more than two levels. To achieve this high level of sophistication, we now turn our attention to the *loglinear model*, a popular methodology for modeling multi-dimensional contingency tables.

Loglinear models describe association patterns among categorical variables without specifying their roles—which is the response and which are explanatory. However, when it is natural to treat one variable as the response and the others as explanatory or independent variables, modifications are simple and the resulting loglinear models are equivalent to the logistic models of Chapter 4. The following are two typical examples that will be used again and again as illustrations throughout this chapter, in addition to Example 1.1 or 2.1.

■ **Example 3.1** The data are from an epidemiologic study following an outbreak of food poisoning that occurred at an outing held for the personnel of an insurance company, taken and slightly modified from Bishop et al. (1975) (see also Korff et al., 1952). Of the food eaten, interest focused on potato salad and crabmeat. The variables are (i) presence or absence of illness, (ii) potato salad eaten or not eaten, and (iii) crabmeat eaten or not eaten.

	Food Eaten			
	Crabmeat Potato Salad		No Crabmeat Potato Salad	
Consumer's Illness	Yes	No	Yes	No
Ill	120	4	22	1
Not Ill	80	31	24	23

■ **Example 3.2** Although cervical cancer is not a major cause of death among American women, it has been suggested that virtually all such deaths are preventable; the basic approach to prevention involves cytologic screening using the Papanicolaou (Pap) test. Yet still many thousands of women die from cervical cancer each year! The data from the 1973 National Health Interview Survey (Kleinman and Kopstein, 1981) are used to examine the relationship between Pap testing and five socioeconomic variables.

				Pap Test	
Age	Residence	Income	Race	No	Yes
25–44	Metropolitan	Poor	White	77	516
			Black	57	344
		Nonpoor	White	476	5796
			Black	47	701
	Nonmetropolitan	Poor	White	63	387
			Black	41	94
		Nonpoor	White	211	3186
45–64	Metropolitan	Poor	White	163	376
			Black	86	172
		Nonpoor	White	859	5646
			Black	121	400

(*Continued*)

Age	Residence	Income	Race	Pap Test	
				No	Yes
45–64	Nonmetropolitan	Poor	White	160	328
			Black	59	68
		Nonpoor	White	491	2361
			Black	44	72
65+	Metropolitan	Poor	White	449	499
			Black	70	86
		Nonpoor	White	903	1544
			Black	63	70
	Nonmetropolitan	Poor	White	396	365
			Black	71	43
		Nonpoor	White	459	698
			Black	21	10

The five variables are (i) age (25–44, 45–64, and 65+), (ii) residence (metropolitan or nonmetropolitan), (iii) income (poor or nonpoor), (iv) race (white or black; data for other ethnic groups are not included), and (v) Pap testing (yes/no).

3.1 LOGLINEAR MODELS FOR TWO-WAY TABLES

Even though we know how to analyze data in two-way tables using the methods of Chapter 2, the aim of this section is to introduce loglinear models for this simple case of two categorical variables. The loglinear model for a two-way table will then be generalized to the case of three-dimensional tables in Section 4.2, which also includes the process of using sample data to fit the models and make inferences. Suppose there is a multinomial sample of size n over the $N = IJ$ cells of an $I \times J$ contingency table; the first factor has I levels represented by rows and the second factor has J levels represented by columns. The cell probabilities π_{ij} for that multinomial distribution form the joint distribution of the two categorical variables. If we define

$$l_{ij} = \ln \pi_{ij}$$

then we can write

$$l_{ij} = \lambda + \lambda_{1(i)} + \lambda_{2(j)} + \lambda_{12(ij)}$$

by an analog with the analysis of variance (ANOVA) model. For this formulation the following condition apply.

(i) The first term, λ, is the grand mean of the logs of the probabilities,

$$\lambda = \frac{l_{++}}{IJ}$$

where the plus sign $(+)$ subscript denotes the total when summing across levels of the corresponding factor.

(ii) $[\lambda + \lambda_{1(i)}]$ is the mean of the logs of the probabilities of the first factor when it is at level i,

$$\frac{\sum_j \ln \pi_{ij}}{J}$$

so that $\lambda_{1(i)}$ is the deviation from the grand mean λ:

$$\lambda_{1(i)} = \frac{l_{i+}}{J} - \frac{l_{++}}{IJ}$$

Therefore these I terms satisfy

$$\sum_i \lambda_{1(i)} = 0$$

and are influenced only by the marginal distribution of the first factor. There are $(I-1)$ of these terms and they are often of no intrinsic interest, representing only the main effects of that factor. Similarly, the main effects of the other factor are represented by $\lambda_{2(j)}$, and there are $(J-1)$ of these terms satisfying

$$\sum_j \lambda_{2(j)} = 0$$

(iii) The remaining component,

$$\lambda_{12(ij)} = l_{ij} - \frac{l_{i+}}{J} - \frac{l_{+j}}{I} + \frac{l_{++}}{IJ}$$

can be regarded as measures of departures from the independence of two factors. For example, if $I = J = 2$, then it can be shown that

$$\begin{aligned}\lambda_{12(11)} &= \frac{1}{4}\ln\left(\frac{\pi_{11}\pi_{22}}{\pi_{12}\pi_{21}}\right)\\ &= \tfrac{1}{4}\ln(\text{Odds ratio})\\ &= 0 \Leftrightarrow (\text{Odds ratio} = 0)\end{aligned}$$

Since

$$\sum_{i} \lambda_{12(ij)} = \sum_{j} \lambda_{12(ij)} = 0$$

there are $(I-1)(J-1)$ of these "interaction terms"; the number is called the *degrees of freedom* (df) in testing for the null hypothesis of independence.

In the case of a general two-way table, numerical values of these $\lambda_{12(ij)}$ would indicate where the interaction is strong; the negative or positive sign is not important, only reflecting the arbitrary coding. Of course, the results may be trivial because we can reach the same conclusion by simply inspecting the cell probabilities. However, it is extremely useful after we generalize the models for use with higher dimensional tables.

3.2 LOGLINEAR MODELS FOR THREE-WAY TABLES

In a typical study even if we are interested only in the relationship between a response and an explanatory variable, we still have to control for at least one confounder that can influence the relationship under investigation. Therefore we end up studying at least three factors simultaneously; and when dealing with three factors, the relationships among them are far more complicated than the case of two factors. For example, we may want to see if:

(i) the three factors are mutually independent; that is, whether

$$\Pr(X_1 = i, X_2 = j, X_3 = k) = \Pr(X_1 = i)\Pr(X_2 = j)\Pr(X_3 = k)$$

or if

(ii) one factor, say, X_3, is jointly independent of the other two factors; that is, whether

$$\Pr(X_1 = i, X_2 = j, X_3 = k) = \Pr(X_1 = i, X_2 = j)\Pr(X_3 = k)$$

or if

(iii) two factors, say, X_1 and X_2, are conditionally independent given the third factor; that is, whether

$$\Pr(X_1 = i, X_2 = j | X_3 = k) = \Pr(X_1 = i | X_3 = k)\Pr(X_2 = j | X_3 = k)$$

The last form—the concept of conditional independence—is very important; it is often the major aim of an epidemiological study. For example, in Example 1.1 of Chapter 1, a case–control study was undertaken to identify reasons for the exceptionally high rate of lung cancer among male residents of coastal Georgia. The exposure under investigation—shipbuilding—refers to employment in shipyards during World War II; smoking is only a confounder that we want to control for. Specifically, we want to know whether or not shipbuilding and lung cancer are related (i) among smokers and (ii) among nonsmokers. The underlying question concerns conditional independence. For Example 3.1 of this chapter, we want to know if eating crabmeat and food poisoning are related considering separately the people who ate potato salad and the people who did not eat potato salad. Again, it is a question of conditional independence. It is interesting and important to note that it is not the same as the marginal independence, what we learn when investigating two factors at a time. For example, consider the following data for a two-center trial to compare two treatments:

		Treatment	
Center	Outcome	A	B
X	Good	162	80
	Bad	38	20
Y	Good	11	21
	Bad	89	179

We can see that Treatment and Outcome are conditionally independent (odds ratios of 1.07 and 1.05) but not marginally independent (odds ratio of 2.68). This example appears to fit Simpson's paradox (see Chapter 1), where the marginal strong association between Outcome and Treatment (with Treatment B showing more effective results) is in a different direction from their weak partial association (at each center, Treatment A is slightly more effective).

With the *loglinear* approach, we model cell probabilities or, equivalently, cell counts or frequencies in a contingency table in terms of association among the variables. Suppose there is a multinomial sample of size n over the $N = IJK$ cells of an $I \times J \times K$ contingency table (I, J, and K are the number of categories for the three factors involved). Then the loglinear model for the two-way tables of the previous section can be generalized and expressed for three-way tables as follows:

$$l_{ijk} = \lambda + \lambda_{1(i)} + \lambda_{2(j)} + \lambda_{3(k)} + \lambda_{12(ij)} + \lambda_{13(ik)} + \lambda_{23(jk)} + \lambda_{123(ijk)}$$

subjected to similar constraints (i.e., summing across indices to zero):

$$\sum_i \lambda_{1(i)} = \sum_j \lambda_{2(j)} = \sum_k \lambda_{3(k)} = 0$$

$$\sum_i \lambda_{12(ij)} = \sum_j \lambda_{12(ij)} = 0$$

$$\sum_i \lambda_{13(ik)} = \sum_k \lambda_{13(ik)} = 0$$
$$\sum_j \lambda_{23(jk)} = \sum_k \lambda_{23(jk)} = 0$$
$$\sum_i \lambda_{123(ijk)} = \sum_j \lambda_{123(ijk)} = \sum_k \lambda_{123(ijk)} = 0$$

The (full or saturated) model decomposes the log of cell probability π_{ijk} (or equivalently log of cell frequency, m_{ijk}, because the expected cell frequency can be expressed as $m_{ijk} = n\pi_{ijk}$) into the following:

(i) A constant λ.

(ii) Terms representing main effects, $\lambda_{1(i)}$, $\lambda_{2(j)}$, and $\lambda_{3(k)}$. There are $(I-1)$, $(J-1)$, and $(K-1)$ of these main effects terms for the three factors; they are only influenced by the marginal distributions of the three factors and are therefore often of no intrinsic interest.

(iii) Terms representing two-factor interactions, $\lambda_{12(ij)}$, $\lambda_{13(ik)}$, and $\lambda_{23(jk)}$; there are $(I-1)(J-1)$, $(I-1)(J-1)$, and $(J-1)(K-1)$ of these.

(iv) Terms used as measures of three-factor interaction, $\lambda_{123(ijk)}$. There are $(I-1)(J-1)(K-1)$ of these three-factor interaction terms.

If the terms in the last group are not zero (i.e., three-factor interaction is present), the presence or absence of a factor would modify the relationship between the other two factors. For example, in the context of Example 1.1 of Chapter 1, if the three-factor interaction is present, the odds ratio relating lung cancer to shipbuilding calculated from nonsmokers would be different from the odds ratio relating lung cancer to shipbuilding calculated from smokers. Similarly, the relationship between lung cancer and smoking varies from one level of shipbuilding to another. The Mantel--Haenszel method of Chapter 2 assumes that these three-factor interaction terms are absent.

3.2.1 The Models of Independence

Starting with the saturated model, we can translate a null hypothesis (or a concept of independence) into a loglinear model by setting certain groups of the above λ-terms equal to zero.

The various concepts of independence for three-way tables can be grouped as follows:

Mutual Independence:

$$(X_1, X_2, X_3) \Leftrightarrow l_{ijk} = \lambda + \lambda_{1(i)} + \lambda_{2(j)} + \lambda_{3(k)}$$

Joint Independence of Two Factors:

$$(X_1, X_2X_3) \Leftrightarrow l_{ijk} = \lambda + \lambda_{1(i)} + \lambda_{2(j)} + \lambda_{3(k)} + \lambda_{23(jk)}$$
$$(X_2, X_1X_3) \Leftrightarrow l_{ijk} = \lambda + \lambda_{1(i)} + \lambda_{2(j)} + \lambda_{3(k)} + \lambda_{13(ik)}$$
$$(X_3, X_1X_2) \Leftrightarrow l_{ijk} = \lambda + \lambda_{1(i)} + \lambda_{2(j)} + \lambda_{3(k)} + \lambda_{12(ij)}$$

Conditional Independence:

$$(X_1X_3, X_2X_3) \Leftrightarrow l_{ijk} = \lambda + \lambda_{1(i)} + \lambda_{2(j)} + \lambda_{3(k)} + \lambda_{13(ik)} + \lambda_{23(jk)}$$
$$(X_1X_2, X_2X_3) \Leftrightarrow l_{ijk} = \lambda + \lambda_{1(i)} + \lambda_{2(j)} + \lambda_{3(k)} + \lambda_{12(ij)} + \lambda_{23(jk)}$$
$$(X_1X_2, X_1X_3) \Leftrightarrow l_{ijk} = \lambda + \lambda_{1(i)} + \lambda_{2(j)} + \lambda_{3(k)} + \lambda_{12(ij)} + \lambda_{12(ij)}$$

No Three-Factor Interaction:

$$(X_1X_2, X_1X_3, X_2X_3) \Leftrightarrow l_{ijk} = \lambda + \lambda_{1(i)} + \lambda_{2(j)} + \lambda_{3(k)} + \lambda_{12(ij)} + \lambda_{13(ik)} + \lambda_{23(jk)}$$

For example, referring to the lung cancer data of Example 1.1, the null hypothesis that lung cancer (L) and shipbuilding (B) are conditionally independent given smoking (S) is expressed as

$$H_0\colon l_{ijk} = \lambda + \lambda_{L(i)} + \lambda_{B(j)} + \lambda_{S(k)} + \lambda_{LS(ij)} + \lambda_{BS(jk)}$$

or, by dropping the subscripts, it could be abbreviated simply as

$$H_0\colon l = \lambda + \lambda_L + \lambda_B + \lambda_S + \lambda_{LS} + \lambda_{BS}$$

As for the food poisoning data of Example 3.1, the null hypothesis that both crabmeat (C) and potato salad (P) have nothing to do with the illness (I) is expressed as

$$H_0\colon l = \lambda + \lambda_I + \lambda_C + \lambda_P + \lambda_{CP}$$

3.2.2 Relationships Between Terms and Hierarchy of Models

Loglinear model term λ_A is a lower order relative of term λ_B if set A is a subset of set B. For example,

(i) λ_2 is a lower order relative of λ_{23} because $\{2\}$ is a subset of $\{2, 3\}$.
(ii) λ_{12} is a lower order relative of λ_{123} because $\{1, 2\}$ is a subset of $\{1, 2, 3\}$.

If term λ_A is a lower order relative of term λ_B, then term λ_B is a higher order relative of term λ_A. For example,

(i) λ_{23} is a higher order relative of λ_2.
(ii) λ_{123} is a higher order relative of λ_{12} because $\{1, 2\}$ is a subset of $\{1, 2, 3\}$.

A loglinear model is a *hierarchical model* under the following conditions:

(i) If a λ-term is zero, then all of its higher order relatives are zero.
(ii) If a λ-term is not zero, then all of its lower order relatives are not zero.

For example, all eight models of independence for the three-way tables of the previous section are hierarchical. However, we have not considered all possible variants of our loglinear models. For example, we have not considered the model

$$l_{ijk} = \lambda + \lambda_{1(i)} + \lambda_{2(j)} + \lambda_{3(k)} + \lambda_{123(ijk)}$$

partly because it does not have any meaningful interpretation, partly because it is a nonhierarchical model. Statistical methodology (e.g., the estimation of expected frequencies) and corresponding computer software are currently valid or available only for hierarchical models. Therefore we will consider only hierarchical models for higher dimensional tables in the remaining parts of this chapter.

3.2.3 Testing a Specific Model

Given the data in a three-way table, we have two different types of statistical inferences:

(i) To test for a specific model; for example, we may want to know whether two specified factors are conditionally independent given the third factor.
(ii) To search for a model that can best explain the relationship(s) found in the observed data.

This section deals with the first issue, where the loglinear model under investigation is the result of a process starting with a research question which leads to a relationship between factors, and finally a null hypothesis.

Expected Frequencies

Expected frequencies are cell counts obtained under the null hypothesis. Given the null hypothesis, these expected frequencies (only for hierarchical models) can be easily estimated. For example, if we consider the model of conditional independence between X_1 and X_2, that is, the model we denoted by (X_1X_3, X_2X_3), or expressed in terms of probabilities as

$$\Pr(X_1 = i, X_2 = j|X_3 = k) = \Pr(X_1 = i|X_3 = k)\Pr(X_2 = j|X_3 = k)$$

then it can be shown that

$$e_{ijk} = \frac{x_{i+k}x_{+jk}}{x_{++k}}$$

where the x's are observed cell counts and the plus sign subscript indicates a summation across the index. However, in practice, these tedious jobs should be left to packaged computer programs, such as SAS; we will show some samples of computer programs in subsequent examples.

Test Statistic

When measuring goodness-of-fit in contingency tables, we often rely on Pearson's chi-squared statistic:

$$X^2 = \sum_{i,j,k} \frac{(x_{ijk} - e_{ijk})^2}{e_{ijk}}$$

However, for a technical reason, the so-called "partition of chi-squares" as seen later, we compare the observed frequencies and the expected frequencies in higher dimensional tables using the likelihood ratio chi-squared statistic:

$$G^2 = 2\sum_{i,j,k} x_{ijk} \ln \frac{x_{ijk}}{e_{ijk}}$$

Degree of Freedom

The degree of freedom for the above likelihood ratio chi-squared statistic is equal to the number of λ-terms that are set equal to zero in the model being tested. For example, if we want to test the model of conditional independence between X_1 and X_2, that is, the model we denoted by (X_1X_3, X_2X_3), since we set all λ_{12} and all λ_{123} terms equal to zero, the degree of freedom is

$$df = (I-1)(J-1) + (I-1)(J-1)(K-1)$$

Similarly, if we want to test for the model of no three-factor interaction, since we set all λ_{12}, λ_{13}, λ_{23}, and λ_{123} terms equal to zero, the degree of freedom is

$$df = (I-1)(J-1)(K-1)$$

(this last model is needed in testing the assumption of the Mantel–Haenszel method).

■ **Example 3.3** Refer to the lung cancer data of Example 1.1. A case–control study was undertaken to identify reasons for the exceptionally high rate of lung cancer among male residents of coastal Georgia (Blot et al., 1978). Cases (of lung cancer) were identified from these sources:

(i) Diagnoses since 1970 at the single large hospital in Brunswick
(ii) Diagnoses during 1975 and 1976 at three major hospitals in Savannah
(iii) Death certificates for the period 1970–1974 in the area.

Controls (or control subjects) were selected from admissions to the four hospitals and from death certificates in the same period for diagnoses other than lung cancer, bladder cancer, or chronic lung cancer. Data are tabulated separately for smokers and nonsmokers as follows:

Smoking	Shipbuilding	Cases	Controls
No	Yes	11	35
	No	50	203
Yes	Yes	84	45
	No	313	270

The exposure under investigation—shipbuilding—refers to employment in shipyards during World War II. Let us suppose we are interested in the null hypothesis that lung cancer (L) and shipbuilding (B) are conditionally independent (smoking is only a confounder):

$$H_0\colon\ l = \lambda + \lambda_{\mathrm{L}} + \lambda_{\mathrm{B}} + \lambda_{\mathrm{S}} + \lambda_{\mathrm{LS}} + \lambda_{\mathrm{BS}}$$

Fitting the above model, we have the result

$$G^2 = 6.09 \text{ with 2 degress of freedom; } p\text{-value} = 0.0477 < 0.05$$

indicating that those two factors are related: workers *are* more likely to have lung cancer.

Note: A SAS program would include these instructions:

```
INPUT CANCER SHIP SMOKING COUNT;
CARDS;
1 1 1 203
1 1 2 270
1 2 1 35
1 2 2 45
2 1 1 50
2 1 2 313
2 2 1 11
2 2 2 84:
PROC CATMOD:
WEIGHT COUNT;
MODEL CANCER*SHIP*SMOKING = RESPONSE/NODESIGN;
LOGLIN CANCER SHIP SMOKING CANCER*SMOKING SHIP*SMOKING;
```

■ **Example 3.4** Refer to the food poisoning data of Example 3.1; data are from an epidemiologic study following an outbreak of food poisoning that occurred at an outing held for the personnel of an insurance company, taken and slightly modified from Bishop et al. (1975) (also see Korff et al., 1952). Of the food eaten, interest focused on potato salad and crabmeat. The variables are (i) presence or absence of illness, (ii) potato salad eaten or not eaten, and (iii) crabmeat eaten or not eaten.

	Food Eaten			
	Crabmeat Potato Salad		No Crabmeat Potato Salad	
Consumer's Illness	Yes	No	Yes	No
Ill	120	4	22	1
Not Ill	80	31	24	23

Let us suppose we are interested in the null hypothesis that both crabmeat (C) and potato salad (P) have nothing to do with the illness (I):

$$H_0\colon\ l = \lambda + \lambda_{\mathrm{I}} + \lambda_{\mathrm{C}} + \lambda_{\mathrm{P}} + \lambda_{\mathrm{CP}}$$

Fitting the above model, we have a highly significant result:

$$G^2 = 62.05 \text{ with 23 degrees of freedom;}\quad p\text{-value} < 0.0001$$

indicating that at least one of the two items, maybe both, are related to the food poisoning outbreak.

Measures of Association

Consider the case of three categorical variables X_1, X_2, and X_3 with data presented in the form of a three-way table. Let X_1 be the response variable with two levels, $I=2$ (say, 1 if event of interest is observed, e.g., being a case; 2 if event of interest is not observed, e.g., being a control), X_2 be the explanatory factor (maybe with more than two categories), and X_3 be the confounder. Suppose we are interested in comparing the effect of X_2 (on the response X_1) when X_2 is at level i to the effect of the same factor when X_2 is at level 1 (chosen to be the baseline level). For example, in the context of Example 1.1, we want to compare the odds of having lung cancer (X_1) for those employed in the shipbuilding industry (level i of X_2) to the odds of having lung cancer (X_1) for those not employed in the shipbuilding industry (level 1 of X_2). Of course, we must control for smoking (X_3).

Let us assume that X_3 is at level k. The above odds ratio is given by the formula

$$\log OR_{ivs1} = 2(\lambda_{12(1i)} - \lambda_{12(11)}) + 2(\lambda_{123(1ik)} - \lambda_{123(11k)})$$

In the special case of no effect modification, there is no three-term interaction; the above formula is simplified to

$$\log OR_{ivs1} = 2(\lambda_{12(1i)} - \lambda_{12(11)})$$

In addition, if the explanatory factor has only two levels (1 if exposed and 2 if not exposed), then we have the odds ratio associated with the exposure:

$$\log OR = 4\lambda_{12(11)}$$

because of the constraint for the model,

$$\lambda_{12(1i)} + \lambda_{12(11)}) = 0$$

■ **Example 3.5** Refer to the food poisoning data of Example 3.1 (or 3.4) and suppose that we are interested in the effect of potato salad (eaten vs. not eaten) on the illness. Consider testing the model

$$H_0\colon l = \lambda + \lambda_{\mathrm{I}} + \lambda_{\mathrm{C}} + \lambda_{\mathrm{P}} + \lambda_{\mathrm{CP}} + \lambda_{\mathrm{IP}}$$

This is the model of conditional independence between crabmeat (C) and illness (I), and the result,

$$G^2 = 3.30 \text{ with 2 degrees of freedom; } p\text{-value} = 0.192$$

indicates a good fit (i.e., the relationship between crabmeat and illness—conditional on potato salad—is not statistically significant). In addition, we obtained from fitting this model

$$\lambda_{\mathrm{IP}(11)} = 0.6727$$

leading to an odds ratio, representing the strength of the relationship between potato salad and illness, of

$$\begin{aligned} OR &= \exp\{(4)(.6727)\} \\ &= 14.74 \end{aligned}$$

This result means that the odds of having the illness for those who ate potato salad are about 14.74 times the odds of having the illness for those who did not eat potato salad. It can be seen that the above result is quite similar to that obtained using the Mantel–Haenszel method. An application of the Mantel–Haenszel procedure would proceed as follows.

(i) From the data for people who ate crabmeat,

Consumer's Illness	Potato Salad		Total
	Yes	No	
Ill	120	4	124
Not Ill	80	31	111
Total	200	35	235

we have

$$\frac{ad}{n} = \frac{(120)(31)}{235}$$
$$\frac{bc}{n} = \frac{(80)(4)}{235}$$

(ii) From the data for people who did not eat crabmeat,

Consumer's Illness	Potato Salad		Total
	Yes	No	
Ill	22	1	23
Not Ill	24	23	47
Total	46	24	70

we have

$$\frac{ad}{n} = \frac{(22)(23)}{70}$$
$$\frac{bc}{n} = \frac{(24)(1)}{70}$$

Therefore

$$OR_{\mathrm{MH}} = \frac{\frac{(120)(31)}{235} + \frac{(22)(23)}{70}}{\frac{(80)(4)}{235} + \frac{(24)(1)}{70}}$$
$$= 13.53$$

■ **Example 3.6** The following table provides data relating infant mortality to the duration of gestation and the smoking status of the mother (Wermuth, 1976):

	Gestation (G; days)			
	260 or Less Smoking (S)		More than 260 Smoking (S)	
Infant Death (D)	Yes	No	Yes	No
Yes	13	91	7	1
No	11	462	583	5606

Suppose we are interested in the effect of premature delivery on infant mortality after controlling for the mother's smoking status. The model of conditional independence to be tested is

$$H_0\colon\ l = \lambda + \lambda_{\mathrm{G}} + \lambda_{\mathrm{S}} + \lambda_{\mathrm{D}} + \lambda_{\mathrm{GS}} + \lambda_{\mathrm{DS}}$$

This is the model of conditional independence between crabmeat and illness, and the result,

$$G^2 = 2.26 \text{ with 2 degrees of freedom; } p\text{-value} = 0.3231$$

indicates a good fit (i.e., the relationship between premature delivery (G) and infant mortality (D), after controlling for the mother's smoking status, is not statistically significant). In addition, we obtained from fitting this model that

$$\lambda_{\mathrm{DS}(11)} = 0.8320$$

leading to an odds ratio, representing the strength of the relationship between premature delivery (G) and infant mortality (D), of

$$\begin{aligned} OR &= \exp\{(4)(0.8320)\} \\ &= 27.88 \end{aligned}$$

This result means that the odds of having an infant death among premature deliveries are about 27.88 times the odds of having an infant death among full-term deliveries. Again, it can seen that the above result is quite similar to that obtained using the Mantel–Haenszel method. An application of the Mantel–Haenszel procedure would proceed as follows.

(i) From the data for the smokers,

Infant Death (D)	Gestation (G) Less than 260 days	More than 260 days	Total
Yes	13	7	20
No	11	583	594
Total	24	590	614

we have

$$\frac{ad}{n} = \frac{(13)(583)}{614}$$
$$\frac{bc}{n} = \frac{(11)(7)}{614}$$

(ii) From the data for nonsmokers,

Infant Death (D)	Gestation (G) Less than 260 days	More than 260 days	Total
Yes	91	38	129
No	462	5606	6068
Total	553	5644	6197

we have

$$\frac{ad}{n} = \frac{(91)(5606)}{6197}$$
$$\frac{bc}{n} = \frac{(462)(38)}{6197}$$

Therefore

$$OR_{\text{MH}} = \frac{\frac{(13)(583)}{614} + \frac{(91)(5606)}{6197}}{\frac{(11)(7)}{614} + \frac{(462)(38)}{6197}}$$
$$= 31.99$$

We should also note that the result on the odds ratio depends on the fitted model. It is conventional to use the smallest good-fit model that contains the interaction term under investigation. For example, the above conditional independence model contains an insignificant term, the interaction between gestation (G) and smoking ($p = 0.4640$). Hence we prefer to estimate the above odds ratio measuring the effect of premature delivery on infant mortality by fitting the smaller model:

$$H_0\colon\ l = \lambda + \lambda_{\mathrm{G}} + \lambda_{\mathrm{S}} + \lambda_{\mathrm{D}} + \lambda_{\mathrm{DS}}$$

even though the results may be very close.

3.2.4 Searching for the Best Model

As mentioned earlier, given the data in a three-way table, we have two different types of statistical inferences:

(i) To test for a specific model; for example, we may want to know whether two specified factors are conditionally independent given the third factor.
(ii) To search for a model that can best explain the relationship(s) found in the observed data.

Section 3.2.3 deals with the first issue of fitting and testing a specific model, which is the result of a process starting with a research question that leads to a relationship between factors, and finally a null hypothesis. This section covers the second issue, where we do not have a specific research question nor a specific model under investigation; the aim here is to search for a model that can best explain the relationships found or observed in a data set.

A loglinear model H_2 is nested in a model H_1 if every nonzero λ-term in H_2 is also contained in H_1 and is denoted by $H_2 < H_1$. For example, if we have

$$\begin{aligned} H_1 &= \lambda + \lambda_1 + \lambda_2 + \lambda_3 + \lambda_{12} + \lambda_{13} + \lambda_{23} \\ H_2 &= \lambda + \lambda_1 + \lambda_2 + \lambda_3 + \lambda_{13} + \lambda_{23} \\ H_3 &= \lambda + \lambda_1 + \lambda_2 + \lambda_3 + \lambda_{23} \end{aligned}$$

then we can say that H_3 is nested in H_2 ($H_3 < H_2$) and H_2 is nested in H_1 ($H_2 < H_1$). In this nested hierarchy of models, it can be shown that if $H_3 < H_2$, then the likelihood ratio chi-squared statistic satisfies the reversed inequality; that is, the larger model has smaller chi-squared statistic value,

$$G_2^2 < G_3^2$$

(One of the reasons we do not often use Pearson's goodness-of-fit with loglinear models is that this property does not necessarily hold for every set of nested models.)

Furthermore, it can be shown that if a model H_2 is nested in H_1 ($H_2 < H_1$), then

$$G^2_{2|1} = G^2_2 - G^2_1$$

is distributed as chi-squared with

$$df_{2|1} = df_2 - df_1$$

(This also does not necessarily hold for Pearson's chi-squared.)

This result has at least two important applications, the second of which provides the necessary framework for selecting the best model.

Application 1 If

(i) a model H_1 fits (i.e., $p \gg 0.05$), and
(ii) $\{H_1 - H_2\}$ consists of only one term, say, λ_θ, we can use $G^2_{2|1}$ as a test statistic for

$$H_0\text{: } \lambda_\theta = 0$$

■ **Example 3.7** Refer to the food poisoning data of Example 3.1 (or 3.4) and suppose we are interested only in the effect of potato salad (eaten vs. not eaten) on the illness, that is, testing against

$$H_0\text{: } \lambda_{\mathrm{IP}} = 0$$

We found the following:

(i) The model of no three-term interaction,

$$H_1 = \lambda + \lambda_{\mathrm{I}} + \lambda_{\mathrm{C}} + \lambda_{\mathrm{P}} + \lambda_{\mathrm{IC}} + \lambda_{\mathrm{IP}} + \lambda_{\mathrm{CP}}$$

fits ($G^2 = 0.27$, 1 degree of freedom).

(ii) Let H_2 be the model of conditional independence between illness and potato salad:

$$H_2 = \lambda + \lambda_{\mathrm{I}} + \lambda_{\mathrm{C}} + \lambda_{\mathrm{P}} + \lambda_{\mathrm{IC}} + \lambda_{\mathrm{CP}}$$

which is only one term smaller than H_1, the term under investigation (λ_{IP}). By fitting H_2, we obtained $G^2 = 47.62$ with 2 degrees of freedom. The test for

$$H_0\text{: } \lambda_{\mathrm{IP}} = 0$$

corresponds to $G^2 = 47.62 - 0.27 = 47.35$ with 1 degree of freedom, which is highly significant.

It can seen that the above result is similar to that obtained using the Mantel–Haenszel method. An application of the Mantel–Haenszel procedure would proceed as follows.

(i) From the data for people who ate crabmeat,

Consumer's Illness	Potato Salad Yes	No	Total
Ill	120	4	124
Not Ill	80	31	111
Total	200	35	235

we have for the top left cell

$$\text{Observed value} = 120$$

$$\text{Mean} = \frac{(200)(124)}{235} = 105.5$$

$$\text{Variance} = \frac{(200)(35)(124)(111)}{(235)^2(235-1)} = 7.46$$

(ii) From the data for people who did not eat crabmeat,

Consumer's Illness	Potato Salad Yes	No	Total
Ill	22	1	23
Not Ill	24	23	47
Total	46	24	70

we have for the top left cell

$$\text{Observed value} = 22$$

$$\text{Mean} = \frac{(46)(23)}{70} = 15.1$$

$$\text{Variance} = \frac{(46)(24)(23)(47)}{(70)^2(70-1)} = 3.53$$

Therefore by putting together the results from the two strata, we have

$$X^2_{\mathrm{MH}} = \frac{[(120-105.5)+(22-15.1)]^2}{7.46+3.53} = 41.67$$

The big difference is that the Mantel–Haenszel method assumes that H_1 (the model of no three-term interaction) fits, whereas in this loglinear model approach, we know that H_1 fits (after testing it).

Application 2 If

(i) a model H_2 is nested in H_1 so that $\{H_1 - H_2\}$ consists of only one term, say, λ_θ;

(ii) both models H_1 and H_2 fit; and

(iii) the test for

$$H_0\colon\ \lambda_\theta = 0$$

is not significant, then H_2 is defined as "better than" or "preferred over" H_1.

The best model is the most preferred model among the hierarchy of good-fit models. In other words, the best model fits and contains *only* significant terms. It should be noted that, unfortunately, more than one best model exists because there may be more than one hierarchy of good-fit models.

■ **Example 3.8** Refer to the food poisoning data of Example 3.1 (or 3.4). By fitting all eight models of independence to this three-way table, we obtain

Mutual Independence:

$$H_4\colon (\mathrm{I,C,P}) = \lambda + \lambda_{\mathrm{I}} + \lambda_{\mathrm{C}} + \lambda_{\mathrm{P}};\ \ G^2 = 68.16, 4\ \mathrm{df}$$

Joint Independence of Two Factors:

$$H_{3a}\colon (\mathrm{P,IC}) = \lambda + \lambda_{\mathrm{I}} + \lambda_{\mathrm{C}} + \lambda_{\mathrm{P}} + \lambda_{\mathrm{IC}};\ G^2 = 59.42, 3\ \mathrm{df}$$
$$H_{3b}\colon (\mathrm{C,IP}) = \lambda + \lambda_{\mathrm{I}} + \lambda_{\mathrm{C}} + \lambda_{\mathrm{P}} + \lambda_{\mathrm{IP}};\ G^2 = 15.10, 3\ \mathrm{df}$$
$$H_{3c}\colon (\mathrm{I,CP}) = \lambda + \lambda_{\mathrm{I}} + \lambda_{\mathrm{C}} + \lambda_{\mathrm{P}} + \lambda_{\mathrm{CP}};\ G^2 = 56.36, 3\ \mathrm{df}$$

Conditional Independence:

$$H_{2a}\colon (\mathrm{P,IC,IP}) = \lambda + \lambda_{\mathrm{I}} + \lambda_{\mathrm{C}} + \lambda_{\mathrm{P}} + \lambda_{\mathrm{IC}} + \lambda_{\mathrm{IP}};\ G^2 = 6.36, 2\ \mathrm{df}$$
$$H_{2b}\colon (\mathrm{C,IC,CP}) = \lambda + \lambda_{\mathrm{I}} + \lambda_{\mathrm{C}} + \lambda_{\mathrm{P}} + \lambda_{\mathrm{IC}} + \lambda_{\mathrm{CP}};\ G^2 = 47.62, 2\ \mathrm{df}$$
$$H_{2c}\colon (\mathrm{I,IP,CP}) = \lambda + \lambda_{\mathrm{I}} + \lambda_{\mathrm{C}} + \lambda_{\mathrm{P}} + \lambda_{\mathrm{IP}} + \lambda_{\mathrm{CP}};\ \mathrm{G}^2 = 3.30, 2\ \mathrm{df}$$

No Three-Factor Interaction:

$$H_1\text{: (I, IC, IP, CP)} = \lambda + \lambda_{\mathrm{I}} + \lambda_{\mathrm{C}} + \lambda_{\mathrm{P}} + \lambda_{\mathrm{IC}} + \lambda_{\mathrm{IP}} + \lambda_{\mathrm{CP}};\ G^2 = 0.27,\ 1\ \text{df}$$

There is one hierarchy consisting of two nested good models: $H_2 < H_1$. Since the single term separating them, IC, is not significant ($G^2 = 3.30 - 0.27 = 3.03$, with $2 - 1 = 1$ df), model H_{2c} is better and therefore the best model to explain the food poisoning data set.

3.2.5 Collapsing Tables

Consider the case of three categorical variables X_1, X_2, and X_3 with data presented in the form of a three-way table and suppose that we are interested only in the relationship between X_1 and X_2. If we collapse the cross-classification over X_3, yielding a two-way table, we would like to know whether, for example, the odds ratio obtained from this marginal table is correct: that is, the same that would result from the loglinear model method of Section 3.2.3. By an analogy with the method for correlation coefficients for continuous data, it can be shown that (Bishop et al., 1975) "in a three-dimensional table the interaction between two variables may be measured from the marginal table by collapsing the tables over the third variable if and only if the third variable is independent of at least one of the two variables exhibiting the interaction."

■ **Example 3.9** Refer to the food poisoning data of Example 3.1 (or 3.4). Since the best model we found in Example 3.8 is

$$H_{2c}\text{: (I, IP, CP)} = \lambda + \lambda_{\mathrm{I}} + \lambda_{\mathrm{C}} + \lambda_{\mathrm{P}} + \lambda_{\mathrm{IP}} + \lambda_{\mathrm{CP}};\ G^2 = 3.30,\ 2\ \text{df}$$

indicating that only crabmeat and illness are independent:

(i) we can collapse on crabmeat in studying the relationship between potato salad and illness, but

(ii) we *cannot* collapse on potato salad in order to study the relationship between crabmeat and illness.

However, this seems an unnecessary question: in order to answer it, we need to perform a thorough loglinear model analysis; and after performing the needed loglinear model analysis, we no longer need to collapse the table!

3.3 LOGLINEAR MODELS FOR HIGHER-DIMENSIONAL TABLES

Although all the models and methods discussed in Section 3.2 have been in the context of three-way tables, the extensions to higher-dimensional tables (four-way tables, five-way tables, etc.) are relatively straightforward.

Given the data in a cross-classified table, regardless of the dimension, we still have two different types of statistical inferences:

(i) To test for a specific model; for example, we may want to know whether two specified factors are conditionally independent given all the other factors.
(ii) To search for a model that can best explain the relationship(s) found in the observed data (i.e., the best model).

3.3.1 Testing a Specific Model

The process is still the same as that in Section 3.2.3 for three-way tables: estimation of expected frequencies under the null hypothesis, calculation of goodness-of-fit statistic (likelihood ratio chi-squared, and determination of the degree of freedom. Of course, we should implement these steps using the same computer package program, SAS PROC CATMOD. The key and only needed step is to describe the loglinear model to be tested starting with a research question, which leads to a relationship between factors, and finally a null hypothesis. We are also limited to consider only hierarchical models.

Suppose we consider the cervical screening data of Example 3.2 with all five variables: age (A), residence (M; yes/no), income (I), race (R), and Pap testing (P).

				Pap Test	
Age	Residence	Income	Race	No	Yes
25–44	Metropolitan	Poor	White	77	516
			Black	57	344
		Nonpoor	White	476	5796
			Black	47	701
	Nonmetropolitan	Poor	White	63	387
			Black	41	94
		Nonpoor	White	211	3186
45–64	Metropolitan	Poor	White	163	376
			Black	86	172
		Nonpoor	White	859	5646
			Black	121	400
	Nonmetropolitan	Poor	White	160	328
			Black	59	68
		Nonpoor	White	491	2361
			Black	44	72
65+	Metropolitan	Poor	White	449	499
			Black	70	86
		Nonpoor	White	903	1544
			Black	63	70

(*Continued*)

Age	Residence	Income	Race	Pap Test No	Pap Test Yes
	Nonmetropolitan	Poor	White	396	365
			Black	71	43
		Nonpoor	White	459	698
			Black	21	10

The null hypothesis of no income effects, for example, can be investigated in two different ways:

(i) *Unconditional approach*: Here we test the hierarchical model H_0 not containing the term under investigation, λ_{IP}.
(ii) *Conditional approach*: This is a two-step procedure where we first fit the same model H_0 as in the unconditional approach, and we than then, we fit another model, H_1, by adding in the term λ_{IP}, and use the process for nested models to focus on $\{G^2(H_0) - G^2(H_1)\}$.

The conditional approach is like an extension of the Mantel–Haenszel method. In order to use this approach, H_1 must fit because we assume that there are no other factors that would modify the effects of income on Pap testing (terms such as λ_{AIP}; otherwise, look for a good-fit model containing λ_{IP}).

■ **Example 3.10** Refer to the cervical screening data of Example 3.2 with data as in the above table, and suppose we are only interested in the effects of income (I) on Pap testing (P)–that is, we focus on the term λ_{IP}. With five variables (P, A, I, M, R), there are:

(i) 10 two-term interactions: PA, PI, PM, PR, AI, AM, AR, IM, IR, MR;
(ii) 10 three-term interactions: PAI, PAM, PAR, PIM, PIR, PMR, AIM, AIR, AMR, IMR;
(iii) 5 four-term interactions: PAIM, PAIR, PAMR, PIMR, AIMR; and
(iv) 1 five-term interaction: PAIMR.

With the unconditional approach, we would fit the model without term IP; using only hierarchical models, this also rules out the following higher order relatives: {PAI, PIM, PIR, PAIM, PAIR, PIMR, and PAIMR}. By fitting the model with the remaining terms $H_0 = \{$PA, PM, PR, AI, AM, AR, IM, IR, MR, PAM, PAR, PMR, AIM, AIR, AMR, IMR, PAMR, AIMR$\}$, we obtain $G^2 = 317.28$ with 12 degrees of freedom ($p < 0.0001$). In addition, model H_1 (same as model H_0 plus interaction term IP) does not fit, which rules out the conditional approach.

3.3.2 Searching for the Best Model

As mentioned earlier, given the data in a contingency table, we have two different types of statistical inferences: (i) to test for a specific model, and (ii) to search for a model that can best explain the relationship(s) found in the observed data (i.e. the best model).

The test for a specific model is conducted in the same way regardless of the dimension of the table. However, the search for the best model is different for higher-dimensional tables. In the case of three-way tables, we fit all eight possible models of independence; there is usually one hierarchy of nested models from which to pick the best model. For higher-dimensional tables, it is impossible to do that; for a four-way table there are 113 possible hierarchical models, and for a ten-way table the number of hierarchical models is 3,475, 978! Therefore we need a more efficient strategy.

The most commonly used procedure is a *stepwise procedure*. The description here is for four-way tables, for simplicity, but can be extended to tables with five or more dimensions. This is, in fact, a two-step process.

Step 1: We look for a "starting model," one that is close to the best model. We begin by choosing a significant level, say 0.05, and then we test for the goodness-of-fit of all models of uniform orders. A "model of uniform orders" contains all interaction terms involving the same number of factors. For example, there are three models of uniform orders for a four-way table:

$$H_1\colon\ \lambda+\lambda_1+\lambda_2+\lambda_3+\lambda_4$$
$$H_2\colon\ \lambda+\lambda_1+\lambda_2+\lambda_3+\lambda_4+\lambda_{12}+\lambda_{13}+\lambda_{14}+\lambda_{23}+\lambda_{24}+\lambda_{34}$$
$$H_3\colon\ \lambda+\lambda_1+\lambda_2+\lambda_3+\lambda_4+\lambda_{12}+\lambda_{13}+\lambda_{14}+\lambda_{23}+\lambda_{24}+\lambda_{34} +\lambda_{123}+\lambda_{124}+\lambda_{134}+\lambda_{234}$$

For example, if model H_3 does not fit the data we stop; the best model is the saturated model. If model H_3 fits (there is no four-term interaction) but model H_2 does not, the best model is somewhere between them—one that contains some but not all three-factor interactions. We can choose as the starting model either H_2 (and try to "add" terms) or H_3 (and try to "remove" terms). If both models H_2 and H_3 fit but model H_1 does not, the best model is somewhere between H_1 and H_2—one that contains some but not all two-factor interactions. In this case, we can use either H_1 or H_2 as the starting model.

Step 2: We apply the stepwise procedure to the starting model found in Step 1 in order to reach the best model. For example, if model H_3 fits but model H_2 does not, then:

(i) we can choose H_2 as a starting model and do a "forward addition" by adding in one three-factor interaction term at a time, or

(ii) we can choose H_3 as a starting model and do a "backward elimination" by deleting one three-factor interaction term at a time. The process continues until no term can be added or deleted.

Since PROC CATMOD of SAS does not have an automatic stepwise option, it is much easier to perform a backward elimination. When we fit the larger model H_3, the computer output also includes significance levels for each three-factor interaction term; we can usually eliminate all nonsignificant terms and reach the best model in one step. We illustrate this process first with a four-way table then a five-way table using the cervical screening data of Example 3.2.

Example 3.11 The following data are from a study of the survival of breast cancer patients (taken from Bishop et al., 1975). The main factor is 3-year survival status (yes/no). Other factors are age (under 50, 50–69, 70 or over), treatment center (Tokyo, Boston, and Glamorgan), and two histological criteria—nuclear grade (malignant appearance/benign appearance) and degree of chronic inflammatory reaction (minimal/moderate–severe). The last two factors were interrelated and together could be regarded as a description of the disease state. Let us call this inflammation with four categories: (a) minimal with malignant appearance, (b) minimal with benign appearance, (c) moderate–severe with malignant appearance, and (d) moderate-severe with benign appearance. The four factors are age (A), center (C), Inflammation (I), and survival status (S); the aim is to search for the best model showing the relationship contained in the observed data.

Fitting the above three models of uniform orders—H_1 (all main effects), H_2 (all two-factor interaction terms), and H_3 (all three-factor interaction terms),

$$H_1\colon\ \lambda + \lambda_{\rm A} + \lambda_{\rm C} + \lambda_{\rm I} + \lambda_{\rm S}$$

$$H_2\colon\ \lambda + \lambda_{\rm A} + \lambda_{\rm C} + \lambda_{\rm I} + \lambda_{\rm S} + \lambda_{\rm AC} + \lambda_{\rm AI} + \lambda_{\rm AS} + \lambda_{\rm CI} + \lambda_{\rm CS} + \lambda_{\rm IS}$$

$$H_3\colon\ \lambda + \lambda_{\rm A} + \lambda_{\rm C} + \lambda_{\rm I} + \lambda_{\rm S} + \lambda_{\rm AC} + \lambda_{\rm AI} + \lambda_{\rm AS} + \lambda_{\rm CI} + \lambda_{\rm CS} + \lambda_{\rm IS} + \lambda_{\rm ACI} + \lambda_{\rm ACS} + \lambda_{\rm AIS} + \lambda_{\rm CIS}$$

we found that H_2 fits ($p = 0.2392$; hence there is no need to fit H_3 in this case) but model H_1 does not ($p < 0.0001$). Therefore we can use H_2 as a starting model and proceed with a backward elimination in the second step.

			Inflammation			
Center	Age	Survival	(a)	(b)	(c)	(d)
Tokyo	Under 50	No	9	7	4	3
		Yes	26	68	25	9
	50–69	No	9	9	11	2
		Yes	20	46	18	5
	70 or over	No	2	3	1	0
		Yes	1	6	5	1
Boston	Under 50	No	6	7	6	0
		Yes	11	24	4	0

(*Continued*)

			Inflammation			
Center	Age	Survival	(a)	(b)	(c)	(d)
	50–69	No	8	20	3	2
		Yes	18	58	10	3
	70 or over	No	9	18	3	0
		Yes	15	26	1	1
Glamorgan	Under 50	No	16	7	3	0
		Yes	16	20	8	1
	50–69	No	16	12	3	0
		Yes	27	39	10	4
	70 or over	No	3	7	3	0
		Yes	12	11	4	1

From fitting model H_2, we obtain

Term	df	G^2	p-Value
C * A	4	58.92	<0.0001
C * S	2	8.51	0.0142
C * I	6	31.49	<0.0001
A * S	2	4.17	0.1245
A * I	6	1.55	0.9560
S * I	3	10.12	0.0176

The results indicate that among the six three-factor interaction terms, two of them are not significant at the 5% level: A * S (age and survival; $p = 0.1245$) and A * I (age and inflammation; $p = 0.9560$). By deleting these two terms, we obtain the best model:

$$H\colon\ \lambda + \lambda_{\mathrm{A}} + \lambda_{\mathrm{C}} + \lambda_{\mathrm{I}} + \lambda_{\mathrm{S}} + \lambda_{\mathrm{AC}} + \lambda_{\mathrm{CI}} + \lambda_{\mathrm{CS}} + \lambda_{\mathrm{IS}}$$

($G^2 = 44.12$, 41 degrees of freedom, $p = 0.3411$). As far as the survival of patients is concerned, there are two significant factors: degree of inflammation (I) and treatment center (C). Recall the condition for collapsing tables in Section 3.2.5; it can be seen that we cannot collapse to two-way tables in order to study the effects of each factor (age, inflammation, and center) on breast cancer survival:

(i) If we are interested in the effects of age (A), we cannot collapse on center (C) because of the significant terms A * C and C * S.

(ii) If we are interested in the effects of inflammation (I), we cannot collapse on center (C) because of the significant terms C * I and C * S.

(iii) If we are interested in the differences between centers, we cannot collapse on inflammation (I) because of the significant terms C * I and I * S.

■ **Example 3.12** Refer to the cervical screening data of Example 3.2 with all five variables: age (A), residence (M; yes/no), income (I), race (R), and Pap testing (P).

				Pap Test	
Age	Residence	Income	Race	No	Yes
25–44	Metropolitan	Poor	White	77	516
			Black	57	344
		Nonpoor	White	476	5796
			Black	47	701
	Nonmetropolitan	Poor	White	63	387
			Black	41	94
		Nonpoor	White	211	3186
45–64	Metropolitan	Poor	White	163	376
			Black	86	172
		Nonpoor	White	859	5646
			Black	121	400
	Nonmetropolitan	Poor	White	160	328
			Black	59	68
		Nonpoor	White	491	2361
			Black	44	72
65 +	Metropolitan	Poor	White	449	499
			Black	70	86
		Nonpoor	White	903	1544
			Black	63	70
	Nonmetropolitan	Poor	White	396	365
			Black	71	43
		Nonpoor	White	459	698
			Black	21	10

Fitting the four models of uniform orders,

H_1: all main effects
H_2: all two-factor interaction terms
H_3: all three-factor interaction terms
H_4: all four-factor interaction terms

we found that H_3 fits ($p = 0.1171$; therefore there is no need to fit H_4 in this case) but model H_2 does not ($p < 0.0001$). Therefore we can use H_3 as a starting model and proceed with a backward elimination.

Term	df	G^2	p-Value
A * M * R	2	2.36	0.3077
A * M * I	2	5.96	0.0508
A * M * P	2	3.13	0.2096
A * R * I	2	10.87	0.0044
A * R * P	2	3.36	0.1861
A * I * P	2	37.61	<0.0001
M * R * I	1	4.42	0.0356
M * R * P	1	24.82	<0.0001
M * I * P	1	0.27	0.6015
R * I * P	1	10.19	0.0014

From fitting model H_3, we found that among the ten three-factor interaction terms, five of them are not significant at the 5% level:

A * M * R ($p = 0.3077$)
A * M * I ($p = 0.0508$)
A * M * P ($p = 0.2096$)
A * R * P ($p = 0.1861$)
M * I * P ($p = 0.6015$)

By deleting these five terms, we obtain the best model. In other words, all four factors (age, income, residence, and race) are related to Pap testing. There are also three significant three-factor interaction terms:

(i) Age and income modify the effects of each other.
(ii) Income and race modify the effects of each other.
(iii) Residence and race modify the effects of each other.

For example, the degree of difference between poor women and nonpoor women, in terms of testing or not testing for cervical cancer, is different when calculating among whites from that when calculating among blacks. This would be much easier to illustrate using the odds ratios, as formulated in the next section.

3.3.3 Measures of Association with an Effect Modification

Consider the case of three categorical variables X_1, X_2, and X_3 with a significant three-factor interaction term $X_1 * X_2 * X_3$. Let X_1 be the response variable with two levels (1 if event of interest is observed, e.g., being not tested; 2 if event of interest is not observed, e.g., being tested in the context of the cervical cancer screening problem),

X_2 be the explanatory factor (maybe with more than two categories), and X_3 be a confounder, which is also an effect modifier. Suppose we are interested in comparing the effect of X_2 (on the response X_1) when X_2 is at level i to the effect of the same factor when X_2 is at level 1. For example, in the context of Example 3.2, we want to compare the odds of not testing for cervical cancer (P) for those aged 70 or over (level 3 of A) to the odds of not testing for cervical cancer (P) for those aged 50 or under (level 1 of A). Of course, we must control for income (I) because age and income modify the effects of each other. Assuming that X_3 is at level k, the above odds ratio is given by the formula

$$\log OR_{ivs1} = 2(\lambda_{12(1i)} - \lambda_{12(11)}) + 2(\lambda_{123(1ik)} - \lambda_{123(11k)})$$

The second term is induced by the effect modification; thus, there are two such terms if there are two effect modifiers. In the special case of no effect modification, there is no three-term interaction and the above formula is simplified to

$$\log OR_{ivs1} = 2(\lambda_{12(1i)} - \lambda_{12(11)})$$

In addition, if the explanatory factor has only two levels (1 if exposed and 2 if not exposed), then we have the odds ratio associated with the exposure:

$$\log OR = 4\lambda_{12(11)}$$

because of the constraint for the model:

$$\lambda_{12(1i)} + \lambda_{12(11)}) = 0$$

■ **Example 3.13** Refer to the breast cancer survival data of Example 3.11, where we found the best model was

$$H\colon \lambda + \lambda_{\mathrm{A}} + \lambda_{\mathrm{C}} + \lambda_{\mathrm{I}} + \lambda_{\mathrm{S}} + \lambda_{\mathrm{AC}} + \lambda_{\mathrm{CI}} + \lambda_{\mathrm{CS}} + \lambda_{\mathrm{IS}}$$

Suppose that we are interested in the difference between two treatment centers, Boston versus Tokyo. By fitting the best model, we obtain the following estimates for λ_{CS}:

Survival	Tokyo	Boston	Glamorgan	Total
Yes	0.2086	−0.1368	−0.0718	0
No	−0.2086	0.1368	0.0718	0
Total	0	0	0	

This leads to

$$\begin{aligned}\log OR_{\text{Boston vs. Tokyo}} &= 2[(-0.1368)-(0.2086)]\\ &= -0.6908\\ OR_{\text{Boston vs. Tokyo}} &= \exp(-0.6908)\\ &= 0.50\end{aligned}$$

In other words, the treatment in Boston is only half as effective as that in Tokyo.

Example 3.14 Refer to the cervical cancer screening data of Example 3.2 where we found, in Example 3.12, the best model with all two-term interactions and five three-term interactions {AIR, MIR, MRP, IRP, and AIP} indicating that:

(i) Age and income modify the effects of each other.
(ii) Income and race modify the effects of each other.
(iii) Residence and race modify the effects of each other.

Suppose we want to compare the older women (65 or over) versus the younger women (25–44). Of course, we have to control for income because λ_{AIP} is significant (age and income modify the effects of each other). By fitting the best model, we obtain the following estimates for λ_{AP}:

	Age			
Pap Test	25–44	45–64	65 and Over	Total
No	−0.5095	0.0019	0.5076	0
Yes	0.5095	−0.0019	0.0718	0
Total	0	0	0	

We also obtain the following estimates for λ_{AIP}:

		Age		
Income	Pap Test	25–44	45–64	65 and Over
Poor	No	0.0392	0.0441	−0.0833
	Yes	−0.0392	−0.0441	0.0833
Nonpoor	No	−0.0392	−0.0441	0.0833
	Yes	0.0392	0.0441	−0.0833

For poor women,

$$\begin{aligned}\log OR_{65+\text{ vs. }25-44} &= 2[(0.5076)-(-0.5095)]+2[(-0.0833)-(0.0392)]\\ &= 1.7892\\ OR_{65+\text{ vs. }25-44} &= \exp(1.7892)\\ &= 5.985\end{aligned}$$

and for nonpoor women

$$\begin{aligned}\log OR_{65+\text{ vs. }25-44} &= 2[(0.5076)-(-0.5095)]+2[(-0.0833)-(-0.0392)]\\ &= 2.2792\\ OR_{65+\text{ vs. }25-44} &= \exp(2.2792)\\ &= 9.770\end{aligned}$$

In other words, the women in age group 65-and-over are much more likely to not be tested for cervical cancer as compared to women in age group 25–44.

(i) The odds ratio is 5.975 if they are poor.
(ii) The odds ratio is 9.770 if they are nonpoor.

■ **Example 3.15** Refer to the cervical cancer screening data of Examples 3.2 and 3.14, but this time we want to compare the black women versus white women. The calculation is more complicated because there are two effect modifiers: terms λ_{MRP} and λ_{IRP} are both significant at the 5% level.

By fitting the best model, we obtain the following estimates for λ_{RP}:

Pap Test	Whites	Blacks	Total
No	−0.1437	0.1437	0
Yes	0.1437	−0.1437	0
Total	0	0	

We also obtain the following estimates for λ_{MRP}:

Residence	Pap Test	Whites	Blacks
Metropolitan	No	0.0698	−0.0698
	Yes	−0.0698	0.0698
Nonmetropolitan	No	−0.0698	0.0698
	Yes	0.0698	−0.0698

and the following estimates for λ_{IRP}:

Income	Pap Test	Whites	Blacks
Poor	No	0.0444	−0.0444
	Yes	−0.0444	0.0444
Nonpoor	No	−0.0444	0.0444
	Yes	0.0444	−0.0444

(i) For poor women in metropolitan areas,

$$\begin{aligned}\log OR_{\text{B vs. W}} &= 2[(0.1437)-(-0.1437)]+2[(-0.0698)-(0.0698)]\\&\quad+2[(-0.0444)-(0.0444)]\\&= 0.118\\OR_{\text{B vs. W}} &= \exp(0.118)\\&= 1.13\end{aligned}$$

(ii) For nonpoor women in metropolitan areas,

$$\begin{aligned}\log OR_{\text{B vs. W}} &= 2[(0.1437)-(-0.1437)]+2[(-0.0698)-(0.0698)]\\&\quad+2[(0.0444)-(-0.0444)]\\&= 0.4732\\OR_{\text{B vs. W}} &= \exp(0.4732)\\&= 1.61\end{aligned}$$

(iii) For poor women in nonmetropolitan areas,

$$\begin{aligned}\log OR_{\text{B vs. W}} &= 2[(0.1437)-(-0.1437)]+2[(0.0698)-(-0.0698)]\\&\quad+2[(-0.0444)-(0.0444)]\\&= 0.6467\\OR_{\text{B vs. W}} &= \exp(0.6467)\\&= 1.97\end{aligned}$$

(iv) For nonpoor women in nonmetropolitan areas,

$$\begin{aligned}\log OR_{\text{B vs. W}} &= 2[(0.1437)-(-0.1437)]+2[(0.0698)-(-0.0698)]\\&\quad +2[(0.0444)-(-0.0444)]\\&= 1.0316\\OR_{\text{B vs. W}} &= \exp(1.0316)\\&= 2.81\end{aligned}$$

In other words, the black women are more likely not tested for cervical cancer as compared to white women:

(i) The odds ratio is 1.13 if the women are poor and from metropolitan areas.
(ii) The odds ratio is 1.61 if the women are nonpoor and from metropolitan areas.
(iii) The odds ratio is 1.97 if the women are poor and from nonmetropolitan areas.
(iv) The odds ratio is 2.81 if the women are nonpoor and from nonmetropolitan areas.

3.3.4 Searching for a Model with a Dependent Variable

Loglinear models describe association patterns among categorical variables without specifying their roles, that is, which one is the response and which ones are explanatory factors. However, when it is natural to treat one variable as the response and the others as explanatory or independent variables, modifications have to be made in order to make the resulting loglinear models equivalent to the logistic regression models of Chapter 4. This can be achieved by simply keeping, in the models to be fitted, all terms relating independent variables with each other; the changes in the numerical results are negligible most times.

■ **Example 3.16** Refer to the breast cancer survival data of Example 3.11 with four factors: age (A), center (C), inflammation (I), and survival status (S); and suppose we are searching for the best model with survival status (S) being identified as the response. Without a response variable, the three models of uniform orders are:

H_1: all main effects
H_2: all two-factor interaction terms
H_3: all three-factor interaction terms

In the presence of the dependent variable (S), these are expanded to include the following terms wherever appropriate: {AC, AI, CI, and ACI}. The resulting best model turns out to be

$$H\!:\ \lambda+\lambda_{\text{A}}+\lambda_{\text{C}}+\lambda_{\text{I}}+\lambda_{\text{S}}+\lambda_{\text{AC}}+\lambda_{\text{CI}}+\lambda_{\text{CS}}+\lambda_{\text{IS}}+\lambda_{\text{AC}}+\lambda_{\text{AC}}+\lambda_{\text{AI}}+\lambda_{\text{ACI}}$$

which appears different from the result in Example 3.11. However, most numerical results of interest are still very similar; for example,

$$\begin{aligned} OR_{\text{Boston vs. Tokyo}} &= \exp(-0.6834) \\ &= 0.501 \end{aligned}$$

as compared to an odds ratio of 0.50 as seen in Example 3.11.

EXERCISES

3.1 Adult male residents of 13 counties of western Washington State, in whom testicular cancer had been diagnosed during 1977–1983, were interviewed over the telephone regarding their history of genital tract conditions, including vasectomy. For comparison, the same interview was given to a sample of men selected from the population of these counties by dialing telephone numbers at random. The following data are tabulated by religious background:

Religion	Vasectomy	Cases	Controls
Protestant	Yes	24	56
	No	205	239
Catholic	Yes	10	6
	No	32	90
Others	Yes	18	39
	No	56	96

(**a**) Test for the conditional independence between testicular cancer and vasectomy.

(**b**) Search for the model that can best explain the relationships found in the observed data; choose to do with or without identifying a dependent variable (i.e., testicular cancer in this exercise).

(**c**) From the best model found in (b), calculate the odds ratio(s) measuring the strength of the relationship between testicular cancer and vasectomy.

3.2 In a seroepidemiologic survey of health workers representing a spectrum of exposure to blood and patients with hepatitis B virus (HBV), it was found that infection increased as a function of contact. The following table provides data for hospital workers with uniform socioeconomic status at an urban teaching hospital in Boston, Massachusetts:

		HBV Test Result	
Personnel	Exposure	Negative	Positive
Physicians	Frequent	64	17
	Infrequent	82	7
Nurses	Frequent	82	22
	Infrequent	115	11

(a) Test for the conditional independence between exposure and test result.

(b) Does "personnel" modify the effect of exposure on the test result?

(c) Search for the model that can best explain the relationships found in the observed data; choose to do with or without identifying a dependent variable (i.e., test result in this exercise).

(d) From the best model found in (c), calculate the odds ratio(s) measuring the strength of the relationship between exposure and test result.

(e) Should we collapse the data across personnel in order to answer question (d)?

3.3 It has been hypothesized that dietary fiber decreases the risk of colon cancer, while meats and fats are thought to increase this risk. A large study was undertaken to confirm these hypotheses. Fiber and fat consumptions are classified as "low" or "high" and data are tabulated separately for males and females as follows ("low" means below median):

	Males		Females	
Diet	Cases	Controls	Cases	Controls
Low fat, high fiber	27	38	23	39
Low fat, low fiber	64	78	82	81
High fat, high fiber	78	61	83	76
High fat, low fiber	36	28	35	27

(a) Test for the conditional independence between colon cancer and diet.

(b) Does gender modify the effect of diet on the disease?

(c) Search for the model that can best explain the relationships found in the observed data; choose to do with or without identifying a dependent variable (i.e., colon cancer in this exercise).

(d) From the best model found in (c), calculate the odds ratio(s) measuring the strength of the relationship between colon cancer and diet.

(e) Should we collapse the data across genders in order to answer question (d)?

3.4 Postneonatal mortality due to respiratory illnesses is known to be inversely related to maternal age, but the role of young motherhood as a risk factor for respiratory morbidity in infants has not been thoroughly explored. A study was conducted in Tucson, Arizona, aimed at the incidence of lower respiratory tract illnesses during the first year of life. In this study, over 1200 infants were enrolled at birth between 1980 and 1984 and the following data are concerned with wheezing lower respiratory tract illnesses (wheezing LRI: no/yes).

	Boys		Girls	
Maternal Age (years)	No	Yes	No	Yes
Below 21	19	8	20	7
21–25	98	40	128	36
26–30	160	45	148	42
Over 30	110	20	116	25

(a) Test for the conditional independence between wheezing LRI and maternal age.

(b) Does gender modify the effect of maternal age on wheezing LRI?

(c) Search for the model that can best explain the relationships found in the observed data; choose to do with or without identifying a dependent variable (i.e., wheezing LRI in this exercise).

(d) From the best model found in (c), calculate the odds ratio(s) measuring the strength of the relationship between wheezing LRI and maternal age.

3.5 Since incidence rates of most cancers rise with age, this must always be considered a confounder. The following are stratified data for an unmatched case–control study; the disease was esophageal cancer among men and the risk factor was alcohol consumption.

		Daily Alcohol Consumption	
Age	Sample	80+ g	0–79 g
25–44	Cases	5	5
	Controls	35	270
45–64	Cases	67	55
	Controls	56	277
65 and over	Cases	24	44
	Controls	18	129

(a) Test for the conditional independence between cancer and alcohol consumption.

(b) Test for the conditional independence between age and alcohol consumption.

(c) Search for the model that can best explain the relationships found in the observed data; choose to do with or without identifying a dependent variable (i.e., esophageal cancer in this exercise).

(d) From the best model found in (c), calculate the odds ratio(s) measuring the strength of the relationship between esophageal cancer and alcohol consumption.

(e) How does the result in (d) compare to that obtained by the Mantel–Haenszel procedure?

3.6 Postmenopausal women who develop endometrial cancer are, on average, heavier than women who do not develop the disease. One possible explanation is that heavy women are more exposed to endogenous estrogens, which are produced in postmenopausal women by conversion of steroid precursors to active estrogens in peripheral fat. In the face of varying levels of endogenous estrogen production, one might ask whether the carcinogenic potential of exogenous estrogens would be the same in all women. A case–control study has been conducted to examine the relation between weight, replacement estrogen therapy, and endometrial cancer; results are as follows:

		Estrogen Replacement	
Weight (kg)	Sample	Yes	No
<57	Cases	20	12
	Controls	61	183
57–75	Cases	37	45
	Controls	113	378
>75	Cases	9	42
	Controls	23	140

(a) Test for the conditional independence between endometrial cancer and estrogen replacement.

(b) Does weight modify the effect of estrogen replacement on endometrial cancer?

(c) Search for the model that can best explain the relationships found in the observed data; choose to do with or without identifying a dependent variable (i.e., endometrial cancer in this exercise).

(d) From the best model found in (c), calculate the odds ratio(s) measuring the strength of the relationship between endometrial cancer and estrogen replacement.

3.7 Data taken from a study to investigate the effects of smoking on cervical cancer are stratified by the number of sexual partners. Results are as follows:

		Cervical Cancer	
Number of Partners	Smoking	Yes	No
Zero or one	Yes	12	21
	No	25	118
Two or more	Yes	96	142
	No	92	150

(a) Test for the conditional independence between cervical cancer and number of partners.

(b) Does smoking modify the effect of the number of partners on cancer risk?

(c) Search for the model that can best explain the relationships found in the observed data; choose to do with or without identifying a dependent variable (i.e., cervical cancer in this exercise).

(d) From the best model found in (c), calculate the odds ratio(s) measuring the strength of the relationship between cervical cancer and number of partners.

3.8 Cases of poliomyelitis were classified by age, paralytic status, and whether or not the subject had been injected with the Salk vaccine, as in the following table; the three variables are age (six levels), Salk vaccine (yes/no), and paralysis (yes/no).

(a) Test for the conditional independence between paralysis and Salk vaccine.

(b) Does the vaccine effect vary with age? (Putting it in a different way, could we investigate the vaccine effect using the method of Mantel–Haenszel?)

(c) Search for the model that can best explain the relationships found in the observed data; choose to do with or without identifying a dependent variable (i.e., paralysis in this exercise).

(d) From the best model found in (c), calculate the odds ratio(s) measuring the strength of the relationship between paralysis and Salk vaccine.

Age	Salk Vaccine	Paralysis	
		No	Yes
0–4	Yes	20	14
	No	10	24
5–9	Yes	15	12
	No	3	15
10–14	Yes	3	2
	No	3	2
15–19	Yes	7	4
	No	1	6
20–29	Yes	12	3
	No	7	5
30 +	Yes	1	0
	No	3	2

3.9 The following are data on smoking from a survey of seventh graders (age: $1 = 12$ or younger, $2 = 13$ or older):

Family Structure	Race	Sex	Age	Smoking	
				None	Some
Both parents	Black	Male	1	27	2
			2	12	2
		Female	1	23	4
			2	7	1
	White	Male	1	394	32
			2	142	19
		Female	1	421	38
			2	94	11
Mother only	Black	Male	1	18	1
			2	13	1
		Female	1	24	0
			2	4	3
	White	Male	1	48	6
			2	25	4
		Female	1	55	15
			2	13	4

(a) Test for the conditional independence between smoking and family structure.

(b) Test to see if there is a difference in smoking rates between boys and girls.

(c) Test to see if race or sex alters the effect of family structure on smoking.

(d) Search for the model that can best explain the relationships found in the observed data; choose to do with or without identifying a dependent variable (i.e., smoking in this exercise).

(e) From the best model found in (d), calculate the odds ratio(s) measuring the strength of the relationship between smoking and family structure.

(f) From the best model found in (d), calculate the odds ratio(s) measuring the strength of the relationship between smoking and sex.

3.10 The following are data on drinking from the same survey of seventh graders of the previous data set, Exercise 3.9 (age: $1 = 12$ or younger, $2 = 13$ or older):

	Mother's			Drinking	
Family Structure	Occupation	Sex	Age	None	Some
Both parents	Other	Male	1	405	44
			2	152	23
		Famele	1	454	26
			2	102	8
	White collar	Male	1	392	39
			2	130	17
		Female	1	434	39
			2	91	11
Mother only	Other	Male	1	61	11
			2	37	4
		Female	1	83	8
			2	21	3
	White collar	Male	1	104	17
			2	40	9
		Female	1	140	20
			2	32	6

(a) Test for the conditional independence between drinking and family structure.

(b) Test to see if there is a difference in drinking rates between boys and girls.

(c) Test to see if race or sex alters the effect of family structure on drinking.

(d) Search for the model that can best explain the relationships found in the observed data; choose to do with or without identifying a dependent variable (i.e., drinking in this exercise).

(e) From the best model found in (d), calculate the odds ratio(s) measuring the strength of the relationship between drinking and family structure.

(f) From the best model found in (d), calculate the odds ratio(s) measuring the strength of the relationship between drinking and sex.

4

LOGISTIC REGRESSION MODELS

In many research studies, the main focus could be the *difference* between populations or subpopulations, where we aim at the comparison of population means or population proportions. In many other studies, however, the purpose of the research is to assess *relationships* among a set of variables. For example, the sample consists of pairs of values, say, a mother's weight and her newborn's weight measured from each of n sets of mother-and-baby, and the research objective is concerned with the *association* between these weights—to see if one is related to the other. Regression analysis is the technique for investigating relationships between variables; it can be used for assessment of association as well as prediction. Consider, for example, an analysis of whether or not a woman's age is predictive of her systolic blood pressure. As another example, the research question could be whether or not a leukemia patient's white blood cell count is predictive of his/her survival time. Research designs may be classified as experimental or observational. *Regression analyses* are applicable to both types; yet the confidence one has in the results of a study can vary with the research type. In most cases, one variable is usually taken to be the *response* or *dependent variable,* which is the variable to be predicted from or explained by other variables. The other variable or variables are called *predictor* or *predictors*, or explanatory variables or *independent variables.* The above examples, and others, show a wide range of applications in which the dependent variable is a continuous measurement. Such a variable is often assumed to be normally

Applied Categorical Data Analysis and Translational Research, Second Edition, By Chap T. Le

distributed and a *model* is formulated to express the *mean* of this normal distribution as a function of potential independent variables under investigation. The dependent variable is denoted by Y and the study often involves a number of *risk factors* or *predictor variables*: $X_1, X_2, \ldots, X_k$ (it's conventional that we use capital letters for variable names and lowercase letters for their values).

Consider the simplest case where only one predictor or independent variable is available for predicting the response of interest. The regression model describes how possible values of Y are distributed to give a fixed value x of X:

$$Y_i = \beta_0 + \beta_1 x_i + \varepsilon_i$$

where

(i) x_i is a given fixed value of the independent variable X and Y_i is the (random) value of the response or dependent variable Y from the ith .subject

(ii) β_0 and β_1 are the two fixed but unknown parameters; $(\beta_0 + \beta_1 x_i)$ is the mean of Y_i as stipulated by the model

(iii) ε_i is a random error term that is distributed as normal with mean zero and variance s^2; s^2 is also the variance of Y_i because $(\beta_0 + \beta_1 x_i)$ is fixed

The above model is referred to as the *simple linear regression model*. It is simple because it contains only one independent variable. It is *linear* because the independent variable appears only in the first power; if we graph the mean of Y versus X, the graph is a *straight line* with *intercept* β_0 and *slope* β_1.

Choosing an appropriate model and analytical technique depends on the type of variable under investigation. As mentioned, in a variety of applications, the dependent variable of interest is a continuous variable that we can assume, perhaps after an appropriate transformation, to be normally distributed. The *regression model* describes the *mean* of that normally distributed dependent variable Y as a function of the value of the predictor or predictors. In a variety of other applications, however, the dependent variable of interest may have only two possible outcomes. Consider, for example, an analysis of whether or not business firms have a daycare facility, according to the number of female employees. The dependent variable in this study was defined to have two possible outcomes: (i) the firm has a daycare facility and (ii) the firm does not have a daycare facility. As another example, consider a study of drug use among middle school children as a function of gender and age of child, family structure (e.g., who is the head of household), and family income. In this study, the dependent variable Y was defined to have two possible outcomes: (i) child uses drug and (ii) child does not use drug. The following is a typical example in health sciences research.

■ **Example 4.1** When a patient is diagnosed as having cancer of the prostate, an important question in deciding on treatment strategy for the patient is whether or not the cancer has spread to the neighboring lymph nodes. The question is so critical in prognosis and treatment that it is customary to operate on the patient (i.e., perform a laparotomy) for the sole purpose of examining the nodes and removing tissue samples to examine under the microscope for evidence of cancer. However, certain variables that can be measured without surgery are predictive of nodal involvement; and the purpose of the study was to examine the data for 53 prostate cancer patients receiving surgery, to determine which of five preoperative variables are predictive of nodal involvement. In particular, the principal investigator was interested in the predictive value of the level of acid phosphatase in blood serum. Table 4.1 presents the complete data set.

For each of the 53 patients, there are two continuous independent variables—age at diagnosis and level of serum acid phosphatase (x100; called "Acid")—and three binary variables—X-ray reading, pathology reading (grade) of a biopsy of the tumor obtained by needle before surgery, and a rough measure of the size and location of the tumor (stage) obtained by palpation with the fingers via the rectum. For these three binary independent variables a value of 1 signifies a positive or more serious state and a value of 0 denotes a negative or less serious finding. In addition, the sixth column presents the finding at surgery—the

TABLE 4.1. Prostate Cancer Data

Point	X-Ray	Grade	Stage	Age	Acid	Node	Point	X-Ray	Grade	Stage	Age	Acid	Node
1	0	1	1	64	40	0	**29**	0	0	0	52	83	0
2	0	0	1	63	40	0	**30**	0	0	1	67	95	0
4	0	1	0	67	47	0	**31**	0	0	0	56	98	0
5	0	0	0	66	48	0	**32**	0	0	1	61	102	0
6	0	1	1	65	48	0	**33**	0	0	0	64	187	0
7	0	0	0	60	49	0	**34**	1	0	1	58	48	1
8	0	0	0	51	49	0	**35**	0	0	1	65	49	1
8	0	0	0	66	50	0	**36**	1	1	1	57	51	1
10	0	0	0	58	50	0	**37**	0	1	0	50	56	1
11	0	1	0	56	50	0	**38**	1	1	0	67	67	1
12	0	0	1	61	50	0	**39**	0	0	1	67	67	1
13	0	1	1	64	50	0	**40**	0	1	1	57	67	1
14	0	0	0	56	52	0	**41**	0	1	1	45	70	1
15	0	0	0	67	52	0	**42**	0	0	1	46	70	1
16	1	0	0	49	55	0	**43**	1	0	1	51	72	1
17	0	1	1	52	55	0	**44**	1	1	1	60	76	1
18	0	0	0	68	56	0	**45**	1	1	1	56	78	1
19	0	1	1	66	59	0	**46**	1	1	1	50	81	1
20	1	0	0	60	62	0	**47**	0	0	0	56	82	1
21	0	0	0	61	62	0	**48**	0	0	1	63	82	1
22	1	1	1	59	63	0	**49**	1	1	1	65	84	1
23	0	0	0	51	65	0	**50**	1	0	1	64	89	1
24	0	1	1	53	66	0	**51**	0	1	0	59	99	1
25	0	0	0	58	71	0	**52**	1	1	1	68	126	1
26	0	0	0	63	75	0	**53**	1	0	0	61	136	1
27	0	0	1	53	76	0							
28	0	0	0	60	78	0							

primary binary response or dependent variable Y, a value of 1 denoting nodal involvement, and a value of 0 denoting no nodal involvement. A careful reading of the data reveals, for example, that a positive X-ray or an elevated acid phosphatase level, in general, seems likely to be associated with nodal involvement found at surgery. However, predictive values of other variables are not clear and to answer the question, for example, concerning the usefulness of acid phosphatase as a prognostic variable, we need a more detailed analysis before a conclusion can be made.

Example 4.1 shows a wide range of applications in which the dependent variable Y is binary dichotomous, and hence may be represented by a variable taking the value 1 with probability π and the value 0 with probability $(1 - \pi)$. Such a variable is a Bernoulli (or "point-binomial") variable and π is the mean of the dependent variable Y.

$$\begin{aligned} E(Y) &= \Pr(Y = 1) \\ &= \pi \end{aligned}$$

To keep the conventional approach of the *regression modeling,* which describes the *mean* of the dependent variable Y as a function of the value of the predictor or predictors, we need to "model" this probability π as a function of potential independent variables under investigation. The most popular result is the *logistic regression model*, which has been used extensively and successfully in the health sciences to describe the probability (or risk) of developing a condition—say, a disease—over a specified time period as a function of certain risk factors.

4.1 MODELING A PROBABILITY

Let us first consider the following "indirect assay"; it is an experiment in which each of a number of predetermined levels of a stimulus (e.g., dose of a drug) is applied to n experimental units; r of them respond and $(n - r)$ do not respond. In a direct assay, the amount of stimulus needed to produce the response in each individual subject can be measured at the time the event occurs and is called the *individual effect dose* (or IED). In this indirect assay, we cannot measure IEDs because only one fixed dose is given to a group of n subject. Therefore,

(i) If that dose is below some particular IED, the response does not occur.
(ii) Subjects who respond are those with IEDs below the given fixed dose.

Assume that the design consists of a series of dose levels with subjects completely randomized to the dose levels. At the end, we have data in the form

$$\{(n_i, r_i; x_i = \log(\text{Dose}_i)\} \Rightarrow \{p_i = r_i/n_i; x_i\}$$

We have, for example, the data in Table 4.2.

TABLE 4.2.

Dose	Number (n)	Number Killed	X	p (%)
0.1	47	8	1.000	17.0
0.15	53	14	1.176	26.4
0.2	55	24	1.301	43.6
0.3	52	32	1.477	61.5
0.5	46	38	1.699	82.6
0.7	54	50	1.845	92.6
0.95	52	50	1.978	96.2

■ **Example 4.2** Data in Table 4.2 show the effect of different concentrations of (nicotine sulfate in a 1% saponin solution) on fruit flies. Here $X = \log(100 \times \text{Dose})$, making the numbers easier to read.

We can interpret these data as follows: (i) 17% (8 out of $n = 47$) of the first group respond to a dose of 0.1 g/100 cc ($X = 1.0$); that means 17% of subjects have IEDs less than 0.1; (ii) 26.4% (14 out of $n = 53$) of the second group respond to a dose of 0.15 g/100 cc ($X = 1.176$); that means 26.4% of subjects have IEDs less than 0.15; and so on. In other words, we view each dose D (with $X = \log D$) as an upper endpoint of an interval and p as the cumulative relative frequency. This view and the symmetric sigmoid dose–response curve in Figure 4.1 suggest that it be considered as some cumulative distribution function (cdf).

Let π be the probability of response at a particular dose—where the log (Dose) is X. It is estimated by $p = r/n$. In the context of this cumulative distribution function, the probability π has been transformed into a variable Y on the "linear"

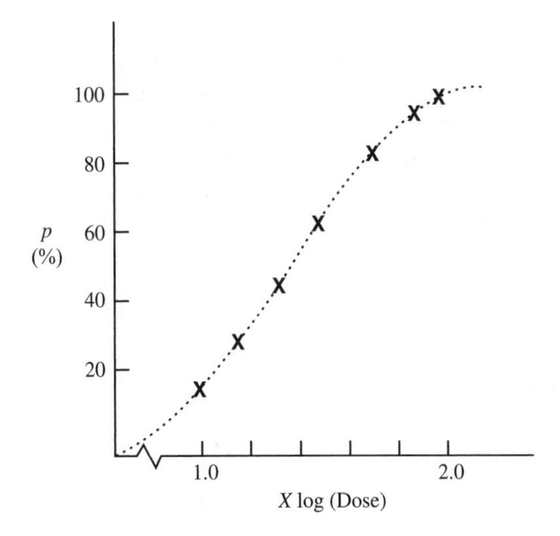

FIGURE 4.1. The symmetric sigmoid dose–response curve.

or continuous scale with unbounded range as

$$\pi = \int_{-\infty}^{Y} f(\theta)d\theta \quad \text{or} \quad \pi = \int_{Y}^{\infty} f(\theta)d\theta$$

The *dose–response relationship*, or any other relationships between a dependent variable Y and an independent variable X, is then stipulated by the *simple linear regression*,

$$Y = \beta_0 + \beta_1 x$$

All we need to implement this modeling process is a probability density function *f*(.). There are many more possibilities.

4.1.1 The Logarithmic Transformation

The simplest choice is the unit exponential probability density function,

$$f(\theta) = e^{-\theta};\ \theta \geq 0$$

leading to

$$\begin{aligned} \pi &= \int_{-\beta_0 - \beta_1 x}^{\infty} e^{-z} dz \\ &= e^{\beta_0 + \beta_1 x};\ \text{or}\ \ln \pi = \beta_0 + \beta_1 x \end{aligned}$$

That is, we model the "log" of the probability as a "linear function" of the value of the independent variable. The advantage of this approach of modeling the log of the probability as a linear function of covariates is the easy interpretation of model parameters and the probability is changed by a multiple constant. Calculations are also simple; however, this approach has the disadvantage that fitted probabilities may exceed 1.0; the unit exponential distribution is defined only on the positive range.

4.1.2 The Probit Transformation

Another choice that had been rather popular for many years is the *standard normal probability density function*,

$$\begin{aligned} f(\theta) &= \frac{1}{\sqrt{2\pi}} \exp\left(-\frac{\theta^2}{2}\right) \\ \pi &= \int_{-\infty}^{\beta_0 + \beta_1 x} \frac{1}{\sqrt{2\pi}} \exp\left(-\frac{\theta^2}{2}\right) d\theta \end{aligned}$$

We call Y the "probit" of π; the word probit is a shortened form of the phrase "PROBability unIT" (but it is not a probability); it is a standard normal variate. There is no closed-form expression that can be used to calculate y from p; as a result, it would be very difficult to check if Y is a linear function of X (e.g., using the scatter diagram).

4.1.3 The Logistic Transformation

The popularity of the probit method is of historical nature and the statistician behind it was D. J. Finney, However, its popularity has since faded. Instead of the standard normal density, one can use the unit logistic probability density leading to a logistic transformation that has many more advantages.

$$f(\theta)=\frac{\exp(\theta)}{[1+\exp(\theta)]^2}$$

$$\pi=\int_{-\infty}^{Y=\beta_0+\beta_1 x}\frac{e^{\theta}}{[1+e^{\theta}]^2}d\theta$$

$$=\frac{\exp(\beta_0+\beta_1 x)}{1+\exp(\beta_0+\beta_1 x)}$$

$$=\frac{1}{1+\exp(-\beta_0-\beta_1 x)}$$

$$=\frac{e^{\beta_0+\beta_1 x}}{1+e^{\beta_0+\beta_1 x}}$$

$$1-\pi=\frac{1}{1+e^{\beta_0+\beta_1 x}}$$

$$\frac{\pi}{1-\pi}=e^{\beta_0+\beta_1 x}$$

$$\log\left(\frac{\pi}{1-\pi}\right)=\beta_0+\beta_1 x$$

The transformation (from observed data) $\log[p/(1-p)]$ is called the *logistic transformation* and the resulting method is the logistic regression.

4.2 SIMPLE REGRESSION ANALYSIS

In this section, we discuss, in greater detail the basic ideas of simple regression analysis when only one predictor or independent variable is available for predicting the response of interest. In the interpretation of the primary parameter of the model, we discuss both scales of measurement, discrete and continuous, even though in practical applications of simple regression, the independent variable under investigation is often on a continuous scale.

4.2.1 The Simple Logistic Regression Model

The usual "regression" analysis goal is to describe the "mean" of a dependent variable Y as a function of a set of predictor variables. We deal exclusively with the case where the basic random variable Y of interest is a dichotomous variable taking the value 1 with probability π and the value 0 with probability $(1-\pi)$. Such a random variable is called a point-binomial or Bernoulli variable, and it has the following simple discrete probability distribution:

$$\Pr(Y=y)=\pi^{y}(1-\pi)^{1-y}$$

In addition, for the simplest case of simple regression, suppose that for the ith individual, $1=i=n$, there is a sample of data $\{(x_i, y_i)\}$. As indicated earlier, there are several possibilities but we will focus only on the simple linear logistic regression model:

$$\begin{aligned}\pi_i &= \frac{1}{1+\exp(-\beta_0-\beta_1 x_i)}\\ &= \frac{\exp(\beta_0+\beta_1 x_i)}{1+\exp(\alpha+\beta_1 x_i)}\\ \log\left(\frac{\pi_i}{1-\pi_i}\right) &= \beta_0+\beta_1 x_i\end{aligned}$$

There are a number of important reasons that make logistic regression popular:

1. The range of the logistic function is between 0 and 1; that makes it suitable for use as a probability model, representing individual risk.
2. The logistic curve has an increasing S-shape with a threshold; that makes it suitable for use as a biological model, representing risk due to exposure.
3. The regression coefficients α and β have interesting interpretation (in terms of disease development).
4. There are strong empirical supports from fields such as pharmacology.

The fit and the origin of empirical supports for the linear logistic model could easily be traced as follows. When a dose y of an agent is applied to a pharmacological system, the fractions f_a and f_u of the system affected and unaffected satisfy the *median effect principle* (Chou, 1976):

$$\frac{f_a}{f_u}=\left\{\frac{y}{ED_{50}}\right\}^{m}$$

where ED_{50} is the *median effective dose* and m is a Hill-type coefficient; $m = 1$ for a first-degree or Michaelis–Menten system. The median effect principle has been investigated thoroughly in pharmacology. If we set $\pi = f_a$, the median effect principle and the logistic regression model are completely identical with a slope $\beta_1 = m$.

Under the above simple logistic regression model, the likelihood function is given by

$$\begin{aligned} L &= \prod_{i=1}^{n} \Pr(Y_i = y_i) \\ &= \prod_{i=1}^{n} \frac{[\exp(\beta_0 + \beta_1 x_i)]^{y_i}}{[1 + \exp(\beta_0 + \beta_1 x_i)]}; \ y_i = 0, 1 \end{aligned}$$

from which we can obtain maximum likelihood estimates of the parameters β_0 and β_1. As mentioned previously, the logistic model has been used extensively and successfully to describe the probability of developing ($Y = 1$) some disease over a specified time period as a function of a risk factor X.

4.2.2 Measure of Association

Regression analysis serves two major purposes: (i) control or intervention, and (ii) prediction. In many studies, such as the one in Example 4.1, one important objective is "measuring" the strength of a statistical relationship between the binary dependent variable and each independent variable or covariate measured from patients. Findings may lead to important decisions in patient management (or public health interventions in other examples). In epidemiological studies, such effects are usually measured by the relative risk or odds ratio; when the logistic model is used, the measure is odds ratio.

For the case of the logistic regression model, the logistic function for the probability π_i can be expressed as a linear model in the log scale:

$$\ln\left(\frac{\pi_i}{1 - \pi_i}\right) = \beta_0 + \beta_1 x_i$$

What we have on the left-hand side of this equation is the "logit" or log of the odds (say, for disease development). Let us first consider the case of a binary covariate with conventional coding:

$$x_i = \begin{cases} 0 & \text{if the } i\text{th subject/patient is not exposed} \\ 1 & \text{if the } i\text{th subject/patient is exposed} \end{cases}$$

Here, the term "exposed" may refer to a risk factor such as smoking, or a patient's characteristic such as race (white/nonwhite) or gender (male/female). It can be seen that, from the above loglinear form of the logistic regression model,

$$\ln(\text{Odds; nonexposed}) = \beta_0$$
$$\ln(\text{Odds; exposed}) = \beta_0 + \beta_1$$

so that the difference (of the two equations) leads to, after an exponentiation (or taking the antilog),

$$e^{\beta_1} = \frac{(\text{Odds; exposed})}{(\text{Odds; nonexposed})}$$

which represents the odds ratio (OR) associated with the exposure—exposed versus nonexposed. In other words, the primary regression coefficient β_1 is the value of the odds ratio on the log scale.

Similarly, we have for a continuous covariate X and any value x of X,

$$\ln(\text{Odds; } X = x) = \beta_0 + \beta_1(x)$$
$$\ln(\text{Odds; } X = x+1) = \beta_0 + \beta_1(x+1)$$

So that the difference (of the two equations) leads to, after an exponentiation (or taking the antilog),

$$e^{\beta_1} = \frac{(\text{Odds; } X = x+1)}{(\text{Odds; } X = x)}$$

which represents the odds ratio (OR) associated with "one unit increase" in the value of X, $X = x + 1$ versus $X = x$. For example, a systolic blood pressure of 114 mmHg versus 113 mmHg. For an m unit increase in the value of X, say, $X = x + m$ versus $X = x$, the corresponding odds ratio is $\exp(m\beta_1)$.

The primary regression coefficients β_1 and β_0 (β_0 is often not needed) can be estimated iteratively using a packaged computer program such as SAS. From the results, we can obtain a point estimate and its 95% confidence interval:

$$\exp[\hat{\beta}_1 + 1.96 SE(\hat{\beta}_1)]$$

The Effect of Measurement Scale

It should be noted that the odds ratio, used as a measure of association between the binary dependent variable and a covariate, depends on the coding scheme for a binary covariate and, for a continuous covariate X, the scale with which to measure X. For example, if we use the following coding for a factor,

$$x_i = \begin{cases} -1 & \text{if the } i\text{th subject/patient is not exposed} \\ 1 & \text{if the } i\text{th subject/patient is exposed} \end{cases}$$

then

$$\begin{aligned} OR &= \frac{(\text{Odds; exposed})}{(\text{Odds; nonexposed})} \\ &= e^{2\beta_1} \end{aligned}$$

and its 95% confidence interval is given by

$$\exp[2[\hat{\beta}_1 + 1.96SE(\hat{\beta}_1)]]$$

Of course, the estimate of β_1—and its standard error—under the new coding scheme is only half of that under the former scheme; therefore the estimate of the OR remains unchanged.

The following example, however, will show the clear effect of measurement scale in the case of a continuous measurement.

■ **Example 4.3** Refer to the data for patients diagnosed as having cancer of the prostate in Example 4.1 and suppose we want to investigate the relationship between nodal involvement found at surgery and the level of acid phosphatase in blood serum in two different ways using either (i) $X =$ Acid or (ii) $X = \log_{10}$(Acid).

(i) For $X =$ Acid, we find $\beta_1 = 0.0204$, from which the odds ratio for (Acid = 100) versus (Acid = 50) would be

$$\begin{aligned} OR &= \exp[(100-50)(0.0204)] \\ &= 2.77 \end{aligned}$$

(ii) For $X = \log_{10}$(Acid), we find $\beta_1 = 5.1683$, from which the odds ratio for (Acid = 100) versus (Acid = 50) would be

$$\begin{aligned} OR &= \exp\{[\log_{10}(100) - \log_{10}(50)][5.1683]\} \\ &= 4.74 \end{aligned}$$

Note: If $X =$ Acid is used, a SAS program would include these instructions:

```
PROC LOGISTIC DESCENDING
DATA = CANCER;
MODEL NODES = ACID;
```

where CANCER is the name assigned to the data set, NODES is the variable name for nodal involvement, and ACID is the variable name for our covariate, the level of acid phosphatase in blood serum. The option DESCENDING is needed because PROC LOGISTIC models $\Pr(Y=0)$ instead of $\Pr(Y=1)$.

The above results are different for two different choices of X and this seems to cause an obvious problem of choosing an appropriate measurement scale. Of course, we assume a "linear model" and one choice of scale for X would fit better than the other. However, it is very difficult to compare different scales unless there were replicated data at each level of X; if such replications are available, one can simply graph a scatter diagram of log(Odds) versus X-value and check for linearity of each choice of scale of measurement for X.

4.2.3 Tests of Association

The last two subsections dealt with inferences concerning the primary regression coefficient β_1, including both point and interval estimation of this parameter and the odds ratio. Another aspect of statistical inference concerns the test of significance. The null hypothesis to be considered is

$$H_0:\ \beta_1 = 0$$

The reason for interest in testing whether or not $\beta_1 = 0$ is that $\{\beta_1 = 0\}$ implies there is no relation between the binary dependent variable and the covariate X under investigation. Since the likelihood function is rather simple, one can easily derive, say, the score test for the above null hypothesis; however, nothing would be gained by going through this exercise. We can simply apply a chi-squared test (if the covariate is binary or categorical) or a t-test or Wilcoxon test (if the covariate under investigation is on a continuous scale). Of course, the application of the logistic model is still desirable, at least in the case of a continuous covariate, because it would provide a measure of association.

4.2.4 Use of the Logistic Model for Different Designs

Data for risk determination may come from different sources, with the two fundamental designs being retrospective and prospective. Prospective studies enroll a group or groups of subjects, follow them over certain periods of time (examples include occupational mortality studies and clinical trials), and observe the occurrence of certain events of interest such as a disease or death. Retrospective studies gather past data from selected cases and controls to determine differences, if any, in the exposure to a suspected risk factor. They are commonly referred to as case–control studies. It can be seen that the logistic model fits in very well with the prospective or follow-up studies and has been used successfully to model the "risk" of developing a condition—say, a disease—over a specified time period as a function of certain risk factor. In such applications, after a logistic model has been fitted, one can estimate the individual risks π_i—given the covariate value x—as well as any risks ratio or relative risk,

$$RR = \frac{\pi(x_i)}{\pi(x_j)}$$

As for case–control studies, it can be shown, using the so-called Bayes' theorem—that if the population is

$$\Pr(Y = 1; \text{given } x) = \frac{1}{1 + \exp[-(\beta_0 + \beta_1 x)]}$$

then

$$\Pr(Y = 1; \text{given } x \text{ and the sample}) = \frac{1}{1 + \exp[-(\beta_0^* + \beta_1 x)]}$$

$$\beta_0^* = \beta_0 + \frac{\theta_1}{\theta_0}$$

where θ_1 is the probability that a case was sampled and θ_0 is the probability that a control was sampled. This result indicates the following for a case–control study:

(i) We cannot estimate individual risks and we cannot estimate the relative risk, unless θ_0 and θ_1 are both known, which is unlikely. The value of the "intercept" β_0^* provided by the computer output is meaningless.

(ii) However, since we have the same β_1 as with the population model, we can still estimate the odds ratio and, if the rare disease assumption applies, we can interpret the numerical result as an approximate relative risk.

4.2.5 Overdispersion

Logistic regression is based on the point-binomial or Bernoulli distribution; its mean is π_i and the variance is $(\pi_i)(1 - \pi_i)$. If we use the variance mean ratio as a dispersion parameter then it is 1 in a standard logistic model, less than 1 in an underdispersed model, and greater than 1 in an overdispersed model. Overdispersion is a common phenomenon in practice and it causes concern because the implication is serious; the analysis that assumes the logistic model often underestimates standard error(s) and thus wrongly inflates the level of significance.

Measuring and Monitoring Dispersion
After a logistic regression model is fitted, dispersion is measured by the scaled deviance or scaled Pearson chi-squared statistic; it is the deviance or Pearson

chi-squared statistic divided by the degrees of freedom. The deviance is defined as twice the difference between the maximum achievable log likelihood and the log likelihood at the maximum likelihood estimates of the regression parameters. Suppose that data with replications consist of m subgroups (with identical covariate values); then the Pearson chi-squared statistic and deviance are given by

$$\chi_{\mathrm{P}}^2 = \sum_i \frac{(r_i - n_i p_i)^2}{n_i p_i}$$

$$\chi_{\mathrm{D}}^2 = \sum_i r_i \log \frac{r_i}{n_i p_i}$$

Each of these goodness-of-fit statistics, divided by the appropriate degrees of freedom, called the scaled Pearson chi-squared statistic and scaled deviance, respectively, can be used as a measure for overdispersion (underdispersion, with measures less than 1, occurs much less often in practice). When their values are much larger than 1, the assumption of binomial variability may not be valid and the data are said to exhibit overdispersion. There can be several reasons for overdispersion: for example, outliers in the data, omitting important covariates in the model, and the need to transform some explanatory factors.

The SAS program PROC LOGISTIC has an option called AGGREGATE that can be used to form subgroups. Without such grouping, data may be too sparse, the Pearson chi-squared statistic and deviance do not have a chi-squared distribution and the scaled Pearson chi-squared statistic and scaled deviance cannot be used as indicators of overdispersion. A large difference between the scaled Pearson chi-squared statistic and scaled deviance provides evidence of this situation.

Fitting an Overdispersed Logistic Model

One way of correcting overdispersion is to multiply the covariance matrix by the value of the overdispersion parameter φ, the scaled Pearson chi-squared statistic or scaled deviance (as used in weighted least squares fitting):

$$E(p_i) = \pi_i$$

$$\mathrm{Var}(p_i) = \varphi \pi_i (1 - \pi_i)$$

In this correction process, the parameter estimates are not changed. However, their standard errors are adjusted (increased), affecting their significance levels (reduced).

■ **Example 4.4** In a study of the toxicity of a certain chemical compound, five groups of 20 rats each were fed for 4 weeks by a diet mixed with that compound at five different doses. At the end of the study, their lungs were harvested and subjected to histopathological examinations to observe for sign(s) of toxicity (yes = 1, no = 0). The results were as follows:

Group	Dose (mg)	Number of Rats	Number of Rats with Toxicity
1	5	20	1
2	10	20	3
3	15	20	7
4	20	20	14
5	30	20	10

A routine fit of the simple logistic regression model yields the following results:

Variable	Coefficient	Standard Error	z-Statistic	p-Value
Intercept	−2.3407	0.538	−4.3507	0.0001
Dose	0.1017	0.0277	3.6715	0.0002

In addition, we obtained these results for the monitoring of overdispersion:

Parameter	Chi-squared	df	Scaled Parameter
Pearson	10.9919	3	3.664
Deviance	10.7863	3	3.595

Note: A SAS program would include these instructions:

```
INPUT DOSE N TOXIC;
PROC LOGISTIC DESCENDING;
MODEL NODES TOXIC/N= DOSE/SCALE=NONE;
```

The above results indicate an obvious sign of overdispersion. By fitting an overdispersed model, controlling for the scaled deviance, we have the following:

Variable	Coefficient	Standard Error	z-Statistic	p-Value
Intercept	−2.3407	1.0297	−2.2732	0.0203
Dose	0.1017	0.053	1.9189	0.0548

Compared to the previous results, the point estimates remain the same but the standard errors are larger, and the effect of dose is no longer significant at the 5% level.

Note: A SAS program would include these instructions (using deviance scale):

```
INPUT DOSE N TOXIC;
PROC LOGISTIC DESCENDING;
MODEL NODES TOXIC/N= DOSE/SCALE=D;
```

4.3 MULTIPLE REGRESSION ANALYSIS

The effect of some factor on a dependent or response variable may be influenced by the presence of other factors through effect modifications (i.e., interactions). Therefore in order to provide a more comprehensive analysis, it is very desirable to consider a large number of factors and sort out which ones are most closely related to the dependent variable. In this section, we discuss a multivariate method for risk determination. This method, which is multiple logistic regression analysis, involves a linear combination of the explanatory or independent variables; the variables must be quantitative with particular numerical values for each patient. A covariate or independent variable—such as a patient characteristic—may be dichotomous, polytomous, or continuous (categorical factors will be represented by dummy variables). Examples of dichotomous covariates are gender and presence or absence of certain comorbidity. Polytomous covariates include race and different grades of symptoms; these can be covered by the use of dummy variables. Continuous covariates include patient age and blood pressure; in many cases, data transformations (e.g., taking the logarithm) may be desirable to satisfy the linearity assumption.

4.3.1 Logistic Regression Model with Several Covariates

Suppose we want to consider k covariates simultaneously. The simple logistic model of the previous section can easily be generalized and expressed as

$$\pi_i = \frac{1}{1+\exp\left[-\left(\beta_0 + \sum_{j=1}^{k} \beta_j x_{ji}\right)\right]}$$

$$= \frac{\exp\left(\beta_0 + \sum_{j=1}^{k} \beta_j x_{ji}\right)}{1+\exp\left(\beta_0 + \sum_{j=1}^{k} \beta_j x_{ji}\right)}$$

$$\log\left(\frac{\pi_i}{1-\pi_i}\right) = \beta_0 + \sum_{j=1}^{k} \beta_j x_{ji}$$

Under the above simple logistic regression model, the likelihood function is given by

$$L = \prod_{i=1}^{n} \Pr(Y_i = y_i)$$

$$= \prod_{i=1}^{n} \frac{\left[\exp\left(\beta_0 + \sum_{j=1}^{k} \beta_j x_{ji}\right)\right]^{y_i}}{\left[1 + \exp\left(\beta_0 + \sum_{j=1}^{k} \beta_j x_{ji}\right)\right]}; \; y_i = 0, 1$$

from which parameters can be estimated iteratively using a packaged computer program such as SAS. Similar to the case of the simple logistic regression model, the multiple logistic model has been used extensively and successfully to describe the probability of developing ($Y = 1$) some disease over a specified time period as a function of several risk factors. Also similar to the univariate case, $\exp(\beta_i)$ represents:

(i) the odds ratio associated with an exposure if X_i is binary (exposed $X_i = 1$ vs. unexposed $X_i = 0$), or

(ii) the odds ratio due to one unit increase if X_i is continuous ($X_i = x + 1$ vs. $X_i = x$).

After β_i has been estimated and the standard error of its estimate has been obtained, a 95% confidence interval for the above odds ratio is given by

$$\exp[\hat{\beta}_i + 1.96 SE(\hat{\beta}_i)]$$

These results are necessary in the effort to identify important risk factors for the binary outcome. Of course, before such analyses are done, the problem and the data have to be examined carefully. If some of the variables are highly correlated, then one or fewer of the correlated factors are as likely to be good predictors as all of them; information from other similar studies also has to be incorporated so as to drop some of these correlated explanatory variables. The uses of products, such as $X_1 * X_2$, and higher power terms, such as X_1^2, may be necessary and can improve the goodness-of-fit. It is important to note that we are assuming a *loglinear* regression model in which, for example, the odds ratio due to one unit increase in the value of a continuous X_i ($X_i = x + 1$ vs. $X_i = x$) is independent of x. Therefore if this linearity seems to be violated, the incorporation of powers of X_i should be seriously considered. The use of products will help in the investigation of possible effect modifications. Finally, we mention the messy problem of missing data; most packaged programs would delete a subject if one or more covariate values are missing.

4.3.2 Effect Modifications

Consider the model for the case of two covariates:

$$\pi_i = \frac{1}{1+\exp[-(\beta_0+\beta_1 x_{1i}+\beta_2 x_{2i}+\beta_3 x_{1i}x_{2i})]}$$

$$\log\left(\frac{\pi_i}{1-\pi_i}\right) = \beta_0+\beta_1 x_{1i}+\beta_2 x_{2i}+\beta_3 x_{1i}x_{2i}$$

The meaning of β_1 and β_2 here is not the same as that given earlier because of the cross-product term $\beta_3 x_1 x_2$. Suppose that both X_1 and X_2 are binary:

(i) For $X_2 = 1$ or exposed, we have

$$\ln(\text{Odds; nonexposed to } X_1, X_1 = 0) = \beta_2$$
$$\ln(\text{Odds; exposed to } X_1 = 1) = \beta_1+\beta_2+\beta_3$$

so that the difference (of the two equations) leads to, after an exponentiation (or taking antilog),

$$e^{\beta_1+\beta_3} = \frac{(\text{Odds; exposed to } X_1 | \text{exposed to } X_2)}{(\text{Odds; not exposed to } X_1 | \text{exposed to } X_2)}$$

which represents the odds ratio associated with the exposure to X_1, exposed versus nonexposed in the presence of X_2.

(ii) For $X_2 = 2$ or nonexposed, we have

$$\ln(\text{Odds; nonexposed to } X_1, X_1 = 0) = 0$$
$$\ln(\text{Odds; exposed to } X_1 = 1) = \beta_1$$

so that the difference (of the two equations) leads to, after an exponentiation (or taking antilog),

$$e^{\beta_1} = \frac{(\text{Odds; exposed to } X_1 | \text{not exposed to } X_2)}{(\text{Odds; not exposed to } X_1 | \text{not exposed to } X_2)}$$

which represents the odds ratio associated with the exposure to X_1, exposed versus nonexposed in the absence of X_2.

In other words, the effect of X_1 depends on the level (presence or absence) of X_2 and vice versa. This phenomenon is called *effect modification*: that is, one factor modifies the effect of the other. The cross-product term $x_1 x_2$ is called an interaction term; the use of these products will help in the investigation of possible effect modifications. If $\beta_3 = 0$, the effect of two factors acting together, as measured by the odds ratio, is equal to the combined effects of two factors acting separately, as measured by the product of two odds ratios:

$$\begin{aligned}\text{Odds ratio (exposed to both vs. none)} &= e^{\beta_1+\beta_2} \\ &= e^{\beta_1}e^{\beta_2}\end{aligned}$$

This fits the classic definition of "no interaction" on a multiplicative scale.

4.3.3 Polynomial Regression

Consider the model for the case of one covariate X:

$$\begin{aligned}\pi_i &= \frac{1}{1+\exp[-(\beta_0+\beta_1 x_i+\beta_2 x_i^2)]} \\ \log\left(\frac{\pi_i}{1-\pi_i}\right) &= \beta_0+\beta_1 x_i+\beta_2 x_i^2\end{aligned}$$

where X is a continuous covariate. The meaning of β_1 here is not the same as that given earlier because of the quadratic term $\beta_2 x_i^2$. We have, for example,

$$\begin{aligned}&\ln(\text{Odds};\ X=x)=\beta_0+\beta_1 x+\beta_2 x^2 \\ &\ln(\text{Odds};\ X=x+1)=\beta_0+\beta_1(x+1)+\beta_2(x+1)^2\end{aligned}$$

So that the difference (of the two equations) leads to, after an exponentiation (or taking antilog),

$$\begin{aligned}OR &= \frac{(\text{Odds}; X=x+1)}{(\text{Odds}; X=x)} \\ &= \exp[\beta_1+\beta_2(2x+1)]\end{aligned}$$

which represents the odds ratio (OR) associated with "one unit increase" in the value of X, $X=x+1$ versus $X=x$; in this case, it is a function of x.

Polynomial models with an independent variable present in higher powers than the second are not often used. The second-order or quadratic model has two basic types of uses:

(i) when the true relationship is a second-degree polynomial or when the true relationship is unknown but the second-degree polynomial provides a better fit than a linear one, but

(ii) more often, a quadratic model is fitted for the purpose of establishing the linearity; the key item to look for is whether $\beta_2=0$.

The use of polynomial models is not without drawbacks. The most important drawback is that multicollinearity is unavoidable: if the covariate is restricted to a narrow range, then the degree of multicollinearity can be quite high. Another problem arises when one wants to use the stepwise regression search method. In addition, finding a satisfactory interpretation for the "curvature" effect coefficient β_2 is not easy.

4.3.4 Testing Hypotheses in Multiple Logistic Regression

Once we have fit a multiple logistic regression model and obtained estimates for the various parameters of interest, we want to answer questions about the contributions of various factors to the prediction of the binary response variable. There are three types of such questions:

(i) *An overall test*: Taken collectively, does the entire set of explanatory or independent variables contribute significantly to the prediction of response?
(ii) *Test for the value of a single factor*: Does the addition of one particular variable of interest add significantly to the prediction of response over and above that achieved by other independent variables?
(iii) *Test for contribution of a group of variables*: Does the addition of a group of variables add significantly to the prediction of response over and above that achieved by other independent variables?

Overall Regression Tests We now consider the first question stated above concerning an overall test for a model containing k factors, say,

$$\begin{aligned}\pi_i &= \frac{1}{1+\exp\left[-\left(\beta_0+\sum_{j=1}^{k}\beta_j x_{ji}\right)\right]}\\ &= \frac{\exp\left(\beta_0+\sum_{j=1}^{k}\beta_j x_{ji}\right)}{1+\exp\left(\beta_0+\sum_{j=1}^{k}\beta_j x_{ji}\right)}\\ \log\left(\frac{\pi_i}{1-\pi_i}\right) &= \beta_0+\sum_{j=1}^{k}\beta_j x_{ji}\end{aligned}$$

The null hypothesis for this test may be stated as: "all k independent variables considered together do not explain the variation in the responses." In other words,

$$H_0\colon\ \beta_1=\beta_2=\cdots=\beta_k=0$$

Two likelihood-based statistics can be used to test this "global" null hypothesis; each has an asymptotic chi-squared distribution with k degrees of freedom under H_0.

(i) *Likelihood Ratio Test.*

$$X_{\mathrm{LR}}^2 = 2[\ln L(\hat{\boldsymbol{\beta}}) - \ln L(\mathbf{0})]$$

(ii) *Score Test.*

$$X_{\mathrm{S}}^2 = \left[\frac{\delta \ln L(\mathbf{0})}{\delta \boldsymbol{\beta}}\right]\left[-\frac{\delta^2 \ln L(\mathbf{0})}{\delta \boldsymbol{\beta}^2}\right]^{-1}\left[\frac{\delta \ln L(\mathbf{0})}{\delta \boldsymbol{\beta}}\right]$$

Both statistics are provided by most standard computer programs such as SAS and they are asymptotically equivalent yielding identical statistical decisions most of the time.

■ **Example 4.5** Refer to the data set on prostate cancer in Example 4.1 with all five covariates; we have the following test statistics for the global null hypothesis:

(i) *Likelihood Ratio Test.*

$$X_{\mathrm{LR}}^2 = 22.126 \text{ with 5 degrees of freedom; } p = 0.0005$$

(ii) *Score Test.*

$$X_{\mathrm{S}}^2 = 19.451 \text{ with 5 degrees of freedom; } p = 0.0016$$

Note: A SAS program would include these instructions:

```
PROC LOGISTIC DESCENDING
DATA = CANCER;
MODEL NODES = X-RAY, GRADE, STAGE, AGE, ACID;
```

where CANCER is the name assigned to the data set, NODES is the variable name for nodal involvement, and X-RAY, GRADE, STAGE, AGE, and ACID are the variable names assigned to the five covariates.

Tests for a Single Variable Let us assume that we now wish to test whether the addition of one particular independent variable of interest adds significantly to the prediction of the response over and above that achieved by other factors already present in the model. The null hypothesis for this test may be stated as: "factor X_i does not have any value added to the prediction of the response given that other factors are already included in the model." In other words,

$$H_0\colon\ \beta_i = 0$$

To test such a null hypothesis, one can perform a likelihood ratio chi-squared test, with 1 degree of freedom, similar to that for the above global hypothesis:

$$X^2_{\text{LR}} = 2[\ln L(\hat{\boldsymbol{\beta}}; \text{all } X\text{'s}) - \ln L(\hat{\boldsymbol{\beta}}; \text{all other } X\text{'s with } X_i \text{ deleted})]$$

A much easier alternative method is to use

$$z = \frac{\hat{\beta}_i}{SE(\hat{\beta}_i)}$$

In performing this test, we refer the value of the z-statistic to percentiles of the standard normal distribution.

■ **Example 4.6** Refer to the data set on prostate cancer in Example 4.1 with all five covariates:

Variable	Coefficient	Standard Error	z-Statistic	p-Value
Intercept	0.0618	3.4599	0.018	0.9857
X-ray	2.0453	0.8072	2.534	0.0113
Stage	1.5641	0.774	2.021	0.0433
Grade	0.7614	0.7708	0.988	0.3232
Age	−0.0693	0.0579	−1.197	0.2314
Acid	0.0243	0.0132	1.85	0.0643

The effects of X-ray and Stage are significant at the 5% level whereas the effect of Acid is marginally significant ($p = 0.0643$).

Note: Use the same SAS program as in the previous example.

Given a continuous variable of interest, one can fit a polynomial model and use this type of test to check for linearity. It can also be used to check for a single product representing an effect modification.

■ **Example 4.7** Refer to the data set on prostate cancer in Example 4.1 but this time we investigate only one covariate, the level of acid phosphatase (Acid).

After fitting the second-degree polynomial model,

$$\log\left(\frac{\pi_i}{1-\pi_i}\right) = \beta_0 + \beta_1(\text{Acid})_i + \beta_2(\text{Acid})_i^2$$

we obtained the following results indicating that the curvature effect should not be ignored ($p = 0.0437$).

Variable	Coefficient	Standard Error	z-Statistic	p-Value
Intercept	−7.32	2.6229	−2.791	0.0053
Acid	0.1489	0.0609	2.445	0.0145
Acid^2	−0.0007	0.0003	−2.017	0.0437

Contribution of a Group of Variables This testing procedure addresses the more general problem of assessing the additional contribution of two or more factors to the prediction of the response over and above that made by other variables already in the regression model. In other words, the null hypothesis is of the form

$$H_0\colon\ \beta_1 = \beta_2 = \cdots = \beta_m = 0$$

where $1 < m < k$

To test such a null hypothesis, one can perform a likelihood ratio chi-squared test, with m degree of freedom:

$$X^2_{\text{LR}} = 2[\ln L(\hat{\boldsymbol{\beta}}; \text{all } X\text{'s}) - \ln L(\hat{\boldsymbol{\beta}}; \text{all other } X\text{'s with } m\, X\text{'s under investigation deleted})]$$

As with the above z-test for a single variable, this procedure is very useful for assessing the importance of potential explanatory variables. In particular, it is often used to test whether a similar group of variables, such as demographic characteristics, is important for the prediction of the response; these variables have some trait in common. Another application would be a collection of powers and/or product terms (referred to as interaction variables). It is often of interest to assess the interaction effects collectively before trying to consider individual interaction terms in a model as previously suggested. In fact, such use reduces the total number of tests to be performed and this, in turn, helps to provide better control of overall Type I error rates, which may be inflated due to multiple testing.

■ **Example 4.8** Refer to the data set on prostate cancer in Example 4.1 with all five covariates and we consider, collectively, these four interaction terms: Acid ∗ X-ray, Acid ∗ Stage, Acid ∗ Grade, and Acid ∗ Age. The basic idea is to see if any of the other variables would modify the effect of the level of acid phosphatase on the response.

(i) With the original five variables, we obtain $\ln L = -24.063$.

(ii) With all nine variables, five original plus four products, we obtain $\ln L = -20.378$.

Therefore

$$\begin{aligned} X^2_{\text{LR}} &= 2[\ln L(\hat{\boldsymbol{\beta}}; \text{all 9 variables}) - \ln L(\hat{\boldsymbol{\beta}}; \text{ 5 variables})] \\ &= 2[(-20.378)-(-24.063)] \\ &= 7.370 \\ df &= 4 \\ p\text{-value} &= 0.11758 \end{aligned}$$

In other words, all four interaction terms, considered together, are not yet statistically significant; however, there may be some weak effect modification and the effect of acid phosphatase on the response may be somewhat stronger for a certain combination of levels of the other four variables.

Stepwise Regression

In many applications (e.g. a case–control study on a specific disease), our major interest is to identify important risk factors. In other words, we wish to identify from many available factors a small subset of factors that relate significantly to the outcome (e.g., the disease under investigation). In that identification process, of course, we wish to avoid a large Type I (false-positive) error. In a regression analysis, a Type I error corresponds to including a predictor that has no real relationship to the outcome; such an inclusion can greatly confuse the interpretation of the regression results. In a standard multiple regression analysis, this goal can be achieved by using a strategy that adds into or removes from a regression model one factor at a time according to a certain order of relative importance. Therefore the two important steps are:

1. Specify a criterion or criteria for selecting a model.
2. Specify a strategy for applying the chosen criterion or criteria.

Strategies This is concerned with specifying the strategy for selecting variables. Traditionally, such a strategy is concerned with which and whether a particular variable should be added to a model or whether any variable should be deleted from a model at a particular stage of the process. As computers became more accessible and more powerful, these practices became more popular.

FORWARD SELECTION PROCEDURE In the forward selection procedure, we proceed as follows:

Step 1: Fit a simple logistic linear regression model to each factor, one at a time.

Step 2: Select the most important factor according to a certain predetermined criterion.

Step 3: Test for the significance of the factor selected in Step 2 and determine, according to a certain predetermined criterion, whether or not to add this factor to the model.

Step 4: Repeat Steps 2 and 3 for those variables not yet in the model. At any subsequent step, if none meets the criterion in Step 3, no more variables are included in the model and the process is terminated.

BACKWARD ELIMINATION PROCEDURE In the backward elimination procedure, we proceed as follows:

Step 1: Fit the multiple logistic regression model containing all available independent variables.

Step 2: Select the least important factor according to a certain predetermined criterion; this is done by considering one factor at a time and treating it as though it were the last variable to enter.

Step 3: Test for the significance of the factor selected in Step 2 and determine, according to a certain predetermined criterion, whether or not to delete this factor from the model.

Step 4: Repeat Steps 2 and 3 for those variables still in the model. At any subsequent step, if none meets the criterion in Step 3, no more variables are removed in the model and the process is terminated.

STEPWISE REGRESSION PROCEDURE Stepwise regression is a modified version of forward regression that permits reexamination, at every step, of the variables incorporated in the model in previous steps. A variable entered at an early stage may become superfluous at a later stage because of its relationship with other variables now in the model; the information it provides becomes redundant. That variable may be removed, if meeting the elimination criterion, and the model is refitted with the remaining variables, and the forward process goes on. The whole process, one step forward followed by one step backward, continues until no more variables can be added or removed.

Criteria For the first step of the forward selection procedure, decisions are based on individual score test results (chi-squared, 1 df). In subsequent steps, both forward and backward, the ordering of levels of importance (Step 2) and the selection (test in Step 3) are based on the likelihood ratio chi-squared statistic:

$$X_{\text{LR}}^2 = 2[\ln L(\hat{\boldsymbol{\beta}}; \text{all } X\text{'s}) - \ln L(\hat{\boldsymbol{\beta}}; \text{all other } X\text{'s with one } X \text{ deleted})]$$

■ **Example 4.9** Refer to the data set on prostate cancer in Example 4.1 with all five covariates: X-ray, Stage, Grade, Age, and Acid. This time we perform a stepwise regression analysis in which we specify that a variable has to be significant at the 0.10 level before it can enter into the model and that a variable in the model has to be significant at the 0.15 level for it to remain in the model (most standard computer programs allow users to make these selections; default values are available). First, we get these individual score test results for all variables:

Variable	Coefficient	Standard Error	z-Statistic	p-Value
Intercept	0.0618	3.4599	0.018	0.9857
X-ray	2.0453	0.8072	2.534	0.0113
Stage	1.5641	0.774	2.021	0.0433
Grade	0.7614	0.7708	0.988	0.3232
Age	−0.0693	0.0579	−1.197	0.2314
Acid	0.0243	0.0132	1.85	0.0643

These indicate that X-ray is the most significant variable ($p = 0.0113$). Thus we proceed as follows.

Step 1: Variable X-ray is entered.

Analysis of Variables Not in the Model

Variable	LR X^2	p-Value
Stage	5.6394	0.0176
Grade	2.371	0.1236
Age	1.3523	0.2449
Acid	2.0733	0.1499

These indicate that Stage is the next most significant variable ($p = 0.0176$).
Step 2: Variable Stage is entered.

Analysis of Variables in the Model Consisting of X-Ray and Stage

Variable	Coefficient	Standard Error	z-Statistic	p-Value
Intercept	−2.0446	0.6100	−3.352	0.0008
X-ray	2.1194	0.7468	2.838	0.0045
Stage	1.5883	0.7000	2.269	0.0233

Neither variable is removed because no p-value is less than 0.15.

Analysis of Variables Not in the Model

Variable	LR X^2	p-Value
Grade	0.5839	0.4448
Age	1.2678	0.2602
Acid	3.0917	0.0787

These indicate that Acid is the next most significant variable ($p = 0.0787$).
Step 3: Variable Acid is entered.

Analysis of Three Variables in the Model

Variable	Coefficient	Standard Error	z-Statistic	p-Value
Intercept	−3.5756	1.1812	−3.027	0.0025
X-ray	2.0618	0.7777	2.651	0.0080
Stage	1.7556	0.7391	2.375	0.0175
Acid	0.0206	0.0126	1.631	0.1029

The results indicate that none of the variables should be removed and the analysis of variables not in the model indicates that no additional variables meet the 0.1 level required for entry into the model:

Variable	LR X^2	p-Value
Grade	1.0650	0.3020
Age	1.5549	0.2124

Note: A SAS program would include these instructions:

```
PROC LOGISTIC DESCENDING;
DATA = CANCER;
MODEL NODES = XRAY, GRADE, STAGE, AGE, ACID
/SELECTION = STEPWISE SLE=.10 SLS=.15 DETAILS;
```

where CANCER is the name assigned to the data set, NODES is the variable name for nodal involvement, and XRAY, GRADE, STAGE, AGE, and ACID are the variable names assigned to the five covariates. The option DETAILS provides step-by-step detailed results; without specifying it, we would have only the final fitted model (which is just fine in practical applications). The default values for SLE(entry) and SLS (stay) probabilities are 0.05 and 0.10, respectively.

4.3.5 Measures of Goodness-of-Fit

In usual (i.e., Gaussian) regression analyses, R^2 gives the proportional reduction in variation in comparing the conditional variation of the response to the marginal variation. It describes the strength of the association between the response and the set of independent variables considered together; for example, with $R^2 = 1$ we can predict the response perfectly. We present here three measures of goodness-of-fit for the logistic regression model; however, none has all the advantages of the coefficient of determination, R^2.

(i) The first option is a likelihood-based measure proposed by McFadden (1974) defined by

$$D = \frac{\ln L(\mathbf{0}) - \ln L(\hat{\boldsymbol{\beta}})}{\ln L(\mathbf{0})}$$

where $\ln L(\boldsymbol{\beta})$ denotes the maximized log likelihood for the fitted model and $\ln L(\mathbf{0})$ the maximized log likelihood for the "null" model containing only an intercept term. Both of these numbers, $\ln L(\boldsymbol{\beta})$ and $\ln L(\mathbf{0})$, are provided by most packaged computer programs such as SAS.

(ii) The second measure is a member of a family developed by Efron (1978). Basically, this measure compares the binary responses y_i's and their corresponding fitted probability values π_i's. It represents the proportional reduction in error obtained by using π instead of the average as a predictor of y_i. This gives

$$E = 1 - \frac{\sum (y_i - \hat{\pi}_i)^2}{\sum (y_i - \bar{y})^2}$$

This measure, E, and the previous measure, D, are not easy to interpret.

(iii) The last measure we present is obtained by an application of a receiver--operator characteristic (ROC) curve analysis; more details are given in Chapter 8. After fitting a logistic regression model, each subject's fitted response probability, π_i, is calculated as in the previous measure, E. Using these probabilities as values of a separator, we can construct a nonparametric ROC curve tracing sensitivities against the estimated false positivities for various cut-points. Such an ROC curve not only makes it easy to determine an optimal cut-point but it also shows the overall performance of the fitted logistic regression model; the better the performance the further away the curve is from the diagonal. The area, C, under this ROC curve can be used as a measure of goodness-of-fit. Measure C represents the separation power of the logistic model under consideration; for example, with $C = 1$, the fitted response probabilities for subjects with $y = 1$ and the fitted response probabilities for subjects with $y = 0$ are completely separated. A function of C can also be used, for example,

$$C^* = \frac{C - 0.5}{0.5}$$

may have a more desirable range, between 0 and 1. To minimize the risk of optimistic bias that occurs when a fitted model is applied to the same data to which it was fit, a Jacknife approach should be used. In this approach, each subject's fitted response probability was calculated from a regression analysis generated by excluding his or her data when fitting the model. Of course, the computing is very tedious and there are no packaged computer programs for the task (it is possible to use PROC LOGISTIC of SAS with the addition of a macrofile).

■ **Example 4.10** Refer to the data set on prostate cancer in Example 4.1 with all five covariates and the fitted results shown in Example 4.6. We have $\ln L(\mathbf{0}) = -35.126$ and $\ln L(\boldsymbol{\beta}) = -24.063$, leading to

$$\begin{aligned} D &= \frac{\ln L(\mathbf{0}) - \ln L(\hat{\boldsymbol{\beta}})}{\ln L(\mathbf{0})} \\ &= \frac{(-35.126)-(-24.063)}{(-35.126)} \\ &= 0.315 \end{aligned}$$

Using the estimated regression parameters obtained from Example 4.6 (where the model was fitted once), we have $E = 0.362$ and $C = 0.845$ ($C^* = 0.690$).

Note: The area under the ROC curve (i.e., measure C) is provided by SAS PROC LOGISTIC. The computer output also includes another measure by Cox and Snell (1989).

Since the measure of goodness-of-fit A has a meaningful interpretation and increases when we add an explanatory variable to the model, it can be used as a criterion in performing stepwise logistic regression instead of the p-value, which is easily influenced by the sample size. For example, in the forward selection procedure, we proceed as follows:

Step 1: Fit a simple logistic linear regression model to each factor, one at a time.

Step 2: Select the most important factor defined as the one with the largest value of the measure of goodness-of-fit A.

Step 3: Compare this value of A for the factor selected in Step 2 and determine, according to a predetermined criterion, whether or not to add this factor to the model—say, increase A by 5% over its value when no factor is considered.

Step 4: Repeat Steps 2 and 3 for those variables not yet in the model. At any subsequent step, if none meets the criterion in Step 3—say, increase the separation power by 5%—no more variables are included in the model and the process is terminated.

4.4 ORDINAL LOGISTIC MODEL

As seen in the previous two sections, logistic regression is most often used for a dichotomous response where it models the "logit" (log of the odds) of the probability π of having an "event" (e.g., a disease). Suppose we want to consider k covariates simultaneously. The logistic model is expressed as

$$\begin{aligned}\operatorname{logit}(\pi) &= \log\left(\frac{\pi}{1-\pi}\right)\\ &= \beta_0 + \sum_{j=1}^{k} \beta_j x_{ji}\end{aligned}$$

This common logistic model is now generalized for use with an ordinal response having more than two levels. Such a targeted ordinal response has several categories with some natural ordering but perhaps without a defined metric (i.e., no numeric value associated with each category). For example, the condition of a disease can be classified as "normal," "mild," "moderate," and "severe," a four-level ordinal response. For any ordinal response, we can define cumulative logits. For example, if the ordinal response under investigation has k levels with corresponding polynomial probabilities $\pi_1, \pi_2, \ldots, \pi_k$, where

$$\sum_{i=1}^{k} \pi_i = 1$$

then the cumulative logits are defined as

$$F_i = \frac{\sum_{j=1}^{i} \pi_j}{\sum_{j=i+1}^{k} \pi_j}$$

An ordinal logistic regression model describes a relationship between an ordinal response and a set of explanatory or independent variables. Suppose we want to consider k covariates simultaneously. The proportional odds model by McCullagh (1980) assumes that

$$\log(F_i) = \beta_0 + \sum_{j=1}^{k} \beta_j x_{ji}$$

In other words, the proportional odds model assumes that each logit follows a linear model that has a separate intercept parameter but other regression parameters (i.e., the slopes relating the response to each covariate) are constant across all cumulative logits for different levels of the response. This is similar to the parallelism assumption in a bioassay and the proportional hazards model in survival analysis. In addition, since

$$F_i \leq F_{i+1}$$

it follows that there is a constraint on the intercepts:

$$\beta_{0i} \leq \beta_{0(i+1)}$$

■ **Example 4.11** To provide a simple illustration of ordinal logistic regression, let us consider the example in Hanley and McNeil (1983). They studied 112 phantoms that were specially constructed to evaluate the "accuracy" of a computer algorithm used in the image construction for CT. Fifty-eight (58) of these phantoms were of uniform density and were designated as normal, $x = 0$; the remaining 54 contained an area of reduced density to simulate a lesion and were designated as abnormal, $x = 1$. The computer algorithm "reads" each image and rates it on a six-point ordinal scale (Y): 1 = definitely normal, 2 = probably normal, 3 = possibly normal, 4 = possibly abnormal, 5 = probably abnormal, and 6 = definitely abnormal. The results (frequencies) were as follows:

	Computer Reading (Y)					
Status (X)	1	2	3	4	5	6
Normal	12	28	8	6	4	0
Abnormal	1	3	6	13	22	9

Using the computer reading as an ordinal dependent variable, a logistic regression analysis yields the following:

Factor	Coefficient	Standard Error	z-Statistic	p-Value
Intercept 1	−4.7595	0.5201	−9.151	0.0001
Intercept 2	−2.7531	0.3976	−6.924	0.0001
Intercept 3	−1.6175	0.3254	−4.970	0.0001
Intercept 4	−0.7541	0.2767	−2.725	0.0064
Intercept 5	1.3146	0.3156	4.165	0.0001
Status (X)	3.0932	0.4464	6.930	0.0001

The results indicate a level of accuracy of the computer algorithm; the odds that an abnormal phantom would be judged as at or above a certain level of abnormality by the computer algorithm, say, probably or definitely abnormal ($\geq$5), is

$$\begin{aligned} OR &= \exp(3.0932) \\ &= 22.04 \end{aligned}$$

times the odds that a normal phantom would be judged as at or above that same level of abnormality.

Note: A SAS program would include these instructions:

```
DATA PHANTOM;
INPUT EXPLANATORY; RATING COUNT;
CARDS;
0 1 12
0 2 28
0 3 8
0 4 6
0 5 4
0 6 0
1 1 1
1 2 3
1 3 6
1 4 13
1 5 22
1 6 9
;
PROC LOGISTIC DESCENDING;
WEIGHT COUNT;
MODEL RATINGY=YEXPLANATORY;
```

where `EXPLANATORY` is the true abnormality (0 being normal), and `COUNT` is the frequency of each group at each level of reading (`RATING`) by the computer algorithm.

We can also obtain almost the same result using the method of Chapter 2. The computer rating would be dichotomized with a moving division point—say, (6) versus (1, 2, 3, 4, 5), then (5, 6) versus (1, 2, 3, 4), and so on—to form five 2×2 tables. We then assume that the odds ratios are constant across tables (the basic proportional odds assumption in the ordinal logistic model). The common odds ratio is then estimated by

$$\begin{aligned} OR &= \frac{\sum ad}{\sum bc} \\ &= 22.19 \end{aligned}$$

In the following example, taken from Rosner (1982), the response variable has numerical values but we can also treat it as ordinal as well.

■ **Example 4.12** This data set consists of 216 persons aged 20–39 with retinitis pigmentosa (RP). The patients were classified on the basis of a detailed family history into the genetic types of autosomal dominant RP (DOM), autosomal recessive RP (AR), sex-linked RP (SL), and isolate RP (ISO). The outcome is obtained from a routine ocular examination; an eye was considered affected if visual acuity was 20/50 or worse, and normal otherwise. The response variable (Y) is the number of eyes affected (0, 1, or 2) as tabulated in the following table:

	Number of Eyes Affected		
Type	0	1	2
DOM	15	6	7
AR	7	5	9
SL	3	2	14
ISO	67	24	57

Using the number of affected eyes as an ordinal dependent variable and the autosomal dominant RP (DOM) group as the baseline for comparison (each other group is represented by an indicator variable), a logistic regression analysis yields the following:

Factor	Coefficient	Standard Error	z-Statistic	p-Value
Intercept 1	−0.9437	0.3726	−2.532	0.0113
Intercept 2	−0.2184	0.3671	−0.595	0.552
AR	0.7688	0.5495	1.399	0.1618
SL	1.9594	0.6323	3.099	0.0019
ISO	0.4392	0.3967	1.107	0.2682

The results indicate that only the SL group is significantly different from the others.

4.5 QUANTAL BIOASSAYS

Translational research is the component of clinical research that interacts with basic science (T1) or with population research (T2). We often emphasize the first area of translational research (T1)—research efforts and activities needed to bring discoveries in the laboratories to the bedsides. Bench-to-bedside research often necessitates some deep understanding of the underlying basic science. And it is hard to pinpoint precisely the starting point of T1; many believe that translational research starts with *biological assays* or *bioassays*.

Biological assays or bioassays are methods for estimating the potency or strength of an "agent" or "stimulus" by utilizing the "response" (or effect, or reaction) caused by its application to biological material or experimental living "subjects." The three components of a bioassay are:

(i) The subject is usually an animal, a human tissue, or a bacteria culture.
(ii) The agent is usually a drug or a chemical.
(iii) The response is usually a change in a particular characteristic or even the death of a subject; responses can be binary or measured on continuous scale.

For example, "Six aspirin tablets can be fatal to a child." In this example the child is the subject, aspirin is the drug, and fatal reaction is the response. In most applications, we do not usually estimate the potency (or strength) of an agent or drug because such estimates depend on the biological system, the "subjects" that make applications more limited. Instead, we usually can estimate the "relative potency" of an agent or drug by comparing its effect on biological material, such as animals or animal tissue, with those of a standard product. This is very similar to the case of epidemiology; we are interested in studying a certain "risk", and we end up measuring "relative risk." Any risk can be measured but it is more difficult and it does not have any special meaning unless we compare the measured risk to "background risk."

There are deterministic or nonstochastic assays, but they are not subjected to statistical analyses. An assay is stochastic if the relative potency is influenced by factors other than the preparations; that is, extraneous factors that cannot be completely controlled or explained. In other words, the response is subjected to a random error; for example, either the "dose" or the "response" is a random variable—depending on the design. The underlying rationale of biological assays is that if the relationship between stimulus level and response exists (by means of an algebraic expression), then it can be used to study the potency of a dose from the response it produces. For stochastic assays, our only targets, we refer to the relationship between stimulus level and the response it produces (by means of an algebraic expression) as a *regression model*. For a certain class of bioassays, it's the logistic regression models of this chapter.

4.5.1 Types of Bioassays

There are two types of bioassays, both stochastic: (i) direct assays and (ii) indirect assays.

In direct assays, the doses of the standard and test preparations are directly measured for (or until) an event of interest. Response is fixed (binary), dose is random. When an event of interest occurs, for example, the death of the subject, and the variable of interest is the dose required to produce that response/event for each subject, the value is called *individual effect dose* (IED). For example, we can increase the dose until the heartbeat (of an animal) ceases to evaluate the IED.

Direct assays are only applicable when "intrasubject dose escalation" is possible. For technical and ethical reasons, this is not done, for example, in experiments with human beings where indirect assays are applicable. In indirect assays, the doses of the standard and test preparations are applied and we observe the "response" that each dose produces; for example, we measure the tension in a tissue or the hormone level or the blood sugar content. For each subject, the dose is fixed in advance, the variable of interest is not the dose but the response it produces in each subject; the response could be binary or continuous.

In indirect assays, the dose is fixed and the response is random; and that response could be a measurement or the occurrence of an event (whereas the response in direct assays is always binary, the occurrence of an event). Depending on the "measurement scale" for the response (of indirect assays), we have the following:

1. Quantal assays, where the response is binary: whether or not an event (like the death of the subject) occurs.
2. Quantitative assays, where measurements for the response are on a continuous scale. Quantitative assays are further divided into (i) parallel-line bioassays and (ii) slope-ratio bioassays. These bioassays, with a continuous outcome or dependent variable, are not covered in this book.

4.5.2 Quantal Response Bioassays

Quantal response assays belong to the class of qualitative indirect assays. They are characterized by experiments in which each of a number of predetermined levels of a stimulus (e.g., dose of a drug) is applied to n experimental units; r of them respond and $(n-r)$ do not respond. That is a "binary" response (yes/no). The group size n may vary from dose to dose; in theory, some n could be 1 (so that $r=0$ or 1).

Quantal Assays Versus Direct Assays

In direct assays, the doses of the standard and test preparations are *measured* for an "event of interest"; intrapatient adjustment is needed. When an (predetermined) event of interest occurs (e.g., the death of the subject), the variable of interest is the dose required to produce that response/event for each subject. That is, the dose is measured right at the time the event occurs; it is not possible to do it if the dose is fixed in advance (indirect assays). It is assumed that each subject has its own tolerance to a particular preparation. In a direct assay, the amount of stimulus needed to produce the response in each individual subject can be measured, called the IED. In quantal bioassays, we cannot measure IEDs because only one fixed dose is given to a group of n subjects: (i) if that dose is below some particular IED, the response does not occur; and (ii) subjects who responsed are those with IEDs below the given fixed dose.

In biological assays, we usually can only estimate "relative potency" of an agent, not its "potency." Quantal assays are the exceptions to this rule: there is a difference here.

(i) In quantitative assays, the absolute standard, or the yardstick, is not available; therefore we cannot estimate potency (the strength)—we can only estimate the relative potency, or relative strength.

(ii) In quantal assays, the yardstick is available—the event; therefore we can estimate the "potency" and "relative potency" too if we wish.

Measure of Potency

Quantal bioassays are qualitative; we observe occurrences of an event—we do not obtain measurements on a continuous scale. Because the event is well defined, we can estimate the agent's potency. The most popular parameter is the level of the stimulus that results in a response by 50% of individuals in a population. It is often denoted by LD_{50} for median lethal dose, or ED_{50} for median effective dose, or EC_{50} for median effective concentration. However, measures of potency depend on the biological system used; the estimates of LD_{50} for preparations of the same system can be used to form a ratio to estimate the relative potency—which would more likely be independent from the system. The most popular parameter, LD_{50} (for median lethal dose), or ED_{50} (for median effective dose), or EC_{50} (for median effective concentration), is the level of the stimulus that results in a response by 50% of individuals in a population:

(i) It is a measure of the agent's potency, which could be used to calculate relative potency.
(ii) It is chosen by a statistical reason; for any fixed number of subjects, one would attain greater precision as compared to estimating, say, LD_{90} or LD_{10} or any other percentiles.

The Assay Procedure

The usual design consists of a series of dose levels with subjects completely randomized among the dose levels. The experiment may include a standard and a test preparation, or perhaps just the test. The dose levels chosen should range from "very low" (few or no subjects would respond) to "rather high" (most or all subjects would respond). The objective is often to estimate the LD_{50}; the number of observations per preparation depends on the desired level of precision of its estimate. Sample size estimation is a very complicated topic.

Estimation of Potency and Relative Potency

Data, from the above assay procedure, are of the form

$$\{(n_i, r_i; x_i = \log(\text{Dose}_i)\} \Rightarrow \{p_i = r_i/n_i; x_i\}$$

As mentioned previously, there are several possible ways to transform the proportion p into some measurement Y on the continuous scale with an unbounded range. The most solid and popular one is the logistic transformation leading to the logistic regression model, mostly because of its strong empirical supports from the median effect principle:

$$p_i = \frac{1}{1+\exp(-\beta_0 - \beta_1 x_i)}$$
$$= \frac{\exp(\beta_0 + \beta_1 x_i)}{1+\exp(\alpha + \beta_1 x_i)}$$
$$\log\left(\frac{p_i}{1-p_i}\right) = y_i$$
$$= \beta_0 + \beta_1 + x_i$$

The parameters β_0 and β_1 are estimated by their maximum likelihood estimates b_0 and b_1, and the log of the LD_{50} (or ED_{50}) is estimated by

$$m = -\frac{b_0}{b_1}$$

It should be noted that log potency is in the form of a ratio b_0/b_1. One can calculate the variance and standard deviation by the delta method; however, it would be the same old problem when it comes to forming confidence intervals for ratios because the normal approximation may not work well unless we have a large sample. One efficient way to deal with confidence intervals (CIs) for ratios is the use of Fieller's theorem.

■ **Example 4.13** Consider the following small data set from an assay on mice:

Dose	Number of Mice	p (%)
0.2	27	14.8
0.4	30	43.3
0.8	30	60
1.6	30	73.3

We can easily obtain these results using SAS PROC LOGISTIC:

$x = \log(\text{Dose})$	n	p	$Y = \ln[p/(1-p)]$
−0.7	27	0.15	−1.73
−0.4	30	0.43	−0.28
−0.1	30	0.6	0.41
0.2	30	0.73	0.99
		log $LD_{50} = -0.22$; $LD_{50} = 0.61$	
		95% CI for LD_{50}: (0.43,0.85)	

EXERCISES

4.1 Radioactive radon is an inert gas that can migrate from soil and rock and accumulate in enclosed areas such as underground mines and homes. The radioactive decay of trace amounts of uranium in the Earth's crust through radium is the source of radon, or more precisely the isotope radon-222. Radon-222 emits alpha particles; when inhaled, the alpha particles rapidly diffuse across the alveolar membrane of the lung and are transported by the blood to all parts of the body. Due to the relatively high flow rate of blood in bone marrow, this may be a biologically plausible mechanism for the development of leukemia. Table 4.3 provides some data from a case–control study to investigate the association between indoor residential radon exposure and risk of childhood acute myeloid leukemia. The variables are as follows:

- Disease (1 = Case, 2 = Control).
- Some characteristics of the child: Sex (1 = Male, 2 = Female) and Race (1 = White, 2 = Black, 3 = Hispanic, 4 = Asian, and 5 = Other).
- Radon (radon concentration in Bq/m^3).
- Risk factors from the parents: M-smoke (1 = Mother a current smoker, 2 = No, 0 = Unknown); M-drink (1 = Mother a current alcohol drinker, 2 = No, 0 = Unknown); F-smoke (1 = Father a current smoker, 2 = No, 0 = Unknown); and F-drink (1 = Father a current alcohol drinker, 2 = No, 0 = Unknown).
- Down syndrome, a known risk factor for leukemia (1 = No, 2 = yes).

  (a) Taken collectively, do the covariates contribute significantly to the separation of the cases and the controls? Give your interpretation for the measure of goodness-of-fit C.

  (b) Fit the multiple regression model to obtain estimates of individual regression coefficients and their standard errors. Draw your conclusion concerning the conditional contribution of each factor.

  (c) Within the context of the multiple regression model in (b), does sex alter the effect of Down syndrome

  (d) Within the context of the multiple regression model in (b), does Down syndrome alter the effect of radon exposure?

  (e) Within the context of the multiple regression model in (b), taken collectively, do smoking–drinking variables (by the father or mother) relate significantly to the disease of the child?

  (f) Within the context of the multiple regression model in (b), is the effect of radon concentration linear?

  (g) Focusing on radon exposure as the primary factor, taken collectively, was this main effect altered by any other covariates?

TABLE 4.3. Radon–Leukemia Data

Disease	Sex	Race	Radon	M-Smoke	M-Drink	F-Smoke	F-Drink	Down Syndrome
1	2	1	17	1	1	1	2	1
1	2	1	8	1	2	2	0	2
2	2	1	8	1	2	1	2	2
2	1	1	1	2	0	2	0	2
1	1	1	4	2	0	2	0	2
2	1	1	4	1	1	1	1	2
1	2	1	5	2	0	2	0	2
2	1	1	4	2	0	2	0	2
1	2	1	7	1	1	2	0	1
1	1	1	15	1	1	1	2	1
2	1	1	16	2	0	1	1	1
1	1	1	12	2	0	1	2	1
2	1	1	14	1	1	1	1	1
2	2	1	12	2	0	2	0	1
2	1	1	14	1	2	1	2	1
2	2	1	9	1	2	1	1	2
2	1	1	4	2	0	1	1	2
1	1	1	2	2	0	1	1	1
1	2	1	12	2	0	1	1	2
1	2	1	13	2	0	2	0	1
2	2	1	13	2	0	1	2	2
2	1	1	18	2	0	2	0	1
1	1	1	13	1	2	1	1	2
1	2	1	16	2	0	2	0	1
1	2	1	10	1	1	2	0	2
1	1	1	11	2	0	2	0	1
2	2	1	4	1	1	1	1	2
1	2	1	1	1	2	1	1	2
1	1	2	9	2	0	2	0	1
1	2	1	15	1	1	1	2	1
2	2	1	17	2	0	1	2	1
1	1	1	9	2	0	1	2	1
2	1	1	15	2	0	2	0	1
1	1	1	10	1	1	1	1	1
2	1	1	11	1	2	2	0	1
1	2	1	8	2	0	2	0	1
1	1	1	14	1	2	2	0	2
2	2	1	14	1	2	1	2	2
1	2	1	1	2	0	1	1	2
2	1	1	1	2	0	1	1	2
1	2	1	6	1	2	1	2	2
2	1	1	16	2	0	2	0	1
2	2	1	3	2	0	2	0	2
1	2	1	5	2	0	1	2	2
2	1	2	15	2	0	1	0	1

(continued)

TABLE 4.3. (*Continued*)

Disease	Sex	Race	Radon	M-Smoke	M-Drink	F-Smoke	F-Drink	Down Syndrome
1	1	1	17	2	0	1	2	2
2	1	1	17	1	2	1	2	2
1	2	1	3	2	0	2	0	2
2	1	1	11	2	0	2	0	2
2	2	1	14	1	2	1	1	2
1	1	1	17	0	1	2	1	1
2	1	1	1	1	2	2	0	2
2	1	1	10	2	0	2	0	2
1	1	1	14	1	1	1	1	1
2	1	1	4	1	2	1	2	2
1	1	3	12	2	0	2	0	2
1	2	1	9	1	1	2	0	2
2	2	1	7	2	0	1	2	2
1	2	1	5	1	2	1	2	2
1	1	1	8	2	0	2	0	2
2	1	1	9	2	0	2	0	2
2	1	1	15	1	2	1	2	2
1	2	1	10	2	0	2	0	1
2	2	1	10	2	0	2	0	1
2	1	1	1	2	0	2	0	2
2	2	1	1	2	0	1	1	2
1	2	1	9	1	1	2	0	1
1	2	5	14	1	1	2	0	2
1	2	1	8	2	0	1	1	2
2	2	1	7	2	0	2	0	2
2	2	1	13	2	0	1	1	2
1	2	1	1	2	0	2	0	2
2	2	1	1	2	0	1	2	2
1	1	5	12	2	0	1	2	1
2	1	1	11	1	2	2	0	1
1	2	1	2	2	0	1	2	2
2	2	1	3	1	1	1	1	2
2	2	1	6	1	2	1	2	1
2	2	1	3	2	0	1	2	2
1	1	3	1	2	0	2	0	2
2	2	5	2	1	2	1	2	2
1	1	5	14	1	1	1	2	1
1	1	1	1	1	1	1	1	2
2	1	1	12	2	0	1	2	1
2	2	1	13	1	1	1	1	2
1	2	1	11	2	0	2	0	1
2	1	1	11	2	0	2	0	1
1	2	1	16	1	2	2	0	1
2	1	1	3	2	0	1	2	2
1	1	1	13	2	0	2	0	1

TABLE 4.3. (*Continued*)

Disease	Sex	Race	Radon	M-Smoke	M-Drink	F-Smoke	F-Drink	Down Syndrome
2	2	1	12	2	0	2	0	1
2	1	1	12	2	0	2	0	1
1	1	1	3	1	2	2	0	2
2	2	1	5	2	0	1	2	2
1	2	1	7	1	1	1	1	1
2	2	1	7	2	0	1	1	1
1	1	1	2	2	0	1	2	2
2	1	1	2	1	1	1	2	2
1	1	1	2	2	0	1	1	2
2	1	1	2	2	0	2	0	2
2	2	1	2	1	1	1	1	2
1	2	1	3	1	2	2	0	2
2	2	1	3	2	0	2	0	2
2	1	1	14	2	0	2	0	1
2	2	1	15	1	1	1	2	1
1	2	1	1	2	0	1	1	2
2	2	1	1	2	0	1	2	2
2	1	1	1	2	0	1	2	2
2	2	3	10	2	0	1	2	1
2	1	3	9	2	0	1	1	1
2	2	1	14	2	0	2	0	2
2	2	1	9	1	1	1	1	1
2	2	1	9	1	1	1	1	2
1	1	1	17	2	0	1	2	1
2	2	1	3	1	1	1	1	2
2	1	1	5	2	0	1	1	2
1	1	1	15	1	2	1	2	1
2	1	1	14	1	1	2	0	2
2	2	1	5	1	2	1	2	2
2	2	3	13	2	0	1	1	1
2	1	1	15	2	0	1	1	2
2	1	1	12	2	0	1	1	1
1	1	1	1	2	0	1	2	2
2	2	1	2	2	0	2	0	2
2	2	3	1	2	0	1	2	2
2	2	5	3	2	0	2	0	2
2	2	1	15	2	0	2	0	1
1	1	2	6	2	0	2	0	1
2	1	1	5	1	2	1	1	2
1	1	1	2	2	0	2	0	2
1	1	1	15	1	1	2	0	1
1	1	2	5	2	0	1	1	2
1	1	1	2	1	1	1	1	2
2	2	1	2	1	2	1	2	2
1	2	1	17	2	0	1	2	2

(*continued*)

TABLE 4.3. (*Continued*)

Disease	Sex	Race	Radon	M-Smoke	M-Drink	F-Smoke	F-Drink	Down Syndrome
2	1	1	8	2	0	2	0	2
2	2	1	6	2	0	1	1	2
1	2	1	1	1	2	1	1	2
1	1	1	13	2	0	1	2	1
2	2	1	12	2	0	1	1	1
2	1	1	9	2	0	2	0	1
1	1	1	10	1	1	2	0	1
2	2	1	15	1	1	2	0	1
1	2	1	3	2	0	1	2	2
2	1	1	2	1	2	1	1	2
1	2	1	11	1	2	1	2	2
2	2	1	11	2	0	1	2	2
2	1	1	4	2	0	2	0	2
1	2	1	2	2	0	2	0	2
1	2	1	12	2	0	1	1	1
1	2	1	1	2	0	1	2	2
1	1	1	15	1	1	1	2	2
2	2	1	8	2	0	2	0	2
2	2	1	1	1	2	1	1	2
2	2	1	2	1	1	1	1	2
1	1	1	8	1	2	1	1	2
2	1	1	2	2	0	1	2	2
1	1	1	1	1	1	1	1	2
1	2	1	12	2	0	2	0	1
2	1	1	12	1	2	1	1	1
1	2	1	1	2	0	1	1	2
2	1	1	12	2	0	2	0	2
1	2	1	16	2	0	2	0	2
2	2	1	9	2	0	2	0	1
2	2	3	7	2	0	2	0	2
1	2	4	2	2	0	1	2	2
2	1	1	1	2	0	2	0	2
1	2	1	11	1	2	1	2	2
2	2	1	13	1	2	1	2	1
2	2	1	7	1	1	1	1	2
2	1	1	13	1	2	1	2	1
1	2	1	14	1	2	1	2	2
2	1	1	2	2	0	2	0	2
1	2	1	1	1	1	2	0	2
2	2	1	2	2	0	2	0	2
1	2	1	2	2	0	2	0	2
2	2	1	1	1	2	2	0	2
2	1	5	1	2	0	1	1	2
2	1	1	9	1	1	1	2	1
1	1	1	13	1	2	2	0	2
2	1	1	12	2	0	2	0	1

TABLE 4.3. (*Continued*)

Disease	Sex	Race	Radon	M-Smoke	M-Drink	F-Smoke	F-Drink	Down Syndrome
1	2	1	17	2	0	2	0	1
2	2	1	10	2	0	2	0	1
2	1	1	11	1	2	1	2	1
2	2	1	5	2	0	2	0	2
2	2	1	15	2	0	2	0	1
1	2	1	2	2	0	1	1	2
2	2	1	2	1	1	2	0	2
2	2	1	11	1	1	1	1	2
1	1	1	10	2	0	1	0	1
1	1	1	5	1	1	1	2	2
2	1	1	4	1	2	2	0	2
2	2	1	1	1	1	1	1	2
2	2	1	1	2	0	2	0	2
1	2	1	2	2	0	2	0	2
1	2	1	10	1	2	2	0	2
2	2	1	9	2	0	2	0	1
1	1	1	13	2	0	2	0	2
2	1	1	12	2	0	2	0	1
2	2	1	14	2	0	1	1	2
1	1	1	3	1	2	2	0	2
1	1	5	12	2	0	2	0	2
1	1	1	2	2	0	1	2	2
2	1	1	3	1	1	2	0	2
2	1	1	11	1	1	1	1	2
1	1	1	15	1	1	1	1	2
2	2	1	15	1	1	1	0	1
1	2	1	9	1	2	1	1	1
2	2	1	10	1	1	1	1	1
2	2	1	2	2	0	1	1	2
2	1	1	2	2	0	2	0	2
2	1	1	13	2	0	2	0	2
2	2	1	10	2	0	2	0	2
2	1	1	9	2	0	2	0	2
1	2	1	12	2	0	1	1	1
2	2	1	12	1	2	2	0	2
1	1	1	8	1	2	1	1	1
2	1	1	7	2	0	2	0	2
1	2	1	1	2	0	2	0	2
2	2	1	14	1	2	1	2	1
1	2	1	9	2	0	1	2	2
2	2	1	9	1	2	1	2	1
1	1	5	15	2	0	1	2	1
2	2	1	14	2	0	2	0	2
1	1	1	12	2	0	1	1	1
1	2	1	1	2	0	2	0	2

(*continued*)

TABLE 4.3. (*Continued*)

Disease	Sex	Race	Radon	M-Smoke	M-Drink	F-Smoke	F-Drink	Down Syndrome
2	2	1	2	2	0	1	2	2
2	2	3	2	2	0	1	2	2
1	2	1	4	2	0	2	0	2
1	1	1	0	2	0	2	0	2
2	2	1	0	2	0	2	0	2
2	1	1	4	2	0	2	0	2
2	1	1	2	1	2	1	1	2
1	2	3	16	2	0	1	2	2
2	1	1	2	2	0	1	2	2

4.2 Recall the following exercise from Chapter 2 (Exercise 2.20): Prematurity, which ranks as the major cause of neonatal morbidity and mortality, has traditionally been defined on the basis of a birth weight under 2500 g. But this definition encompasses two distinct types of infants: infants who are small because they are born early, and infants who are born at or near term but are small because their growth was retarded. "Prematurity" has now been replaced by (i) low birth weight to describe the second type and (ii) preterm to characterize the first type (babies born before 37 weeks of gestation).

A case–control study of the epidemiology of preterm delivery was undertaken at Yale–New Haven Hospital in Connecticut during 1977. The study population consisted of 175 mothers of singleton preterm infants and 303 mothers of singleton full-term infants. The following table gives the distribution of socioeconomic status:

Socioeconomic Level	Cases	Controls
Upper	11	40
Upper-middle	14	45
Middle	33	64
Lower-middle	59	91
Lower	53	58

(a) Investigate the relationship between socioeconomic status and the disease.

(b) Is it true that the poorer the mother the higher the risk? Measure the strength of this relationship by an odds ratio.

(c) Reanalyze this problem using an appropriate ordinal logistic regression model.

4.3 Use the following data set to estimate the LD_{50}, both point estimate and 95% confidence interval.

Dose	n	r	$p(\%)$
0.1	47	8	17
0.15	53	14	26.4
0.2	55	24	43.6
0.3	52	32	61.5
0.5	46	38	82.6
0.7	54	50	92.6
0.95	52	50	96.2

4.4 Use the following data set to estimate the LD_{50}, both point estimate and 95% confidence interval, and test for goodness-of-fit.

Dose	Mice (n)	$p(\%)$
0.2	27	14.8
0.4	30	43.3
0.8	30	60
1.6	30	73.3

4.5 Use the following data set to estimate the relative potency.

Preparation	Dose	n	r
Standard	3.4	33	0
	5.2	32	5
	7	38	11
	8.5	37	14
	13	37	21
	18	31	23
	28	30	27
Test	6.5	40	2
	10	30	10
	14	40	18
	21.5	35	21
	29	37	27

5

METHODS FOR MATCHED DATA

Case–control studies have been perhaps the most popular form of research design in epidemiology. They generally can be carried out in a much shorter period of time than cohort studies and are cost effective. As a technique for controlling effects of confounders, randomization and stratification are possible solutions at the design stage and statistical adjustments can be made at the analysis stage. Statistical adjustments are done using regression methods, such as logistic regression discussed in Chapter 4. Stratification is more often introduced at the analysis stage too, and methods such as the Mantel–Haenszel method are available to complete the task.

Stratification can also be introduced at the design stage; its advantage is that one can avoid inefficiencies resulting from having some strata with a gross imbalance of cases and controls. A popular form of stratified design occurs when each case is individually matched with one or more controls chosen to have similar characteristics (i.e., values of confounding or matching variables). Matched designs have several advantages. They make it possible to control for confounding variables that are difficult to measure directly and, therefore, difficult to adjust at the analysis stage. For example, subject can be matched using area of residence so as to control for environmental exposure. Matching also provides more adequate control of confounding than can adjustment in analysis using regression because matching does not need any specific assumptions on functional form, which may be needed in regression models. Of course, matching has disadvantages too. Matches for individuals with unusual characteristics are hard to find. In addition, when cases and controls are matched on a specific characteristic, the influence of that characteristic on the disease can no longer be studied. Finally, a sample of

Applied Categorical Data Analysis and Translational Research, Second Edition, By Chap T. Le

matched cases and controls is not usually representative of any specific population, which may reduce our ability to generalize the analysis results.

Before coming back to the analysis of matched case–control studies, let us first start with a simple use of matched design to measure agreement.

5.1 MEASURING AGREEMENT

Many research studies rely on an observer's judgment to determine whether a disease, a trait, or an attribute is present or absent. For example, results of ear examinations will have effects on a comparison of competing treatments for ear infection. Of course, the basic concern is the issue of reliability. In order to judge a method's validity, an exact method for classification, or a "gold standard," must be available for the calculation of sensitivity and specificity (see Chapter 8 for more details). When an exact method is not available, reliability can only be judged indirectly in terms of reproducibility; the most common way for doing that is measuring the agreement between examiners.

For simplicity, assume that each of two observers independently assigns each of n items or subjects to one of two categories. The sample may then be enumerated in a 2×2 table as follows:

	Observer 2		
Observer 1	Category 1	Category 2	Total
Category 1	n_{11}	n_{12}	n_{1+}
Category 2	n_{21}	n_{22}	n_{2+}
Total	n_{+1}	n_{+2}	$n = n_{++}$

or, in terms of the cell probabilities,

	Observer 2		
Observer 1	Category 1	Category 2	Total
Category 1	p_{11}	p_{12}	p_{1+}
Category 2	p_{21}	p_{22}	p_{2+}
Total	p_{+1}	p_{+2}	$1 = p_{++}$

Using the frequencies in the first table, we can define the following:

(i) An overall proportion of *concordance*:

$$C = \frac{n_{11} + n_{22}}{n}$$

(ii) Category-specific proportions of concordance:

$$C_1 = \frac{2n_{11}}{2n_{11} + n_{12} + n_{21}}$$

$$C_2 = \frac{2n_{22}}{2n_{22} + n_{12} + n_{21}}$$

The distinction between concordance and association is that for two responses to be perfectly associated we only require that we can predict the category on one response from the category of the other response, while for two responses to have a perfect concordance, they must fall into the identical category. However, the proportions of concordance, overall or category-specific, do not measure agreement. Among other reasons, they are affected by the marginal totals. Cohen (1960) and others suggest comparing the overall concordance,

$$\theta_1 = \sum_i p_{ii}$$

where p's are the proportions in the above second 2×2 table, with the chance concordance,

$$\theta_2 = \sum_i p_{i+} p_{+i}$$

that occurs if the row variable is independent of the column variable, because if two events are independent, the probability of their joint occurrence is the product of their individual or marginal probabilities. This leads to a measure of agreement,

$$\kappa = \frac{\theta_1 - \theta_2}{1 - \theta_2}$$

called the *kappa statistic*, $0 = \kappa = 1$.

For the case of two categories, kappa and its standard error are given by

$$\kappa = \frac{2(n_{11}n_{22} - n_{12}n_{21})}{n_{1+}n_{+2} + n_{+1}n_{2+}}$$

$$SE(\kappa) = \frac{2\sqrt{(n_{1+}n_{+1}n_{2+}n_{+2})/2}}{n^2 - (n_{1+}n_{+2} + n_{+1}n_{2+})}$$

and the following are guidelines for the evaluation of kappa in clinical research:

$0.75 < \kappa \leq 1.0$: *Excellent reproducibility.*

$0.40 \leq \kappa \leq 0.75$: *Good reproducibility.*

$0 \leq \kappa < 0.40$: *Marginal/poor reproducibility.*

In general, when the reproducibility is not good, it indicates the need for multiple assessments.

■ **Example 5.1** Two nurses perform ear examinations focusing on the color of the eardrum (tympanic membrane). Each nurse independently assigns each of 100 ears to one of two categories: (i) normal or gray, or (ii) not normal (white, pink, orange, or red). The data were as follows:

	Nurse 2		
Nurse 1	Normal	Not Normal	Total
Normal	35	10	45
Not normal	20	35	55
Total	55	45	100

The results are

$$\kappa = 0.406$$

$$SE(\kappa) = 0.098$$

which indicate that the agreement is barely acceptable.

Note: A SAS program would include these instructions:

```
INPUT N11 N12 N21 N22;
N10 = N11 + N12; N20 = N21 + N22; N01 = N11 + N21; N02 = N12+N22;
N = N10 + N20; M = N10*N01 + N20*N02;
KAPPA = 2*(N11*N22 - N12*N21)/(N10*N02 + N01*N20);
SE = 2*SQRT((N10*N01*N20*N02)/N)/(N*N - M);
CARDS; 35 10 20 35;
PROC PRINT; KAPPA SE;
(The newer version of SAS, PROC FREQ, has an option for kappa.)
```

The following should also be pointed out.

1. The kappa statistic, as a measure for agreement, can also be used when there are more than two categories for classification:

$$\kappa = \frac{\sum p_{ii} - \sum_i p_{i+}p_{+i}}{1 - \sum_i p_{i+}p_{+i}}$$

2. We can form category-specific kappa statistics; for example, with two categories, we have

$$\kappa_1 = \frac{p_{11} - p_{1+}p_{+1}}{1 - p_{1+}p_{+1}}$$

$$\kappa_2 = \frac{p_{22} - p_{2+}p_{+2}}{1 - p_{2+}p_{+2}}$$

3. Kappa can be used with stratification. For example, if there are m strata (e.g., male and female subjects, or even m pairs of observers), let κ_i and $v_i = \text{Var}(\kappa_i)$ be the kappa and its estimated variance for the ith stratum, and $w_i = 1/v_i$. Defining the weighted average as

$$\bar{\kappa} = \frac{\sum_i w_i \kappa_i}{\sum_i w_i}$$

we then have

$$SE(\bar{\kappa}) = \sqrt{\frac{1}{\sum_i w_i}}$$

The chi-squared statistic,

$$X^2 = \sum_i w_i (\kappa_i - \bar{\kappa})^2$$

can be used to test for homogeneity across strata, with $(m - 1)$ degrees of freedom.

4. The major problem with kappa is that it approaches zero (even with high degree of agreement) if the prevalence is near 0 or 1.

5.2 PAIR-MATCHED CASE–CONTROL STUDIES

One-to-one matching is a cost-effective design and is perhaps the most popular form used in practice. It is conceptually easy and usually leads to a simple analysis.

5.2.1 The Model

Consider a case–control design and suppose that each individual in a large population has been classified as exposed or not exposed to a certain factor, and as having or not having some disease. The population may then be enumerated in a 2×2 table as follows, with entries being the proportions of the total population:

	Case		
Control	Exposed	Nonexposed	Total
Exposed	P_1	P_3	$P_1 + P_3$
Nonexposed	P_2	P_4	$P_2 + P_4$
Total	$P_1 + P_2$	$P_3 + P_4$	1

Using these proportions, the association (if any) between the factor and the disease could be measured by the ratio of risks (or relative risk) of being disease positive for those with or without the factor:

$$\begin{aligned}\text{Relative risk} &= \frac{P_1}{P_1 + P_3} \div \frac{P_2}{P_2 + P_4} \\ &= \frac{P_1(P_2 + P_4)}{P_2(P_1 + P_3)}\end{aligned}$$

In many (although not all) situations, the proportions of subjects classified as disease positive will be small: that is, P_1 is small in comparison with P_3, and P_2 is small in comparison with P_4. In such a case, the relative risk is almost equal to

$$\begin{aligned}OR &= \frac{P_1 P_4}{P_2 P_3} \\ &= \frac{P_1/P_3}{P_2/P_4}\end{aligned}$$

the odds ratio of being disease positive, or

$$\begin{aligned}OR &= \frac{P_1 P_4}{P_2 P_3} \\ &= \frac{P_1/P_2}{P_3/P_4}\end{aligned}$$

the odds ratio of being exposed. This justifies the use of odds ratio to measure differences, if any, in the exposure to a suspected risk factor.

As a technique to control confounding factors in a designed study, individual cases are matched, often one-to-one, to a set of controls chosen to have similar values for the important confounding variables. The simplest example of pair-matched data occurs with a single binary exposure (e.g., smoking versus nonsmoking). The data for outcomes can be represented by a 2×2 table where $(+, -)$ denotes (exposed, unexposed).

Control	Case		Total
	Exposed	Unexposed	
Exposed	n_{11}	n_{01}	$n_{11} + n_{01}$
Unexposed	n_{10}	n_{00}	n_{2+}
Total	$n_{11} + n_{10}$	$n_{01} + n_{00}$	n

For example, n_{10} denotes the number of pairs where the case is exposed but the matched control is unexposed. The most suitable statistical model for making inferences about the odds ratio θ is to use the conditional probability of the number of exposed cases among the discordant pairs. Given $(n_{10} + n_{01})$ being fixed, it can be seen that n_{10} follows the binomial distribution $B(n_{10} + n_{01}, \pi)$ where

$$\pi = \frac{OR}{1+OR}$$

5.2.2 The Analysis

Use the above binomial model, with the likelihood function

$$\left\{\frac{OR}{1+OR}\right\}^{n_{10}} \left\{\frac{1}{1+OR}\right\}^{n_{01}}$$

From this simple likelihood function, it is straightforward to estimate the odds ratio. The results are (we omit the "hat" over terms for simplicity; OR, e.g., is its estimate):

$$OR = \frac{n_{10}}{n_{01}}$$

$$SE(OR) = \sqrt{\frac{n_{10}(n_{10}+n_{01})}{n_{01}^3}}$$

For example, with large samples, a 95% confidence interval for the odds ratio is given by

$$OR \pm 1.96\, SE(OR)$$

The null hypothesis of no risk effect can be tested using the score procedure where the z-statistic,

$$z = \frac{(n_{10} - n_{01})}{\sqrt{n_{10}+n_{01}}}$$

is compared to percentiles of the standard normal distribution. The corresponding two-tailed procedure, based on

$$X^2 = \frac{(n_{10} - n_{01})^2}{n_{10} + n_{01}}$$

is often called the McNemar chi-squared test (at 1 degree of freedom). It is interesting to note that if we treat a matched pair as a group, or a level of a confounder, and present the data in the form of a 2×2 table,

	Disease Classification		
Exposure	Cases	Controls	Total
Yes	a_i	b_i	$a_i + b_i$
No	c_i	d_i	$c_i + d_i$
Total	1	1	2

then the Mantel–Haenszel method of Section 2.4 would yield the same estimate for the odds ratio:

$$\begin{aligned} OR_{\text{MH}} &= OR \\ &= \frac{n_{10}}{n_{01}} \end{aligned}$$

An alternative large-sample method is based on

$$\text{Var}(\log OR) = \frac{1}{n_{10}} + \frac{1}{n_{01}}$$

leading to a 95% confidence interval of

$$\frac{n_{10}}{n_{01}} \exp\left\{ \pm 1.96 \sqrt{\frac{1}{n_{10}} + \frac{1}{n_{01}}} \right\}$$

This works better when the sample size is not very large.

The etiologic fraction or population attributable risk is a measure of the impact of an exposure on the population (whereas the relative risk measures the impact of the exposure on the exposed subpopulation). The etiologic fraction is defined as the proportion of disease cases attributable to the risk factor and is expressible as

$$\lambda = p_{1e}(1 - 1/RR)$$

where p_{1e} is the exposure rate of the subpopulation of cases and RR is the relative risk. If the disease is considered as rare (most diseases are rare!), the relative risk can be approximated by the odds ratio OR. In addition, p_{1e} can be estimated using the sample of n cases (also the number of case–control pairs):

$$p_{1e} = \frac{n_{11} + n_{10}}{n}$$

Putting these results together, we have

$$\lambda = \frac{(n_{11} + n_{10})(n_{10} - n_{01})}{nn_{10}}$$

$$SE(\lambda) = \frac{1}{nn_{10}} \sqrt{n_{11}(n_{10} - n_{01})^2 + \frac{(n_{10}^2 + n_{11}n_{01})^2}{n_{10}} + n_{01}(n_{11} + n_{10})^2 - \frac{(n_{11} + n_{10})(n_{10} - n_{01})^2}{n}}$$

For example, with large samples, a 95% confidence interval for the attributable risk or etiologic fraction is given by $\lambda \pm 1.96\ SE(\lambda)$.

■ **Example 5.2** Breslow and Day (1980) used the data on endometrial cancer, which were taken from Mack et al. (1976). The investigators identified 63 cases of endometrial cancer occurring in a retirement community near Los Angeles, California, from 1971 to 1975. Each disease individual was matched with $R = 4$ controls who were alive and living in the community at the time the case was diagnosed, who were born within 1 year of the case, who were the same marital status, and who had entered the community at approximately the same time. The risk factor was previous use of estrogen (yes/no) and the data in the following table were obtained from the first-found matched control; the complete data set with 4 matched controls will be given in Section 5.3:

	Case		
Control	Exposed	Nonexposed	Total
Exposed	27	3	30
Nonexposed	29	4	33
Total	56	7	63

An application of the above methods yields

$$\begin{aligned} OR &= \frac{n_{10}}{n_{01}} \\ &= \frac{29}{3} \\ &= 9.67 \end{aligned}$$

$$SE(OR) = \sqrt{\frac{n_{10}(n_{10}+n_{01})}{n_{01}^3}}$$
$$= \sqrt{\frac{(29)(29+3)}{3^3}}$$
$$= 5.86$$

and a 95% confidence interval for OR is

$$\frac{n_{10}}{n_{01}}\exp\left\{\pm 1.96\sqrt{\frac{1}{n_{10}}+\frac{1}{n_{01}}}\right\} = \frac{29}{3}\exp\left\{\pm 1.96\sqrt{\frac{1}{29}+\frac{1}{3}}\right\}$$
$$= (2.96,\ 31.74)$$

Similarly, we have

$$\lambda = \frac{(n_{11}+n_{10})(n_{10}-n_{01})}{nn_{10}}$$
$$= \frac{(27+29)(29-3)}{(63)(29)}$$
$$= 0.797$$
$$SE(\lambda) = \frac{1}{nn_{10}}\sqrt{n_{11}(n_{10}-n_{01})^2 + \frac{(n_{10}^2+n_{11}n_{01})^2}{n_{10}} + n_{01}(n_{11}+n_{10})^2 - \frac{(n_{11}+n_{10})(n_{10}-n_{01})^2}{n}}$$
$$= \frac{1}{(63)(29)}\sqrt{(27)(29-3)^2 + \frac{[(29)^2+(27)(3)]^2}{29} + (3)(27+29)^2 - \frac{(27+29)(29-3)^2}{63}}$$
$$= 0.130$$

5.2.3 The Case of Small Samples

Methods in the previous section are likelihood-based asymptotic methods; they work well when sample sizes are large. However, one of the major reasons that investigators prefer matched design is that they do not have many disease cases. For small studies, say, $(n_{10} + n_{01}) = 25$, a 95% confidence interval for OR is obtained as follows:

(i) First, an exact 95% confidence interval for a binomial parameter π is calculated from the tail probabilities of the binomial distribution using the formulas for lower and upper limits:

$$\pi_L = \frac{n_{10}}{n_{10} + (n_{01}+1)F_{0.975}(2n_{01}+2, 2n_{10})}$$
$$\pi_U = \frac{(n_{10}+1)F_{0.975}(2n_{10}+2, 2n_{01})}{n_{01} + (n_{10}+1)F(2n_{10}+2, 2n_{01})}$$

where $F_{0.975}$(df1,df2) is the 97.5 percentile with df1 numerator degrees of freedom and df2 denominator degrees of freedom.

(ii) Once limits for π are found, they are converted into the limit for the OR by using the transformation

$$OR = \frac{\pi}{1-\pi}$$

In addition, the p-value for testing

$$H_0:\ OR = 1$$

against

$$H_A:\ OR > 1$$

is given by the cumulative binomial probability:

$$\sum_{i=n_{10}}^{n_{10}+n_{01}} \binom{n_{10}+n_{01}}{i} \left(\frac{1}{2}\right)^{n_{10}+n_{01}}$$

■ **Example 5.3** Refer to the data in Example 5.2 but use the method for small samples. We have

$$\begin{aligned}
\pi_{\mathrm{L}} &= \frac{n_{10}}{n_{10}+(n_{01}+1)F_{0.975}(2n_{01}+2,\ 2n_{10})} \\
&= \frac{29}{29+(3+1)F_{0.975}(8,\ 58)} \\
&= \frac{29}{29+(4)(2.42)} \\
&= 0.750 \\
\pi_{\mathrm{U}} &= \frac{(n_{10}+1)F_{0.975}(2n_{10}+2,\ 2n_{01})}{n_{01}+(n_{10}+1)F_{0.975}(2n_{10}+2,\ 2n_{01})} \\
&= \frac{(29+1)F_{0.975}(60,\ 6)}{3+(29+1)F_{0.975}(60,\ 6)} \\
&= \frac{(30)(4.96)}{3+(30)(4.96)} \\
&= 0.980
\end{aligned}$$

leading to a 95% confidence interval for the odds ratio of (3.0, 49.6).

5.2.4 Risk Factors with Multiple Categories and Ordinal Risks

This section is presented here only for completeness; its implementation may be hard for readers in applied fields or students at this level because it involves a matrix inversion and packaged computer programs are not readily available.

Consider the pair-matched case–control study with a risk factor having k levels $(k=2)$. Let

$$\theta_{ij} = \text{Odds ratio for level } i \text{ versus level } j$$

then the θ_{ij}'s are subject to the constraint

$$\begin{aligned}\theta_{ij} &= \frac{\theta_{i1}}{\theta_{j1}} \\ &= \frac{\theta_i}{\theta_j},\end{aligned}$$

which is called the *consistency relationship* between odds ratios. If x_{ij} denotes the number of pairs for which the case has been exposed to level and the control level j of the risk factor, then it can be shown that the log-likelihood function is given by

$$\begin{aligned}\ln L &= \sum_{i>j} \{x_{ij}\ln(\theta_i/\theta_j) - n_{ij}\ \ln(1+\theta_i/\theta_j)\} \\ n_{ij} &= x_{ij} + x_{ji}\end{aligned}$$

Consider the null hypothesis

$$H_0\colon\ \theta_1 = \theta_2 = \cdots = \theta_k = 1$$

It can be shown that the efficient score chi-squared test, called *Stuart's test* for marginal homogeneity in square tables (Stuart, 1955), can be expressed in a matrix form as follows:

$$X^2_{\text{ES}} = (\mathbf{O}-\mathbf{E})^T\mathbf{V}^{-1}(\mathbf{O}-\mathbf{E});\ \ df = (k-1)$$

where **0** and **E** denote the $(k-1)$-dimensional vectors of observed and expected values of the x_{i+}'s (only for $1=i=(k-1)$ while **V** is the corresponding $(k-1) \times (k-1)$-dimensional covariance matrix. The elements of **E** and **V** are given by

$$\begin{aligned}E(x_{i+}) &= e_{i+} \\ &= (x_{i+} + x_{+i})/2 \\ \text{Var}(x_{i+}) &= (x_{i+} + x_{+i})/4 - x_{ii}/2 \\ \text{Cov}(x_{i+}) &= -n_{ii}/2\ \ \text{for } i \neq j\end{aligned}$$

If the k levels are ordered so that we can assign scores $z_1, z_2, \ldots, z_k$ to these levels, then assuming an additional loglinear model,

$$\theta_i = \exp\{\beta(z_i - z_1)\}$$

the log-likelihood function becomes

$$\ln L = \sum_{i>j} \{\beta x_{ij}(z_i - z_j) - n_{ij}\ln[1 + \exp\{\beta(z_i - z_j)\}]\}$$

and a test for a *linear trend* (i.e., H_0: $\beta = 0$) can be shown to be expressed as

$$X^2_{\text{ES}} = \frac{\left[\sum_{i<j}(x_{ij} - x_{ji})(z_i - z_j)\right]^2}{\sum_{i<j} n_{ij}(z_i - z_j)^2}; \quad df = 1$$

■ **Example 5.4** In the study of endometrial cancer in Example 5.2, using the first control and the following four levels of exposure to conjugated estrogen—(1) none, (2) 0.1 to 0.299, (3) 0.3 to 0.625, and (4) more than 0.625 mg—the data for the 59 matched pairs are as follows:

	Control Dose				
Case Dose	(1)	(2)	(3)	(4)	Total
(1)	6	2	3	1	12
(2)	9	4	2	1	16
(3)	9	2	3	1	15
(4)	12	1	2	1	16
Total	30	9	10	4	53

We obtain the test statistic,

$$X^2_{\text{ES}} = [36-24, \; 9-12.5 \;\; 10-12.5] \begin{bmatrix} 9 & -2.75 & -3 \\ -2.75 & 4.25 & -1 \\ -3 & -1 & -4.75 \end{bmatrix}^{-1} \begin{bmatrix} 36-24 \\ 9-12.5 \\ 10-12.5 \end{bmatrix}$$

$$= 16.96$$

which is highly significant by reference to tables of chi-squared with 3 degrees of freedom. Assigning scores $z = (1, 2, 3, 4)$ to the four exposure levels, we have a 1 degree of freedom highly significant test statistic for the trend $X^2 = 14.71$.

5.3 MULTIPLE MATCHING

One-to-one matching is a cost-effective design. However, an increase in the number of controls may give the study more power. In epidemiologic studies, there are typically a small number of cases and a large number of potential controls to select from. When the controls are more easily available than cases, it is more efficient and effective to select more controls for each case. Breslow and Day (1980) proved that the efficiency of an M-to-one control–case ratio for estimating a relative risk relative to having complete information on the control population (i.e., $M = 8$) is $M/(M + 1)$. Hence one-to-one matching is 50% efficient, four-to-one matching is 80% efficient, five-to-one matching is 83% efficient, and so on. The gain in efficiency is rapidly diminishing for designs with $M = 5$.

5.3.1 The Conditional Approach

The analysis of one-to-one matching design was conditional on the number of pairs showing differences in the exposure history, the $(-, +)$ and $(+, -)$ cells. Similarly, considering an M-to-one matching design, we use a conditional approach, fixing the number m of exposed individuals in a matched set; and the sets with $m = 0$ or $m = (M + 1)$ (similar to $(-,-)$ and $(+,+)$ cells in the one-to-one matching design) are ignored.

If we fix the number of exposed individuals in each stratum, then it can be shown straightforwardly that

$$\Pr(\text{case exposed} \mid m \text{ members of stratum exposed}) = \frac{(m)(OR)}{(m)(OR) + M - m + 1}$$

where OR is the odds ratio representing the effect of exposure. The result for pair-matched design in Section 5.2.1 is a special case where $M = m = 1$.

For the strata, or matched sets, with m ($m = 1, 2, \ldots, M$) exposed individuals, let

$$\begin{aligned} n_{1,m-1} &= \text{number of sets with an exposed case} \\ n_{0,m} &= \text{number of sets with an unexposed case} \\ n_m &= n_{1,m-1} + n_{0,m} \end{aligned}$$

Then given n_m being fixed, $n_{1,m}$ is distributed as the binomial $B(n_m, p_m)$, where

$$p_m = \frac{(m)(OR)}{(m)(OR) + M - m + 1}$$

5.3.2 Estimation of the Odds Ratio

From the joint (conditional) likelihood function,

$$L = \prod_{m=1}^{M} \left\{ \frac{(m)(OR)}{(m)(OR) + M - m + 1} \right\}^{n_{1,m-1}} \left\{ \frac{M - m + 1}{(m)(OR) + M - m + 1} \right\}^{n_{0,m}}$$

or

$$\ln L = \sum_{m=1}^{M} \{ n_{1,m-1} \ln[(m)(OR)] + n_{0,m} \ln(M - m + 1) - n_m \ln[(m)(OR) + M - m + 1] \}$$

from which one can obtain the maximum likelihood estimator (MLE) of OR and its standard error. But such a solution requires an iterative procedure and a computer algorithm. We will come back to this topic with a complete solution in Section 5.4.

A simple method for estimating the odds ratio would be to treat a matched set, consisting of one case and M matched controls, as a stratum (i.e., a level of some confounder). We then present data from this stratum in the form of a 2×2 table:

	Disease Classification		
Exposure	Cases	Controls	Total
Yes	a_i	b_i	$a_i + b_i$
No	c_i	d_i	$c_i + d_i$
Total	1	M	$M + 1$

and obtain the Mantel–Haenszel estimate for the odds ratio, which is very simple:

$$\begin{aligned} OR_{\text{MH}} &= \frac{\sum_i \{a_i d_i / (M+1)\}}{\sum_i \{b_i c_i / (M+1)\}} \\ &= \frac{\sum_{m=1}^{M} (M - m + 1) n_{1,m-1}}{\sum_{m=1}^{M} m n_{0,m}} \end{aligned}$$

The Mantel–Haenszel estimate has been used widely in the analysis of case–control studies with multiple matching (see Example 5.5).

5.3.3 Testing for Exposure Effect

From the likelihood function of Section 5.3.2,

$$L = \prod_{m=1}^{M} \left\{ \frac{(m)(OR)}{(m)(OR) + M - m + 1} \right\}^{n_{1,m-1}} \left\{ \frac{M - m + 1}{(m)(OR) + M - m + 1} \right\}^{n_{0,m}}$$

it can be verified that the score test for H_0: $OR = 1$ is given by a chi-squared test with 1 degree of freedom:

$$X_{\text{ES}}^2 = \frac{\left[\sum_m^M \left\{ n_{1,m-1} - \frac{mn_m}{M+1} \right\} \right]^2}{\frac{1}{(M+1)^2} \sum_{m=1}^M mn_m(M - m + 1)}$$

■ **Example 5.5** Breslow and Day (1980) used the data on endometrial cancer, which were taken from Mack et al. (1976). The investigators identified 63 cases of endometrial cancer occurring in a retirement community near Los Angeles, California, from 1971 to 1975. Each disease individual was matched with $M = 4$ controls who were alive and living in the community at the time the case was diagnosed, who were born within 1 year of the case, who were the same marital status, and who had entered the community at approximately the same time. The risk factor was previous use of estrogen (yes/no) and the data in Example 5.2 involve only the first-found matched control. We are now able to analyze the complete data set with 4-to-1 matching:

	Number of Exposed Individuals in Each Matched Set			
Case Status	1	2	3	4
Exposed	4	17	11	9
Unexposed	6	3	1	1
Total	10	20	12	10

Using these 52 sets with 4 matched controls, we have

$$\begin{aligned} X_{\text{ES}}^2 &= \frac{(25)\left[\left(4 - \frac{(1)(10)}{5}\right) + \left(17 - \frac{(2)(20)}{5}\right) + \left(11 - \frac{(3)(12)}{5}\right) + \left(9 - \frac{(4)(10)}{5}\right) \right]^2}{(1)(10)(4) + (2)(20)(3) + (3)(12)(2) + (4)(10)(1)} \\ &= 22.95 \end{aligned}$$

The Mantel–Haenszel estimate for the odds ratio is

$$\begin{aligned} OR_{\mathrm{MH}} &= \frac{(4)(4)+(3)(17)+(2)(11)+(1)(9)}{(1)(6)+(2)(3)+(3)(1)+(4)(1)} \\ &= 5.16 \end{aligned}$$

When the number of controls matched to a case, M, is variable (mostly due to missing data), the test for exposure effects should incorporate data from all strata:

$$X_{\mathrm{ES}}^2 = \frac{\left[\sum_M \sum_m^M \left\{n_{1,m-1} - \frac{mn_m}{M+1}\right\}\right]^2}{\sum_M \sum_{m=1}^M \frac{mn_m(M-m+1)}{(M+1)^2}}$$

The corresponding Mantel–Haenszel estimate for the odds ratio is

$$OR_{\mathrm{MH}} = \frac{\sum_M \sum_{m=1}^M (M-m+1)n_{1,m-1}}{\sum_M \sum_{m=1}^M mn_{0,m}}$$

■ **Example 5.6** Refer to the data on endometrial cancer of Example 5.5. Due to missing data, we have some cases matching to four and some matching to three controls. In addition to the 52 sets with 4-to-1 matching,

	Number of Exposed Individuals in Each Matched Set			
Case Status	1	2	3	4
Exposed	4	17	11	9
Unexposed	6	3	1	1
Total	10	20	12	10

we also have 4 sets of data with 3-to-1 matching,

	Number of Exposed Individuals in Each Matched Set		
Case Status	1	2	3
Exposed	1	3	0
Unexposed	0	0	0
Total	1	3	0

With the inclusion of the 4 sets having 3 matched controls, the result of Example 5.5 becomes

$$X_{\mathrm{ES}}^2 = \frac{\left[\left(4-\frac{(1)(10)}{5}\right)+\left(17-\frac{(2)(20)}{5}\right)+\left(11-\frac{(3)(12)}{5}\right)+\left(9-\frac{(4)(10)}{5}\right)+\left[1-\frac{(1)(1)}{4}\right]+\left[3-\frac{(2)(3)}{4}\right]\right]^2}{\frac{1}{25}\{(1)(10)(4)+(2)(20)(3)+(3)(12)(2)+(4)(10)(1)\}+\frac{1}{16}\{(1)(1)(3)+(2)(3)(2)\}}$$

$$= 27.57$$

The corresponding Mantel–Haenszel estimate for the odds ratio is

$$OR_{\mathrm{MH}} = \frac{[(4)(4)+(3)(17)+(2)(11)+(1)(9)]/(5)+[(3)(1)+(2)(3)+(1)(0)]/(4)}{[(1)(6)+(2)(3)+(3)(1)+(4)(1)]/(5)+[(1)(0)+(2)(0)+(3)(0)]/(4)}$$

$$= 5.75$$

5.3.4 Testing for Homogeneity

This section is concerned with an important aspect of case–control studies—the homogeneity of the odds ratio. To estimate the odds ratio (as seen in Section 5.3.1), we made the assumption that the exposure probabilities of a risk factor are allowed to differ from matched set to matched set (due to confounding effects) but the odds ratio is constant across matched sets. This homogeneity can be tested using the statistic

$$S = \sum_{m=1}^{M} p_m(n_m p_m - n_{1,m-1})$$

where p_m, $1 = m = M$, is the probability that the case has been exposed given that there are m exposed members of a set,

$$p_m = \frac{mOR}{mOR + M - m + 1}$$

The estimated variance of S is given by

$$\mathrm{Var}(S) = \sum_{m=1}^{M} n_m p_m(1-p_m) - \frac{\left\{\sum_{m=1}^{M} n_m p_m(1-p_m)\right\}^2}{\sum_{m=1}^{M} n_m p_m(1-p_m)}$$

The standardized statistic,

$$z = \frac{S}{\sqrt{\mathrm{Var}(S)}}$$

is distributed asymptotically as standard normal (Liang and Self, 1985).

Example 5.7 Refer to the 4-to-1 matched data of Example 5.5 and the Mantel–Haenszel estimate for the odds ratio as in Example 5.5 (= 5.16). We have

$$\begin{aligned} S &= -0.056 \\ \mathrm{Var}(S) &= 4.23 - \frac{(5.58)^2}{7.60} \\ &= 0.133 \\ z &= \frac{-0.056}{\sqrt{0.133}} \\ &= 0.15 \end{aligned}$$

indicating the homogeneity of the odds ratio fits rather well.

5.4 CONDITIONAL LOGISTIC REGRESSION

Recall from Chapter 4 that, in a variety of applications using regression analysis, the dependent variable of interest has only two possible outcomes, and therefore can be represented by an indicator variable taking on values 0 and 1. An important application is the analysis of case–control studies where the dependent variable represents the disease status: 1 for a case and 0 for a control. The methods that have been used widely and successfully for these applications are based on the logistic model. This section also deals with the cases where the dependent variable of interest is binary following a binomial distribution—the same as those using logistic regression analyses, but the data are matched. Again, the term refers to the pairing of one or more controls to each case on the basis of their similarity with respect to selected variables used as matching criteria, as seen in Sections 5.2 and 5.3. Although the primary objective of matching is the elimination of biased comparison between cases and controls, this objective can only be accomplished if matching is followed by an analysis that corresponds to the matched design. Unless the analysis properly accounts for the matching used in the selection phase of a case–control study, the results can be biased. In other words, matching (which refers to the selection process) is only the first step of a two-step process that can be used effectively to control for confounders: (i) matching design followed by (ii) matched analysis.

Suppose the purpose of the research is to assess relationships between the disease and a set of covariates using a matched case–control design. The regression techniques for the statistical analysis of such relationships is based on the conditional logistic model.

Following are two typical examples; the first one is a case–control study of vaginal carcinoma, which involves two binary risk factors. Of course, we can investigate one covariate at a time using the method of Section 5.3.

Example 5.8 Consider the data taken from a study by Herbst et al. (1971); the cases were eight women 15–22 years of age who were diagnosed with vaginal carcinoma between 1966 and 1969. For each case, four controls were found in

TABLE 5.1. Vaginal Carcinoma Data

	Responses (Bleeding, Previous Loss)				
		Control Subject Number			
Set	Case	1	2	3	4
1	(N,Y)	(N,Y)	(N,N)	(N,N)	(N,N)
2	(N,Y)	(N,Y)	(N,N)	(N,N)	(N,N)
3	(Y,N)	(N,Y)	(N,N)	(N,N)	(N,N)
4	(Y,Y)	(N,N)	(N,N)	(N,N)	(N,N)
5	(N,N)	(Y,Y)	(N,N)	(N,N)	(N,N)
6	(Y,Y)	(N,N)	(N,N)	(N,N)	(N,N)
7	(N,Y)	(N,Y)	(N,N)	(N,N)	(N,N)
8	(N,Y)	(N,N)	(N,N)	(N,N)	(N,N)

the birth records of patients having their babies delivered within 5 days of the case in the same hospital.

The risk factors of interest are the mother's bleeding in this pregnancy (N = No, Y = Yes) and any previous pregnancy loss by the mother (N = No, Y = Yes). The data are given in Table 5.1, as used by Holford (1982).

In the next example, one of the four covariates is on a continuous scale.

■ **Example 5.9** The data in Table 5.2 were taken from Hosmer and Lemeshow (1989) in a study of low-birth-weight babies (the cases); only a small portion of the data set is presented here for illustration. For each of the first 15 cases, we retain only the first three matched controls (even though the number of controls per case need not be the same). Four risk factors are under investigation: weight (in pounds) of the mother (M Weight) at the last menstrual period, hypertension, smoking, and uterine irritability (U Irritability); for the last three factors, a value of 1 indicates a "yes" and a value of 0 indicates a "no." The mother's age was used as the matching variable.

5.4.1 Simple Regression Analysis

In this section we discuss the basic ideas of simple regression analysis when only one predictor or independent variable is available for predicting the binary response of interest. We illustrate these for the more simple designs in which each matched set has one case and case i is matched to m_i controls; the number of controls m_i varies from case to case.

The Likelihood Function

In our framework, let x_i be the covariate value for case i and x_{ij} be the covariate value the jth control matched to case i. Then, for the ith matched set, the conditional probability of the observed outcome (that the subject with covariate value x_i be the

TABLE 5.2. Low-Birth-Weight Data

Matched Set	Case (1)/ Control (0)	M Weight	Hypertension	Smoking	U Irritability
1	1	130	0	0	0
	0	112	0	0	0
	0	135	1	0	0
	0	270	0	0	0
2	1	110	0	0	0
	0	103	0	0	0
	0	113	0	0	0
	0	142	0	1	0
3	1	110	1	0	0
	0	100	1	0	0
	0	120	1	0	0
	0	229	0	0	0
4	1	102	0	0	0
	0	182	0	0	0
	0	150	0	0	0
	0	150	0	0	0
5	1	125	0	0	1
	0	120	0	0	1
	0	169	0	0	1
	0	158	0	0	0
6	1	200	0	0	1
	0	108	1	0	1
	0	185	1	0	0
	0	110	1	0	1
7	1	130	1	0	0
	0	95	0	1	0
	0	120	0	1	0
	0	169	0	0	0
8	1	97	0	0	1
	0	128	0	0	0
	0	115	1	0	0
	0	190	0	0	0
9	1	132	0	1	0
	0	90	1	0	0
	0	110	0	0	0
	0	133	0	0	0
10	1	105	0	1	0
	0	118	1	0	0
	0	155	0	0	0
	0	241	0	1	0

TABLE 5.2. (*Continued*)

Matched Set	Case (1)/ Control (0)	M Weight	Hypertension	Smoking	U Irritability
11	1	96	0	0	0
	0	168	1	0	0
	0	160	0	0	0
	0	133	1	0	0
12	1	120	1	0	1
	0	120	1	0	0
	0	167	0	0	0
	0	250	1	0	0
13	1	130	0	0	1
	0	150	0	0	0
	0	135	0	0	0
	0	154	0	0	0
14	1	142	1	0	0
	0	153	0	0	0
	0	110	0	0	0
	0	112	0	0	0
15	1	102	1	0	0
	0	215	1	0	0
	0	120	0	0	0
	0	150	1	0	0

case) given that we have one case per matched set is

$$\frac{\exp(\beta x_i)}{\exp(\beta x_i) + \sum_{j=1}^{m_i} \exp(\beta x_{ij})}$$

If the sample consists of N matched sets, then the conditional likelihood function is the product of the above terms over the N matched sets:

$$L = \prod_{i=1}^{N} \frac{\exp(\beta x_i)}{\exp(\beta x_i) + \sum_{j=1}^{m_i} \exp(\beta x_{ij})}$$

from which we can obtain the maximum likelihood estimate of the parameter β (Breslow, 1982).

Measure of Association

Similar to the case of the logistic model, $\exp(\beta)$ represents:

(i) the odds ratio associated with an exposure if X is binary (exposed $X = 1$ vs. unexposed $X = 0$), or

(ii) the odds ratio due to one unit increase if X is continuous ($X = x + 1$ vs. $X = x$).

After β has been estimated and its standard error has been obtained, a 95% confidence interval for the above odds ratio is given by

$$\exp[\beta \pm 1.96\, SE(\beta)]$$

A Special Case

Consider now the simplest case of a pair matched (i.e., one-to-one matching) with a binary covariate: exposed $X = 1$ versus unexposed $X = 0$. Let the data be summarized and presented as in Section 5.2:

	Case		
Control	Exposed	Unexposed	Total
Exposed	n_{11}	n_{01}	$n_{11} + n_{01}$
Unexposed	n_{10}	n_{00}	n_{2+}
Total	$n_{11} + n_{10}$	$n_{01} + n_{00}$	n

For example, n_{10} denotes the number of pairs where the case is exposed but the matched control is unexposed. The above likelihood function is reduced to

$$L = \left\{\frac{1}{1+1}\right\}^{n_{00}} \left\{\frac{\exp(\beta)}{1+\exp(\beta)}\right\}^{n_{10}} \left\{\frac{1}{1+\exp(\beta)}\right\}^{n_{01}} \left\{\frac{\exp(\beta)}{\exp(\beta)+\exp(\beta)}\right\}^{n_{11}}$$

$$= \frac{\exp(\beta n_{10})}{[1+\exp(\beta)]^{n_{10}+n_{01}}}$$

From this we can obtain a point estimate:

$$\hat{\beta} = \frac{n_{10}}{n_{01}}$$

which is the usual odds ratio estimate from pair-matched data identical to that of Section 5.2.2.

Tests of Association

Another aspect of statistical inference concerns the test of significance; the null hypothesis to be considered is

$$H_0\colon\ \beta = 0$$

The reason for interest in testing whether or not $\beta = 0$ is that $\beta = 0$ implies that there is no relationship between the binary dependent variable and the covariate X under investigation. Since the likelihood function is rather simple, one can easily derive, say, the score test for the above null hypothesis; however, nothing would be gained by going through this exercise. We can simply apply a McNemar chi-squared test (if the covariate is binary or categorical; see Sections 5.2.2 and 5.2.4) or a paired t-test or signed-rank Wilcoxon test (if the covariate under investigation is on a continuous scale). Of course, the application of the conditional logistic model is still desirable, at least in the case of a continuous covariate, because it would provide a measure of association—the odds ratio.

■ **Example 5.10** Refer to the data for low-birth-weight babies in Example 5.9 and suppose we want to investigate the relationship between the low-birth-weight problem, our outcome for the study, and the weight of the mother taken at the last menstrual period. An application of the simple conditional logistic regression analysis yields the following:

Variable	Coefficient	Standard Error	z-Statistic	p-Value
M Weight	−0.0211	0.0112	−1.884	0.0593

The result indicates that the effect of the mother's weight is nearly significant at the 5% level ($p = 0.0593$).

The odds ratio associated with, say, a 10-pound (10-lb) increase in weight is

$$\exp(-0.2114) = 0.809$$

If a mother increases her weight about 10 pounds, the odds of having a low-birth-weight baby is reduced by almost 20%.

Note: A SAS program would include these instructions:

```
INPUT SET CASE MWEIGHT;
DUMMYTIME=2-CASE;
CARDS;
(data);
PROC PHREG DATA = LOWWEIGHT;
MODEL = DUMMYTIME*CASE(0) = MWEIGHT/TIES = DISCRETE;
STRATA = SET;
```

where `LOWWEIGHT` is the name assigned to the data set, `DUMMYTIME` is the name for the make-up time variable defined in the upper part of the program, `CASE` is the case–control status indicator (coded as 1 for a case and 0 for a control), and `MWEIGHT` is the variable name for weight of the mother at the last menstrual period. The matched `SET` number (1 to 15 in this example) is used as the stratification factor.

5.4.2 Multiple Regression Analysis

The effect of some factor on a dependent or response variable may be influenced by the presence of other factors through effect modifications (i.e., interactions). Therefore in order to provide a more comprehensive analysis, it is very desirable to consider a large number of factors and sort out which ones are most closely related to the dependent variable. In this section, we discuss a multivariate method for such a risk determination. This method, which is multiple conditional logistic regression analysis, involves a linear combination of the explanatory or independent variables; the variables must be quantitative with particular numerical values for each patient. A covariate or independent variable—such as a patient characteristic—may be dichotomous, polytomous, or continuous (categorical factors will be represented by dummy variables). Examples of dichotomous covariates are gender and presence or absence of certain comorbidity. Polytomous covariates include race and different grades of symptoms; these can be covered by the use of dummy variables. Continuous covariates include patient age and blood pressure. In many cases, data transformations (e.g., taking the logarithm) may be desirable to satisfy the linearity assumption. We illustrate this process for a very general design in which matched set i $(1=N)$, has n_i cases that were matched to m_i controls; the numbers of cases n_i and controls m_i vary from matched set to matched set.

The Likelihood Function Parameter Estimation

For the general case of n_i cases matched to m_i controls in a set, we have the conditional probability of the observed outcome (that a specific set of n_i subjects are cases) given that the number of cases is n_i (any n_i subjects could be cases):

$$\frac{\exp\left(\sum_{j=1}^{n_i}[\beta^T \mathbf{x}_j]\right)}{\sum_{R(n_i,m_i)} \exp\left(\sum_{j=1}^{n_i}[\beta^T \mathbf{x}_j]\right)}$$

where the sum in the denominator ranges over the collections $R(n_i, m_i)$ of all partitions of the $(n_i + m_i)$ subjects—one of size n_i and one of size m_i. The full conditional likelihood is the product over all matched sets—one probability for each set (Breslow, 1982) that is,

$$L = \prod_{i=1}^{N} \frac{\exp\left(\sum_{j=1}^{n_i}[\beta^T \mathbf{x}_j]\right)}{\sum_{R(n_i,m_i)} \exp\left(\sum_{j=1}^{n_i}[\beta^T \mathbf{x}_j]\right)}$$

From the above result, it follows that the full likelihood of the observed outcome has the same mathematical form as the overall partial likelihood for the proportional hazards survival model with strata: one for each matched set, and one event time for each (more in Chapter 7). This enables us to adapt programs written for proportional hazards model to analyze epidemiologic matched studies as seen in subsequent examples. The essential features of the adaptation are:

(i) Creating matched set numbers and using them as different levels of a stratification factor.
(ii) Assigning a number to each subject; these numbers will be used in place of duration times. These numbers are chosen arbitrarily as long as the number assigned to a case is smaller than the number assigned to a control in the same matched set. This is possible because when there is only one event in each set, the numerical value for the time to event becomes irrelevant.

Similar to the univariate case, $\exp(\beta_i)$ represents

(i) the odds ratio associated with an exposure if X_i is binary (exposed $X_i = 1$ vs. unexposed $X_i = 0$), or
(ii) the odds ratio due to 1 unit increase if X_i is continuous ($X_i = x + 1$ vs. $X_i = x$).

After β_i has been estimated and its standard error has been obtained, a 95% confidence interval for the above odds ratio is given by

$$\exp[\beta \pm 1.96\, SE(\beta_i)]$$

These results are necessary in the effort to identify important risk factors in matched designs. Of course, before such analyses are done, the problem and the data have to be examined carefully. If some of the variables are highly correlated, then one or fewer of the correlated factors are as likely to be good predictors as all of them; information from other similar studies also has to be incorporated so as to drop some of these correlated explanatory variables. The uses of products, such as X_1X_2, and higher power terms, such as X_1^2, may be necessary and can improve the goodness-of-fit (unfortunately, it is very hard to tell!). It is important to note that we are assuming a linear regression model in which, for example, the odds ratio due to one unit increase in the value of a continuous X_i ($X_i = x + 1$ vs. $X_i = x$) is independent of x. Therefore if this linearity seems to be violated (again, it is very hard to tell; the only easy way is fitting a polynomial model as seen in a later example), the incorporation of powers of X_i should be seriously considered. The use of products will help in the investigation of possible effect modifications. Finally, for the messy problem of missing data, most packaged programs would delete a subject if one or more covariate values are missing.

Testing Hypotheses in Multiple Regression

Once we have fit a multiple conditional logistic regression model and obtained estimates for the various parameters of interest, we want to answer questions about the contributions of various factors to the prediction of the binary response variable using matched designs. There are three types of such questions:

(i) An overall test: Taken collectively, does the entire set of explanatory or independent variables contribute significantly to the prediction of response?
(ii) Test for the value of a single factor: Does the addition of one particular variable of interest add significantly to the prediction of response over and above that achieved by other independent variables?
(iii) Test for contribution of a group of variables: Does the addition of a group of variables add significantly to the prediction of response over and above that achieved by other independent variables?

Overall Regression Test We now consider the first question stated above concerning an overall test for a model containing k factors. The null hypothesis for this test may be stated as: "all k independent variables considered together do not explain the variation in the response any more than the size alone." In other words,

$$H_0\colon\ \beta_1 = \beta_2 = \cdots = \beta_k = 0$$

Three statistics can be used to test this global null hypothesis; each has an asymptotic chi-squared distribution with k degrees of freedom under the null hypothesis H_0.

(i) *Likelihood Ratio Test*:

$$X_{\mathrm{LR}}^2 = 2\{\ln L(\boldsymbol{\beta}) - \ln L(\mathbf{0})\}$$

(ii) *Wald's Test*:

$$X_{\mathrm{W}}^2 = \boldsymbol{\beta}^{\mathrm{T}}[\mathbf{V}(\boldsymbol{\beta})]^{-1}\boldsymbol{\beta}$$

(iii) *Score Test*:

$$X_{\mathrm{S}}^2 = \left[\frac{\delta \ln L(\mathbf{0})}{\delta \boldsymbol{\beta}}\right]^T \left[-\frac{\delta^2 \ln L(\mathbf{0})}{\delta \boldsymbol{\beta}^2}\right]^{-1} \left[\frac{\delta \ln L(\mathbf{0})}{\delta \boldsymbol{\beta}}\right]$$

All three statistics are provided by most standard computer programs such as SAS and they are asymptotically equivalent (i.e., for very large sample sizes), yielding identical statistical decisions most of the time; however, Wald's test is much less often used than the other two.

■ **Example 5.11** Refer to the data for low-birth-weight babies in Example 5.9 with all four covariates. We have the following test statistics; each is a chi-squared test with 4 degrees of freedom for the global null hypothesis:

(i) *Likelihood Ratio Test*:

$$X^2_{\text{LR}} = 9.530; \quad p = 0.0491$$

(ii) *Wald's Test*:

$$X^2_{\text{W}} = 6.001; \quad p = 0.1991$$

(iii) *Score Test*:

$$X^2_{\text{S}} = 8.491; \quad p = 0.0752$$

The results indicate a weak combined explanatory power; the Wald's test is not even significant. Very often, this implicitly means that perhaps only one or two covariates are significantly associated with the response of interest (a weak overall correlation).

Test for a Single Variable Let us assume that we now wish to test whether the addition of one particular independent variable of interest adds significantly to the prediction of the response over and above that achieved by other factors already present in the model (usually after seeing a significant result for the above global hypothesis). The null hypothesis for this single-variable test may be stated as: "factor X_i does not have any value added to the prediction of the response given that other factors are already included in the model." In other words,

$$H_0: \beta_i = 0$$

To test such a null hypothesis, one can employ

$$z = \frac{\hat{\beta}}{SE(\hat{\beta})}$$

using the corresponding estimated regression coefficient and its standard error, both of which are printed by standard packaged computer programs such as SAS. In performing this test, we refer the value of the z-score to percentiles of the standard normal distribution; for example, we compare the absolute value of z to 1.96 for a two-sided test at the 5% level.

Example 5.12 Refer to the data for low-birth-weight babies in Example 5.9 with all four covariates; we have the following results. Only the mother's weight ($p = 0.0942$) and uterine irritability ($p = 0.0745$) are marginally significant. In fact, these two variables are highly correlated: if one is deleted from the model, the other would become more significant.

Variable	Coefficient	Standard Error	z-Statistic	p-Value
M Weight	−0.0191	0.0114	−1.673	0.0942
Smoking	−0.0885	0.8618	−0.103	0.9182
Hypertension	0.6325	1.1979	0.528	0.5975
U Irritability	2.1376	1.1985	1.784	0.0745

The overall tests and the tests for single variables are implemented simultaneously sing the same computer program.

Example 5.13 Refer to the data for vaginal carcinoma in Example 5.8. An application of a conditional logistic regression analysis yields the following results; each is a chi-squared test with 2 degrees of freedom for the global null hypothesis:

(i) *Likelihood Test*:

$$X_{\mathrm{LR}}^2 = 9.924; \quad p = 0.0081$$

(ii) *Wald's Test*:

$$X_{\mathrm{W}}^2 = 6.336; \quad p = 0.0027$$

(iii) *Score Test*:

$$X_{\mathrm{S}}^2 = 11.860; \quad p = 0.0421$$

The results for individual covariates are as follows:

Variable	Coefficient	Standard Error	z-Statistic	p-Value
Bleeding	1.6198	1.3689	1.183	0.2367
Pregnancy Loss	1.7319	0.8934	1.938	0.526

In addition to a priori interests in the effects of individual covariates, given a continuous variable of interest, one can fit a polynomial model and use this type of test to check for linearity. It can also be used to check for a single product representing an effect modification.

■ **Example 5.14** Refer to the data for low-birth-weight babies in Example 5.9, but this time we investigate only one covariate, the mother's weight. After fitting the second-degree polynomial model, we obtain a result that indicates that the curvature effect is negligible ($p=0.9131$).

Contribution of a Group of Variables This testing procedure addresses the more general problem of assessing the additional contribution of two or more factors to the prediction of the response over and above that made by other variables already in the regression model. In other words, the null hypothesis is of the form

$$H_0\colon\ \beta_1 = \beta_2 = \cdots = \beta_m = 0$$

To test such a null hypothesis, one can perform a likelihood ratio chi-squared test, with m degrees of freedom:

$$X^2_{\text{LR}} = 2[\ln L(\beta;\text{all covariates}) - \ln L(\beta;\text{all except } m \text{ covariates})]$$

As with the above tests for individual covariates, this multiple contribution procedure is very useful for assessing the importance of potential explanatory variables. In particular, it is often used to test whether a similar group of variables, such as demographic characteristics, is important for the prediction of the response; these variables have some trait in common. Another application would be a collection of powers and/or product terms (referred to as interaction variables). It is often of interest to assess the interaction effects collectively before trying to consider individual interaction terms in a model as previously suggested. In fact, such use reduces the total number of tests to be performed and this, in turn, helps to provide better control of overall Type I error rates, which may be inflated due to multiple testing.

■ **Example 5.15** Refer to the data for low-birth-weight babies in Example 5.9 with all four covariates. We consider, collectively, these three interaction terms: M-Weight * Smoking, M-Weight * Hypertension, and M-Weight * U-Irritability. The basic idea is to see if any of the other variables would modify the effect of the mother's weight on the response (having a low-birth-weight baby).

(i) With the original four variables, we obtain. $\ln L = -16.030$.
(ii) With all seven variables, four original plus three products, we obtain. $\ln L = -14.199$.

Therefore we have

$$\begin{aligned} X^2_{\text{LR}} &= 2[\ln L(\beta;\text{seven variables}) - \ln L(\beta;\text{four original variables})] \\ &= 3.662 \text{ with 2 degrees of freedom};\ \ p \geq 0.10 \end{aligned}$$

indicating a rather weak level of interactions

Stepwise Regression

In many applications, our major interest is to identify important risk factors. In other words, we wish to identify from many available factors a small subset of factors that relate significantly to the outcome (e.g., the disease under investigation). In that identification process, of course, we wish to avoid a large Type I (false-positive) error. In a regression analysis, a Type I error corresponds to including a predictor that has no real relationship to the outcome; such an inclusion can greatly confuse the interpretation of the regression results. In a standard multiple regression analysis, this goal can be achieved by using a strategy that adds into or removes from a regression model one factor at a time according to a certain order of relative importance. Therefore the two important steps are:

1. Specify a criterion or criteria for selecting a model.
2. Specify a strategy for applying the chosen criterion or criteria.

The process follows the same outline of Chapter 5 for logistic regression where we combine the forward selection and backward elimination in the stepwise process and selection at each step is based on the likelihood ratio chi-squared test. The SAS PROC PHREG does have an automatic stepwise option to implement these features.

■ **Example 5.16** Refer to the data for low-birth-weight babies in Example 5.9 with all four covariates: mother's weight, smoking, hypertension, and uterine irritability. This time we perform a stepwise regression analysis in which we specify that a variable has to be significant at the 0.10 level before it can enter into the model and that a variable in the model has to be significant at the 0.15 for it to remain in the model (most standard computer programs allow users to make these selections; default values are available). First, we get the individual score test results for all variables:

Variable	Score X^2	p-Value
M Weight	3.9754	0.0462
Smoking	0.0000	1.0000
Hypertension	0.2857	0.5930
U Irritability	5.5556	0.0184

These indicate that U Irritability is the most significant variable. We thus proceed as follows:

Step 1: Variable U Irritability is entered.

Analysis of Variables Not in the Model

Variable	Score X^2	p-Value
M Weight	2.9401	0.0864
Smoking	0.0027	4.0000
Hypertension	0.2857	0.5930

Step 2: Variable M Weight is entered.

Analysis of Variables in the Model

Variable	Coefficient	Standard Error	z-Statistic	p-Value
M Weight	−0.0192	0.0116	−1.673	0.0978
U Irritability	2.141	1.1983	1.787	0.074

Neither variable is removed.

Analysis of Variables Not in the Model

Variable	Score X^2	p-Value
Smoking	0.0840	0.7720
Hypertension	0.3596	0.5487

No (additional) variables meet the 0.1 level for entry into the model.

Note: A SAS program would include these instructions:

```
PROC PHREG DATA LOWWEIGHT;
MODEL DUMMYTIME*CASE(0) = MWEIGHT SMOKING HYPERT UIRRIT/
SELECTION=STEPWISE SLENTRY=.10 SLSTAY=.15;
STRATA=SET;
```

`HYPERT` and `UIRRIT` are hypertension and uterine irritability. The default values for `SLENTRY` (p-value to enter) and `SLSTAY` (p-value to stay) are 0.05 and 0.10.

EXERCISES

5.1 It has been noted that metal workers have an increased risk for cancer of the internal nose and paranasal sinuses, perhaps as a result of exposure to cutting oils. Therefore a study was conducted to see whether this particular exposure also increases the risk for squamous cell carcinoma of the scrotum (Rousch et al., 1982). Cases included all 45 squamous cell carcinomas of the scrotum diagnosed in Connecticut residents from 1955 to 1973, as obtained from the

Connecticut Tumor Registry. Matched controls were selected for each case based on the age at death (within 8 years), year of death (within 3 years), and number of jobs as obtained from combined death certificate and directory sources. An occupational indicator of metal worker (yes/no) was evaluated as the possible risk factor in this study. Results are as follows:

	Control	
Case	Yes	No
Yes	2	26
No	5	12

(**a**) Find a 95% confidence interval for the odds ratio measuring the strength of the relationship between the disease and the exposure.

(**b**)Test for the independence between the disease and the exposure.

5.2 Ninety-eight heterosexual couples, at least one of whom was HIV infected, were enrolled in an HIV transmission study and interviewed about sexual behavior (Padian, 1990). The following table summarized data reported by heterosexual partners:

	Man	
Woman	Ever	Never
Ever	45	6
Never	7	40

Test to compare the reporting results between men and women.

5.3 A matched case–control study was conducted in order to evaluate the cumulative effects of acrylate and methacrylate vapors on olfactory function (Schwarts et al., 1989). Cases were defined as scoring at or below the 10th percentile on the UPSIT (University of Pennsylvania Smell Identification Test).

	Case	
Control	Exposed	Nonexposed
Exposed	25	22
Nonexposed	9	21

(**a**) Find a 95% confidence interval for the odds ratio measuring the strength of the relationship between the disease and the exposure.

(**b**) Test for the independence between the disease and the exposure.

5.4 A study in Maryland identified 4032 white persons, enumerated in a non-official 1963 census, who became widowed between 1963 and 1974 (Helsing and Szklo, 1981). These people were matched, one-to-one, to married persons on the basis of race, sex, year of birth, and geography of residence. The matched pairs were followed to a second census in 1975. We have the following overall male mortality; test to compare the mortality of widowed men versus married men:

	Married Men	
Widowed Men	Died	Alive
Died	2	292
Alive	210	700

The data for 2828 matched pairs of women were as follows; test to compare the mortality of widowed women versus married women:

	Married Women	
Widowed Women	Died	Alive
Died	1	264
Alive	249	2314

5.5 Table 5.3 provides some data from a matched case–control study to investigate the association between the use of X-ray and risk of childhood acute myeloid leukemia. In each matched set or pair, the case and control(s) were matched by age, race, and the county of residence. The variables are:

- Matched set (or pair).
- Disease (1 = Case, 2 = Control).
- Some characteristics of the child: Sex (1 = Male, 2 = Female),
- Down syndrome (a known risk factor for leukemia; 1 = No, 2 = Yes), and Age.
- Risk factors related to the use of X-rays: M-Xray (Mother ever had X-ray during pregnancy; 1 = No, 2 = Yes), MU-Xray (Mother ever had upper body X-ray during pregnancy; 0 = No, 1 = Yes), ML-Xray (Mother ever had lower body X-ray during pregnancy; 0 = No, 1 = Yes), F-Xray (Father ever had

TABLE 5.3. X-ray–Leukemia Data

Set	Disease	Sex	Down Syndrome	Age	M-Xray	MU-Xray	ML-Xray	F-Xray	C-Xray	CN-Xray
1	1	2	1	0	1	0	0	1	1	1
	2	2	1	0	1	0	0	1	1	1
2	1	1	1	6	1	0	0	1	2	3
	2	1	1	6	1	0	0	1	2	2
3	1	2	1	8	1	0	0	1	1	1
	2	2	1	8	1	0	0	1	1	1
4	1	1	2	1	1	0	0	1	1	1
	2	1	1	1	1	0	0	1	1	1
5	1	1	1	4	2	0	1	1	1	1
	2	1	1	4	1	0	0	1	2	2
6	1	2	1	9	2	1	0	1	1	1
	2	1	1	9	1	0	0	1	1	1
7	1	2	1	17	1	0	0	1	2	2
	2	2	1	17	1	0	0	1	2	2
8	1	2	1	5	1	0	0	1	1	1
	2	1	1	5	1	0	0	1	1	1
9	1	2	2	0	1	0	0	1	1	1
	2	2	1	0	2	1	0	2	1	1
10	1	2	1	7	1	0	0	1	1	1
	2	1	1	7	1	0	0	2	1	1
11	1	1	1	15	1	0	0	1	1	1
	2	1	1	15	1	0	0	1	2	2
12	1	1	1	12	1	0	0	1	2	2
	2	1	1	12	1	0	0	1	1	1
13	1	1	1	4	1	0	0	1	1	1
	2	2	1	4	1	0	0	1	1	1
14	1	1	1	14	1	0	0	1	2	2
	2	2	1	14	1	0	0	1	1	1
15	1	1	1	7	1	0	0	2	1	1
	2	1	1	7	1	0	0	2	2	2
16	1	1	1	8	1	0	0	1	2	2
	2	2	1	8	1	0	0	2	1	1
17	1	1	1	6	1	0	0	2	1	1
	2	1	1	6	1	0	0	1	1	1
18	1	2	1	13	1	0	0	2	2	3
	2	2	1	13	1	0	0	1	2	2
19	1	2	1	17	1	0	0	2	1	1
	2	1	1	17	1	0	0	1	2	2
20	1	2	1	5	2	0	1	1	2	4
	2	2	1	5	1	0	0	1	1	1
21	1	1	1	13	1	0	0	1	2	4
	2	1	1	13	1	0	0	1	1	1
22	1	2	1	16	1	0	0	2	2	2
	2	2	1	16	1	0	0	2	1	1

TABLE 5.3. (*Continued*)

Set	Disease	Sex	Down Syndrome	Age	M-Xray	MU-Xray	ML-Xray	F-Xray	C-Xray	CN-Xray
23	1	2	1	10	1	0	0	2	1	1
	2	2	1	10	1	0	0	1	1	1
24	1	1	1	0	1	0	0	2	1	1
	2	1	1	0	1	0	0	1	1	1
25	1	1	1	1	1	0	0	2	1	1
	2	2	1	1	1	0	0	2	1	1
26	1	2	1	13	1	0	0	2	1	1
	2	1	1	13	2	1	1	1	1	1
27	1	1	1	11	1	0	0	1	1	1
	2	1	1	11	1	0	0	1	2	2
28	1	2	1	4	1	0	0	2	1	1
	2	2	1	4	1	0	0	1	2	2
29	1	2	1	1	1	0	0	2	1	1
	2	1	1	1	1		0	1	1	1
30	1	2	1	15	2	0	1	2	2	3
	2	2	1	15	1	0	0	2	2	2
31	1	1	1	9	1	0	0	1	1	1
	2	2	1	9	1	0	0	1	1	1
32	1	1	1	15	1	0	0	2	2	2
	2	1	1	15	1	0	0	2	2	3
33	1	2	1	5	1	0	0	1	2	3
	2	2	1	5	2	1	0	1	1	1
34	1	1	1	10	2	0	1	2	1	1
	2	1	1	10	1	0	0	2	2	2
35	1	2	1	8	1	0	0	1	2	2
	2	2	1	8	1	0	0	1	1	1
36	1	2	1	15	1	0	0	1	2	4
	2	2	1	15	1	0	0	1	2	2
37	1	2	2	1	1	0	0	1	1	1
	2	2	1	1	2	1	0	2	1	1
38	1	2	1	0	1	0	0	1	1	1
	2	1	1	0	1	0	0	1	1	1
39	1	1	1	6	1	0	0	2	2	2
	2	2	1	6	1	0	0	1	1	1
40	1	1	1	14	1	0	0	2	2	2
	2	2	1	14	1	0	0	1	1	1
41	1	1	1	2	1	0	0	2	1	1
	2	2	1	2	1	0	0	1	1	1
42	1	1	2	1	2	1	0	1	1	1
	2	1	1	1	1	0	0	1	1	1
43	1	2	1	6	1	0	0	1	1	1
	2	2	1	6	1	0	0	1	1	1
44	1	1	1	16	1	0	0	1	1	1
	2	1	1	16	2	1	0	1	1	1

(*continued*)

TABLE 5.3. (*Continued*)

Set	Disease	Sex	Down Syndrome	Age	M-Xray	MU-Xray	ML-Xray	F-Xray	C-Xray	CN-Xray
45	1	1	1	4	1	0	0	1	2	2
	2	2	1	4	1	0	0	1	1	1
46	1	2	1	1	1	0	0	1	1	1
	2	1	1	1	1	0	0	1	1	1
47	1	1	1	0	1	0	0	1	1	1
	2	1	1	0	1	0	0	2	1	1
48	1	2	1	0	1	0	0	1	1	1
	2	2	1	0	1	0	0	2	1	1
49	1	1	1		1	0	0	1	2	4
	2	1	1		1	0	0	1	2	4
50	1	2	1		1	0	0	1	1	1
	2	1	1		1	0	0	2	2	2
51	1	1	1		1	0	0	1	1	1
	2	1	1		1	0	0	1	1	1
52	1	2	1		1	0	0	2	2	2
	2	2	1		1	0	0	1	1	1
53	1	1	1		1	0	0	1	2	1
	2	2	1		1	0	0	2	1	1
54	1	2	1		1	0	0	1	1	1
	2	2	1		2	1	0	1	1	1
55	1	1	1		1	0	0	1	2	2
	2	1	1		1	0	0	1	1	1
56	1	1	1		1	0	0	1	1	1
	2	1	1		2	1	0	1	2	2
57	1	2	1		1	0	0	1	1	1
	2	2	1		1	0	0	2	1	1
58	1	1	1		1	0	0	1	2	4
	2	1	1		2	0	1	1	1	1
59	1	1	1		1	0	0	1	2	2
	2	2	1		2	0	1	1	1	1
60	1	1	1		1	0	0	1	1	1
	2	1	1		1	0	0	1	1	1
61	1	1	1		1	0	0	1	2	3
	2	1	1		1	0	0	2	1	1
62	1	1	1		1	0	0	1	2	2
	2	1	1		1	0	0	1	2	4
63	1	2	1		1	0	0	1	1	1
	2	1	1		1	0	0	1	1	1
64	1	1	1		1	0	0	2	1	1
	2	2	1		1	0	0	1	1	1
65	1	2	1		2	0	1	1	2	2
	2	2	1		1	0	0	1	1	1
66	1	2	1		1	0	0	1	2	4
	2	1	1		1	0	0	2	1	1

TABLE 5.3. (*Continued*)

Set	Disease	Sex	Down Syndrome	Age	M-Xray	MU-Xray	ML-Xray	F-Xray	C-Xray	CN-Xray
67	1	1	1		1	0	0	1	1	1
	2	1	1		1	0	0	1	1	1
68	1	1	1		1	0	0	1	2	4
	2		1		1	0	0	2	1	1
69	1	2	1		1	0	0	1	1	1
	2	2	1		1	0	0	1	1	1
70	1	1	1	3	1	0	0	1	1	1
	2	2	1	3	1	0	0	1	1	1
71	1	1	1	1	1	0	0	1	1	1
	2	1	1	1	1	0	0	1	1	1
72	1	2	1	9	1	0	0	1	1	1
	2	2	1	9	1	0	0	2	1	1
73	1	1	1	1	1	0	0	1	1	1
	2	1	1	1	1	0	0	1	1	1
74	1	2	1	8	1	0	0	1	2	3
	2	2	1	8	1	0	0	2	2	2
75	1	1	1	12	1	0	0	1	1	1
	2	2	1	12	1	0	0	1	1	1
76	1	2	1	1	1	0	0	1	1	1
	2	2	1	1	1	0	0	1	1	1
77	1	2	1	4	1	0	0	1	1	1
	2	2	1	4	1	0	0	2	1	1
78	1	2	1	11	1	0	0	1	2	2
	2	2	1	11	1	0	0	1	1	1
79	1	1	1	2	1	0	0	1	1	1
	2	1	1	2	1	0	0	1	1	1
80	1	2	1	4	1	0	0	1	1	1
	2	2	1	4	1	0	0	1	1	1
81	1	1	1	1	1	0	0	2	1	1
	2	1	1	1	1	0	0	1	1	1
82	1	2	1	0	1	0	0	1	1	1
	2	1	1	0	1	0	0	1	1	1
83	1	1	1	5	1	0	0	1	1	1
	2	1	1	5	1	0	0	2	2	2
84	1	2	2	1	2	0	1	2	1	1
	2	1	1	1	1	0	0	2	1	1
85	1	1	1	12	1	0	0	2	2	2
	2	1	1	12	1	0	0	1	1	1
86	1	1	1	12	2	0	1	2	2	2
	2	1	1	12	1	0	0	2	2	4
87	1	1	1	1	1	0	0	1	1	1
	2	1	1	1	1	0	0	1	1	1
88	1	1	1	9	1	0	0	2	1	1
	2	1	1	9	1	0	0	2	1	1

(*continued*)

TABLE 5.3. (*Continued*)

Set	Disease	Sex	Down Syndrome	Age	M-Xray	MU-Xray	ML-Xray	F-Xray	C-Xray	CN-Xray
89	1	2	1	2	1	0	0	2	1	1
	2	2	1	2	1	0	0	1	1	1
90	1	2	1	1	1	0	0	1	1	1
	2	2	1	1	1	0	0	2	1	1
91	1	2	1	2	1	0	0	1	1	1
	2	2	1	2	1	0	0	1	1	1
92	1	1	1	15	1	0	0	2	1	1
	2	1	1	15	1	0	0	1	1	
93	1	1	1	13	1	0	0	2	1	1
	2	1	1	13	1	0	0	1	1	1
94	1	1	1	6	1	0	0	2	2	4
	2	2	1	6	1	0	0	1	1	1
95	1	2	2	1	1	0	0	1	1	1
	2	2	1	1	1	0	0	1	1	1
96	1	2	1	8	1	0	0	1	1	1
	2	2	1	8	1	0	0	2	1	1
97	1	2	1	4	1	0	0	2	2	4
	2	2	1	4	1	0	0	2	1	1
98	1	1	1	6	1	0	0	1	2	2
	2	1	1	6	1	0	0	2	1	1
99	1	1	1	1	1	0	0	2	1	1
	2	2	1	1	1	0	0	1	1	1
100	1	1	1	14	1	0	0	1	2	4
	2	1	1	14	1	0	0	1	2	3
101	1	1	1	1	1	0	0	2	1	1
	2	1	1	1	1	0	0	1	1	1
102	1	2	2	1	1	0	0	1	1	1
	2	2	1	1	1	0	0	1	1	1
103	1	1	1	13	1	0	0	2	1	1
	2	1	1	13	1	0	0	1	1	1
104	1	2	1	13	1	0	0	1	2	4
	2	2	1	13	1	0	0	1	1	1
105	1	2	1	11	1	0	0	2	2	3
	2	2	1	11	1	0	0	1	1	1
106	1	2	1	13	1	0	0	1	2	3
	2	1	1	13	1	0	0	2	1	1
107	1	2	1	2	1	0	0	2	1	1
	2	2	1	2	1	0	0	1	1	1
108	1	2	1	7	1	0	0	1	1	1
	2	1	1	7	1	0	0	1	2	2
109	1	2	1	16	1	0	0	1	2	3
	2	1	1	16	1	0	0	1	1	1
110	1	1	1	3	1	0	0	1	1	1
	2	1	1	3	1	0	0	1	1	1

TABLE 5.3. (*Continued*)

Set	Disease	Sex	Down Syndrome	Age	M-Xray	MU-Xray	ML-Xray	F-Xray	C-Xray	CN-Xray
111	1	1	1	13	1	0	0	2	2	4
	2	2	1	13	1	0	0	1	1	1
112	1	2	1	6	1	0	0	1	1	1
	2	2	1	6	1	0	0	1	1	1

X-ray; 1 = No, 2 = Yes), C-Xray (Child ever had X-ray; 1 = No, 2 = Yes), and CN-Xray (Child's total number of X-rays; 1 = None, 2 = 1–2, 3 = 3–4, 4 = 5 or more).

(**a**) Taken collectively, do the covariates contribute significantly to the separation of the cases and the controls?

(**b**) Fit the multiple regression model to obtain estimates of individual regression coefficients and their standard errors. Draw your conclusion concerning the conditional contribution of each factor.

(**c**) Within the context of the multiple regression model in (b), does sex alter the effect of Down syndrome?

(**d**) Within the context of the multiple regression model in (b), taken collectively, does the exposure to X-rays (by the father, or mother, or the child) relate significantly to the disease of the child?

(**e**) Within the context of the multiple regression model in (b), is the effect of age linear?

(**f**) Focusing on Down syndrome (DS) as the primary factor, taken collectively, was this main effect altered by any other covariates?

6

METHODS FOR COUNT DATA

Topics in Chapters 2 and 3 focused mainly on contingency tables and Chapter 4 on regression methods for binomially and multinomially distributed responses. This chapter is devoted to a different type of categorical data—count data—and the eventual focus is the Poisson regression model. As usual, the purpose of the research is to assess relationships among a set of variables, one of which is taken to be the response or dependent variable that is a variable to be predicted from or explained by other variables called predictors, explanatory or independent variables. Choosing an appropriate model and analytical technique depends on the type of response variable under investigation. The Poisson regression model applies when the dependent variable follows a Poisson distribution.

6.1 THE POISSON DISTRIBUTION

The binomial distribution of Chapter 2 is used to characterize an experiment when each trial of the experiment has two possible outcomes (often referred to as "failure" and "success"). Let the probabilities of failure and success be, respectively, $(1 - \pi)$ and π, and we "code" these two outcomes as 0 (zero successes) and 1 (one success). The experiment consists of n repeated trials satisfying these assumptions:

(i) The n trials are all independent.
(ii) The parameter π is the same for each trial.

The target for the binomial distribution is the total number X of successes in n trials. The distribution of X is characterized by

$$\Pr(X = x) = \binom{n}{x} \pi^{x} (1-\pi)^{n-x}$$

For $x = 1, 2, \ldots, n$, when n is large, direct calculation of binomial probabilities can involve a prohibitive amount of work. However, it can be shown that the limiting form of the binomial distribution, when n is increasingly large ($n \to \infty$) and when n is increasingly small ($n \to 0$) while $\theta = n\pi$ remains constant, is rather simple. It is given by

$$\begin{aligned}\Pr(X = x) &= \frac{\theta^{x} e^{-\theta}}{x!} \\ &= p(x;\theta)\end{aligned}$$

A random variable having this probability function is said to have the Poisson distribution $P(\theta)$. To illustrate the Poisson approximation of the binomial distribution, consider $n = 48$ and $\pi = 0.05$. Then

$$\begin{aligned}b(x = 5; n, \pi) &= 0.059 \\ p(x = 5; \theta) &= 0.060\end{aligned}$$

The Poisson model is often used when the random variable X is supposed to represent the number of occurrences of some random event in an interval of time or space, or some volume of matter. Numerous applications in health sciences have been documented: for example, the number of viruses in a solution, the number of defective teeth per individual, the number of focal lesions in virology, the number of victims of specific diseases, the number of cancer deaths per household, the number of infant deaths in a certain locality during a given year, among others. The mean and variance of the Poisson distribution are equal:

$$\begin{aligned}\mu &= \theta \\ \sigma^2 &= \theta\end{aligned}$$

Given a sample of counts from the Poisson distribution $P(\theta)$, the sample mean is an unbiased estimator for θ. Its standard error is given by

$$SE(\bar{x}) = \sqrt{\frac{\bar{x}}{n}}$$

■ **Example 6.1** In estimating the infection rates in populations of organisms, sometimes it is impossible to assay each organism individually. Instead, the organisms are randomly divided into a number of pools and each pool is tested as a unit. Let

N = number of insects in the sample.

n = number of pools used in the experiment.

m = number of insects per pool, $N = nm$ (for simplicity, assume that m is the same for every pool).

The random variable X concerned is the number of pools that show a negative test result, that is, none of the insects are infected.

(i) Let λ be the population infection rate. The probability that all m insects in a pool are negative (in order to have a negative pool) is given by

$$\pi = (1-\lambda)^m$$

Designating a negative pool as "success," we have a binomial distribution for X which is $B(n, \pi)$.

(ii) In situations where the infection rate λ is very small, the Poisson distribution could be used as an approximation with $\theta = m\lambda$ being the expected number of infected insects in a pool. The Poisson probability of this number being zero is

$$\pi = \exp(-\theta)$$

and we have the same binomial distribution $B(n, \pi)$ for the experiment. For detailed description, see Bhattacharyya et al. (1979), Walter et al. (1980), or Le (1981) for extension of the above procedure to using pools of variable size.

The U.S. Army tested for syphilis (as well as other very rare diseases) in this way. The method was also adapted to estimate the infectious potential of blood containing human immunodeficiency virus (HIV) (Connett et al., 1990).

■ **Example 6.2** For the year of 1981, the infant mortality rate (IMR) for the United States was 11.9 deaths per 1000 live births. For the same period, the New England states (Connecticut, Maine, Massachusetts, New Hampshire, Rhode Island, and Vermont) had 164,200 live births and 1585 infant deaths (Freeman, 1980). If the national IMR applies, the mean and variance of the number of infant deaths in the New England states would be

$$(164.2)(11.9) = 1954$$

From the z-score,

$$\begin{aligned} z &= \frac{1585-1954}{\sqrt{1954}} \\ &= -8.35 \end{aligned}$$

It is clear that the IMR in the New England states is below the national average.

■ **Example 6.3** Cohort studies are designs in which one enrolls a group of healthy persons and follows them over certain periods of time, for example, occupational mortality studies. The cohort study design focuses attention on a particular exposure rather than a particular disease as in case–control studies. Advantages of a longitudinal approach include the opportunity for more accurate measurement of exposure history and a careful examination of the time relationships between exposure and disease.

The observed mortality of the cohort under investigation often needs to be compared with that expected from the death rates of the national population (serving as standard), with allowance made for age, sex, race, and time period. Rates may be calculated either for total deaths or for separate causes of interest. The statistical method is often referred to as the person-years method. The basis of this method is the comparison of the observed number of deaths, d, from the cohort with the mortality that would have been expected if the group had experienced similar death rates to those of the standard population of which the cohort is a part. The expected number of deaths is usually calculated using published national life tables and the method is similar to that of indirect standardization of death rates.

Each member of the cohort contributes to the calculation of the expected deaths for those years in which he/she was at risk of dying during the study. There are three types of subjects:

(i) Subjects still alive on the analysis date.
(ii) Subjects who died on a known date within the study period.
(iii) Subjects who are lost to follow-up after a certain date. These cases are a potential source of bias; effort should be expended in reducing the number of subjects in this category.

Each subject who entered the study is characterized by the year and age at entry, and is followed until the study date, the date of death, or the date the subject was lost to follow-up. Period and age are divided into 5-year intervals corresponding to the usual availability of referenced death rates. Then a quantity, r, is defined for each individual as the cumulative risk over the follow-up period:

$$r = \sum x\varphi$$

where summation is over all 5-year periods entered by the subject, x is the time spent in a period (5 years, except for the first and the last periods), and φ is the corresponding death rate for the given age–period combination. For the cohort, the individual values of r are added to give the expected number of deaths:

$$m = \sum r$$

For various statistical analyses, the observed number of deaths d may be treated as a Poisson variable with mean $\theta = m\rho$, where ρ is the relative risk of being a member of the cohort as compared to the standard population (Berry, 1983).

6.2 TESTING GOODNESS-OF-FIT

A goodness-of-fit test is used when one wishes to decide if an observed distribution of frequencies is incompatible with some hypothesized distribution. The Poisson is a very special distribution; its mean and its variance are equal. Therefore given a sample of count data

$$\{x_i\}_{i=1}^{n}$$

we often wish to know whether these data provide sufficient evidence to indicate that the sample did not come from a Poisson-distributed population. We have the following hypotheses:

H_0: The sampled population is distributed as Poisson

H_A: The sampled population is not distributed as Poisson

The most frequent violation is an overdispersion—the variance is larger than the mean. The implication is serious; the analysis that assumes the Poisson model often underestimates standard error(s) and thus wrongly inflates the level of significance.

The test is statistic is the familiar Pearson's chi-squared statistic:

$$X^2 = \sum_{i=1}^{k} \frac{(O_i - E_i)^2}{E_i}$$

where O_i and E_i refer to the ith observed and expected frequencies, respectively. In this formula, k is the number of groups for which observed and expected frequencies are available. When the null hypothesis is true, the test statistic is distributed as chi-squared with $(k-2)$ degrees of freedom; 1 degree of freedom was lost because the mean needs to be estimated and another was lost because of the constraint

$$\sum O_i = \sum E_i$$

It is also recommended that adjacent groups at the bottom of the table be combined in order to avoid having any expected frequencies less than 1.

Example 6.4 The purpose of this study was to examine the data for 44 physicians working in the emergency room at a major hospital. The response variable is the number of complaints received during the previous year; other details of the study are given in the next example. For the purpose of testing the goodness-of-fit, the data are summarized in Table 6.1; original data are in the first two columns.

To obtain the expected frequencies, we first obtain relative frequencies by evaluating the Poisson probability for each value of $X = x$; for example,

$$\Pr(X = x) = \frac{\theta^x e^{-\theta}}{x!} \quad \text{and} \quad \hat{\theta} = \frac{\sum x}{44} = 3.34$$

We have

$$\begin{aligned} \Pr(X = 2) &= \frac{3.34^2\, e^{-3.34}}{2!} \\ &= 0.198 \end{aligned}$$

Each of the expected relative frequencies is multiplied by the sample size, 44, to obtain the expected frequencies: for example,

$$\begin{aligned} E_2 &= (44)(0.198) \\ &= 8.71 \end{aligned}$$

TABLE 6.1. Emergency Service Data

Number of Complaints	Observed O_i	Expected E_i
0	1	1.54
1	12	5.19
2	12	8.71
3	5	9.68
4	1	8.10
5	4	5.46
6	2	2.99
7	2	1.45
8	2	0.62
9	1	0.22
10	1	0.09
11	1	0.04

In order to avoid having any expected frequencies less than 1, we combine the last five groups together resulting in eight groups available for testing goodness-of-fit with

$$\begin{aligned} O_8 &= 2+2+1+1+1; \\ &= 7; \\ E_8 &= 1.45+0.62+0.22+0.09+0.04 \\ &= 2.42 \end{aligned}$$

The result is

$$\begin{aligned} X^2 &= \frac{(1-1.59)^2}{1.59} + \frac{(12-5.19)^2}{5.19} + \cdots + \frac{(7-2.42)^2}{2.42} \\ &= 28.24 \end{aligned}$$

with $8-2=6$ degrees of freedom, indicating a significant deviation from the Poisson distribution ($p<0.005$). A simple inspection of Table 6.1 reveals an obvious overdispersion.

6.3 THE POISSON REGRESSION MODEL

As previously mentioned, the Poisson model is often used when the random variable X is supposed to represent the number of occurrences of some random event in an interval of time or space or some volume of matter. Numerous applications in health sciences have been documented. In some of these applications, one may be interested to see if the Poisson-distributed dependent variable can be predicted from or explained by other variables. The other variables are called predictors, or explanatory or independent variables. For example, we may be interested in the number of defective teeth per individual as a function of gender and age of a child, brand of toothpaste, and whether the family has or does not have dental insurance. In this and other examples, the dependent variable Y is assumed to follow a Poisson distribution with mean θ.

The Poisson regression model expresses this mean as a function of certain independent variables $X_1, X_2, \ldots, X_k$, in addition to the *size* of the observation unit from which one obtained the count of interest. For example, if Y is the number of viruses in a solution, then the size is the volume of the solution; or if Y is the number of defective teeth, then the size is the total number of teeth for that same individual. The following is a continuation of Example 6.4 on the emergency room service data; however, data in Table 6.2 also include information on four covariates.

■ **Example 6.5** The purpose of this study was to examine the data for 44 physicians working in the emergency room at a major hospital so as to determine which of the following four variables are related to the number of complaints

TABLE 6.2. Complete Emergency Service Data

Number of Visits	Complaint	Gender	Residency	Revenue	Hours
2014	2	Y	F	263.02	1287.25
3091	3	N	M	334.94	1588.00
879	1	Y	M	206.42	705.25
1780	1	N	M	236.32	1005.50
3646	11	N	M	288.91	1667.25
2690	1	N	M	275.94	1517.75
1864	2	Y	M	295.71	967.00
2782	6	N	M	224.91	1609.25
3071	9	N	F	249.32	1747.75
1502	3	Y	M	269.00	906.25
2438	2	N	F	225.61	1787.75
2278	2	N	M	212.43	1480.50
2458	5	N	M	211.05	1733.50
2269	2	N	F	213.23	1847.25
2431	7	N	M	257.30	1433.00
3010	2	Y	M	326.49	1520.00
2234	5	Y	M	290.53	1404.75
2906	4	N	M	268.73	1608.50
2043	2	Y	M	231.61	1220.00
3022	7	N	M	241.04	1917.25
2123	5	N	F	238.65	1506.25
1029	1	Y	F	287.76	589.00
3003	3	Y	F	280.52	1552.75
2178	2	N	M	237.31	1518.00
2504	1	Y	F	218.70	1793.75
2211	1	N	F	250.01	1548.00
2338	6	Y	M	251.54	1446.00
3060	2	Y	M	270.52	1858.25
2302	1	N	M	247.31	1486.25
1486	1	Y	F	277.78	933.95
1863	1	Y	M	259.68	1168.25
1661	0	N	M	260.92	877.25
2008	2	N	M	240.22	1387.25
2138	2	N	M	217.49	1312.00
2556	5	N	M	250.31	1551.50
1451	3	Y	F	229.43	9.73.75
3328	3	Y	M	313.48	1638.25
2928	8	N	M	293.47	1668.25
2701	8	N	M	275.40	16.52.75
2046	1	Y	M	289.56	1029.75
2548	2	Y	M	305.67	1127.00
2592	1	N	M	252.35	1547.25
2741	1	Y	F	276.86	1499.25
3763	10	Y	M	308.84	1747.50

received during the previous year. In addition to the number of complaints, which serves as the dependent variable, data available consist of the number of visits, which serves as the *size* for the observation unit, the physician, and four covariates. Table 6.2 presents the complete data set. For each of the 44 physicians there are two continuous independent variables—the revenue (dollars per hour) and work load at the emergency room (hours)—and two binary variables—gender (female/male) and residency training in emergency services (no/yes).

6.3.1 Simple Regression Analysis

In this section we discuss the basic ideas of simple regression analysis when only one predictor or independent variable is available for predicting the response of interest.

The Poisson Regression Model

In our framework, the dependent variable Y is assumed to follow a Poisson distribution; its values, y_i's, are available from "observation unit" n, which is also characterized by an independent variable X. For the observation unit i ($i = n$), let s_i be the size and x_i be the covariate value. The Poisson regression model assumes that the relationship between the mean of Y and the covariate X is described by

$$\begin{aligned} E(Y_i) &= s_i \lambda(x_i) \\ &= s_i \exp(\beta_0 + \beta_1 x_i) \end{aligned}$$

where $\lambda(x_i)$ is called the "risk" of/for observation unit i ($1 = i = n$).

Under the assumption that Y_i is distributed as a Poisson distribution with the above mean, the likelihood function is given by

$$L(y;\beta) = \prod_{i=1}^{n} \left\{ \frac{[s_i \lambda(x_i)]^{y_i} \exp[-s_i \lambda(x_i)]}{y_i!} \right\}$$

$$\ln L(y;\beta) = \sum_i \{ y_i \ln s_i - \ln y_i! + y_i[\beta_0 + \beta_1 x_i] - s_i \exp[\beta_0 + \beta_1 x_i] \}$$

from which estimates of the two regression coefficients β_0 and β_1 can be obtained by the maximum likelihood procedure.

Measure of Association

Consider the case of a binary covariate X, say, representing an exposure (1 = exposed, 0 = not exposed). We have the following:

$$\ln \lambda_i = \begin{cases} \beta_0 + \beta_1 + \ln s_i & \text{if exposed (or } x = 1) \\ \beta_0 + \ln s_i & \text{if unexposed (or } x = 0) \end{cases}$$

$$\frac{\lambda_i(\text{exposed})}{\lambda_i(\text{unexposed})} = \exp(\beta_1)$$

This quantity, represented by $\exp(\beta_1)$, is the relative risk associated with the exposure.

Similarly, for a continuous covariate X and any value x of X, we have

$$\ln \lambda_i = \begin{cases} \beta_0 + \beta_1 x + \ln s_i & \text{if } X = x \\ \beta_0 + \beta_1 (x+1) + \ln s_i & \text{if } X = x+1 \end{cases}$$

$$\frac{\lambda_i(X = x+1)}{\lambda_i(X = x)} = \exp(\beta_1)$$

This quantity, represented by $\exp(\beta_1)$, is the relative risk associated with one unit increase in the value of X, from x to $(x + 1)$.

The basic rationale for using the terms "risk" and "relative risk" is the approximation of the binomial distribution by the Poisson distribution. Recall from the previous section that, when n goes to infinity, π tends to 0, and $\theta = n\pi$ remains constant, the binomial distribution $B(n,\pi)$ can be approximated by the Poisson distribution $P(\theta)$. The number n is the size of the observation unit; so the ratio between the mean and the size represents π (or $\lambda(x)$ in the new model), the probability or "risk" and the ratio of risks is the risks ratio or *relative risk*.

■ **Example 6.6** Refer to the emergency service data in Example 6.5 and suppose we want to investigate the relationship between the number of complaints (adjusted for number of visits) and residency training. It may be perceived that by having training in the specialty a physician would perform better and therefore be less likely to provoke complaints. An application of the simple Poisson regression analysis yields the following:

Variable	Coefficient	Standard Error	z-Statistic	p-Value
Intercept	−6.7566	0.1387	−48.714	<0.0001
No Residency	0.3041	0.1725	1.763	0.0779

The result indicates that the common perception is almost true, that the relationship between the number of complaints and no residency training in emergency service is marginally significant ($p = 0.0779$). The relative risk associated with no residency training is

$$\exp(0.3041) = 1.36$$

Those without previous training are 36% more likely to receive the same number of complaints as those who were trained in the specialty.

Note: A SAS program would include these instructions:

```
DATA EMERGENCY;
INPUT VISITS CASES RESIDENCY;
LN = LOG(VISITS);
CARDS;
(Data);
```

```
PROC GENMOD DATA EMERGENCY;
CLASS RESIDENCY;
MODEL CASES = RESIDENCY/ DIST = POISSON LINK = LOG OFFSET = LN;
```

where EMERGENCY is the name assigned to the data set, VISITS is the number of visits, CASES is the number of complaints, and RESIDENCY is the binary covariate indicating whether the physician received residency training in the specialty. The option CLASS is used to declare that the covariate is categorical.

6.3.2 Multiple Regression Analysis

The effect of some factor on a dependent or response variable may be influenced by the presence of other factors through effect modifications (i.e., interactions). Therefore in order to provide a more comprehensive analysis, it is very desirable to consider a large number of factors and sort out which ones are most closely related to the dependent variable. This method, which is multiple Poisson regression analysis, involves a linear combination of the explanatory or independent variables; the variables must be quantitative with particular numerical values for each observation unit. A covariate or independent variable may be dichotomous, polytomous, or continuous; categorical factors will be represented by dummy variables. In many cases, data transformations of continuous measurements (e.g., taking the logarithm) may be desirable so as to satisfy the linearity assumption.

Poisson Regression Model with Several Covariates

Suppose we want to consider k covariates, $X_1, X_2, \ldots, X_k$, simultaneously. The simple Poisson regression model of the previous section can easily be generalized and expressed as

$$\begin{aligned} E(Y_i) &= s_i\lambda(x_{1i}, x_{2i}, \ldots, x_{ki}) \\ &= s_i \exp(\beta_0 + \beta_1 x_{1i} + \beta_2 x_{2i} + \cdots + \beta_1 x_{ki}) \\ &= s_i \exp\left(\beta_0 + \sum_{j=1}^{k} \beta_j x_{ji}\right) \end{aligned}$$

where $\lambda(x_{ji})$ is called the "risk" of/for observation unit i $(1 = i = n)$, and x_{ji} is the value of the covariate X_j measured from subject i.

Under the assumption that Y_i is distributed as a Poisson distribution with the above mean, the likelihood function is given by

$$L(y;\beta) = \prod_{i=1}^{n} \left\{ \frac{[s_i\lambda(x_{ji})]^{y_i} \exp[-s_i\lambda(x_{ji})]}{y_i!} \right\}$$

$$\ln L(y;\beta) = \sum_i \left\{ y_i \ln s_i - \ln y_i! + y_i \left(\beta_0 + \sum_{j=1}^{k} \beta_j x_{ji}\right) - s_i \exp\left(\beta_0 + \sum_{j=1}^{k} \beta_j x_{ji}\right) \right\}$$

from which estimates of the two regression coefficients $\beta_0, \beta_1, \beta_2, \ldots, \beta_k$ can be obtained by the maximum likelihood procedure.

Similar to the simple regression case, $\exp(\beta_i)$ represents:

(i) the relative risk associated with, say, an exposure if X_i is binary (exposed or $X_i = 1$ vs. unexposed or $X_i = 0$), or

(ii) the relative risk due to one unit increase in the value of X_i if X_i is continuous ($X_i = x + 1$ vs. $X_i = x$).

After estimates, b's, of the regression coefficients, β's, and their standard errors have been obtained, a 95% confidence interval for the relative risk associated with the ith factor is given by $b_i \pm SE(b_i)$. These results are necessary to identify important risk factors for the Poisson outcome—the "count." Of course, before such analyses are done, the problem and the data have to be examined carefully. If some of the variables are highly correlated, then one or fewer of the correlated factors are as likely to be good predictors as all of them; information from other similar studies also has to be incorporated so as to drop some of these correlated explanatory variables. The uses of products, such as $x_1 * x_2$, and higher power terms, such as x_1^2, may be necessary and can improve the goodness-of-fit. It is important to note that we are assuming a loglinear regression model in which, for example, the relative risk due to a one unit increase in the value of a continuous X_i ($X_i = x + 1$ vs. $X_i = x$) is independent of x. Therefore if this "linearity" seems to be violated, the incorporation of powers of X_i should be seriously considered. The use of products will help in the investigation of possible effect modifications. Finally, consider the messy problem of missing data: most packaged programs would delete a subject if one or more covariate values are missing.

Testing Hypotheses in Multiple Poisson Regression

Once we have fit a multiple Poisson regression model and obtained estimates for the various parameters of interest, we want to answer questions about the contributions of various factors to the prediction of the Poisson-distributed response variable. There are three types of such questions:

(i) An overall test: Taken collectively, does the entire set of explanatory or independent variables contribute significantly to the prediction of response?

(ii) Test for the value of a single factor: Does the addition of one particular variable of interest add significantly to the prediction of response over and above that achieved by other independent variables?

(iii) Test for contribution of a group of variables: Does the addition of a group of variables add significantly to the prediction of response over and above that achieved by other independent variables?

Overall Regression Test We now consider the first question stated above concerning an overall test for a model containing k factors. The null hypothesis for this test may be stated as: "all k independent variables considered together do not explain the variation in the response any more than the size alone." In other words,

$$H_0\colon\ \beta_1 = \beta_2 = \cdots = \beta_k = 0$$

This can be tested using the likelihood ratio chi-squared test at k degrees of freedom:

$$X^2 = 2\{\ln L_k - \ln L_0\}$$

where $\ln L_k$ is the log-likelihood value for the model containing all k covariates and $\ln L_0$ is the log-likelihood value for the model containing only the intercept. Packaged computer programs, such as SAS PROC GENMOD, provide these log-likelihood values but in separate runs.

■ **Example 6.7** Refer to the data set on emergency service of Example 6.5 with four covariates: gender, residency, revenue, and work load (hours). We have the following:

(i) with all four covariates included, $\ln L_4 = 47.783$, whereas
(ii) with no covariates included, $\ln L_0 = 43.324$,

leading to $X^2 = 8.918$ with 4 degrees of freedom ($p = 0.0632$), indicating that at least one covariate is almost significantly related to the number of complaints.

Note: For model (i), the SAS program would include this instruction:

```
MODEL CASES = GENDER RESIDENCY REVENUE HOURS/ DIST = POISSON LINK
= LOG OFFSET = LN;
```

and for model (ii),

```
MODEL CASES = / DIST = POISSON LINK = LOG OFFSET = LN;
```

(See note after Example 6.6 for other details of the program.)

Test for a Single Variable Let us assume that we now wish to test whether the addition of one particular independent variable of interest adds significantly to the prediction of the response over and above that achieved by other factors already present in the model. The null hypothesis for this test may be stated as: "factor X_i does not have any value added to the prediction of the response given that other factors are already included in the model." In other words,

$$H_0\colon\ \beta_i = 0$$

To test such a null hypothesis, one can use the above likelihood ratio chi-squared test with 1 degree of freedom or, more simply, use the test statistic:

$$z_i = \frac{b_i}{SE(b_i)}$$

where b_i is the corresponding estimated regression coefficient for factor X_i and $SE(b_i)$ is the estimate of the standard error of b_i, both of which are printed by standard packaged computer programs such as SAS. In performing this test, we refer the value of the z-score to percentiles of the standard normal distribution; for example, we compare the absolute value of z to 1.96 for a two-sided test at the 5% level.

■ **Example 6.8** Refer to the data set on emergency service of Example 6.5 with all four covariates. We have the following:

Variable	Coefficient	Standard Error	z-Statistic	p-Value
Intercept	−8.1338	0.9220	−8.822	<0.0001
No Residency	0.2090	0.2012	1.039	0.2988
Female	−0.1954	0.2182	−0.896	0.3703
Revenue	0.0016	0.0028	0.571	0.5775
Hours	0.0007	0.0004	1.750	0.0452

Only the effect of work load (hours) is significant at the 5% level.

Note: Use the same SAS program as in Examples 6.6 and 6.7.

Given a continuous variable of interest, one can fit a polynomial model and use this type of test to check for linearity (see "type 1 analysis" in next section). It can also be used to check for a single product representing an effect modification.

■ **Example 6.9** Refer to the data set on emergency service of Example 6.5, but this time we investigate only one covariate, the work load (hours). After fitting the second-degree polynomial model,

$$\begin{aligned} E(Y_i) &= s_i \lambda(x_i) \\ &= s_i \exp(\beta_0 + \beta_1 Hour_i + \beta_1 Hour_i^2) \end{aligned}$$

we obtain a result that indicates that the curvature effect is negligible ($p = 0.8797$).

Note: A SAS program would include the instruction:

```
MODEL CASES = HOURS HOURS*HOURS/ DIST = POISSON
LINK = LOG OFFSET = LN;
```

The following is another interesting example comparing the incidences of nonmelanoma skin cancer among women from two major metropolitan areas, one in the South and one in the North.

TABLE 6.3. Skin Cancer Data

	Minneapolis–St. Paul		Dallas–Ft. Worth	
Age Group	Cases	Population	Cases	Population
15–24	1	172,675	4	181,343
25–34	16	123,063	38	146,207
35–44	30	96,216	119	121,374
45–54	71	92,051	221	111,353
55–64	102	72,159	259	83,004
65–74	130	54,722	310	55,932
75–84	133	32,185	226	29,007
85 +	40	8,328	65	7,538

■ **Example 6.10** In this example, the dependent variable is the number of cases of skin cancer. Data were obtained from two metropolitan areas: Minneapolis–St. Paul and Dallas–Ft. Worth. The population of each area is divided into eight age groups and the data are shown in Table 6.3 (Kleinbaum et al., 1988).

This problem involves two covariates—age and location; both are categorical. Using seven dummy variables to represent the eight age groups (with "85 + " age group being the baseline) and one for location (with Minneapolis–St. Paul as the baseline), we obtain the results in the following table. These results indicate a clear upward trend of skin cancer incidence with age and, with Minneapolis–St. Paul as the baseline, the relative risk associated with Dallas–Ft. Worth is

$$\begin{aligned} RR &= \exp(0.8043) \\ &= 2.235 \end{aligned}$$

an increase of more than twofold for this southern metropolitan area.

Variable	Coefficient	Standard Error	z-Statistic	p-Value
Intercept	−5.4797	0.1037	−13.498	<0.0001
Age 15–24	−6.1782	0.4577	−13.498	<0.0001
Age 25–34	−3.5480	0.1675	−21.182	<0.0001
Age 35–44	−2.3308	0.1275	−18.281	<0.0001
Age 45–54	−1.5830	0.1138	−13.910	<0.0001
Age 55–64	−1.0909	0.1109	−9.837	<0.0001
Age 65–74	−0.5328	0.1086	−4.906	<0.0001
Age 75–84	−0.1196	0.1109	−1.078	0.2809
Dallas–Ft. Worth	0.8043	0.0522	15.408	<0.0001

Note: A SAS program would include the following instruction:

```
INPUT AGEGROUP CITY \ POP CASES;
LN = LOG(POP);
MODEL CASES = AGEGROUP CITY/ DIST = POISSON
 LINK = LOG OFFSET = LN;
```

Contribution of a Group of Variables This testing procedure addresses the more general problem of assessing the additional contribution of two or more factors to the prediction of the response over and above that made by other variables already in the regression model. In other words, the null hypothesis is of the form

$$H_0\colon\ \beta_{i+1} = \beta_{i+2} = \cdots = \beta_{i+m} = 0$$

To test such a null hypothesis, one can perform a likelihood ratio chi-squared test, with m degrees of freedom,

$$X^2 = 2\{\ln L(\text{All factors included}) - \ln L(\text{Factors under investigation not included})\}$$

As with the above z-test, this multiple contribution procedure is very useful for assessing the importance of potential explanatory variables. In particular, it is often used to test whether a similar group of variables, such as demographic characteristics, is important for the prediction of the response; these variables have some trait in common. Another application would be a collection of powers and/or product terms (referred to as interaction variables). It is often of interest to assess the interaction effects collectively before trying to consider individual interaction terms in a model as previously suggested. In fact, such use reduces the total number of tests to be performed and this, in turn, helps to provide better control of overall Type I error rates, which may be inflated due to multiple testing.

■ **Example 6.11** Refer to the data set on skin cancer of Example 6.10 with all eight covariates. We consider, collectively, the seven dummy variables representing the age; the basic idea is to see if there are any differences without drawing seven separate conclusions comparing each age group versus the baseline.

(i) With all eight variables included, we obtain. $\ln L = 7201.864$.
(ii) When the seven age variables are deleted, we obtain. $\ln L = 5921.076$.

Therefore

$$\begin{aligned}
X^2 &= 2\{\ln L(8 \text{ variables}) - \ln L(\text{only Location})\} \\
&= 2(7201.864 - 5921.076) \\
&= 2561.568 \\
df &= 7 \\
p\text{-value} &< 0.0001
\end{aligned}$$

In other words, the difference between the age group is highly significant; in fact, it is more so than the difference between the cities.

Main Effects
The z-tests for single variables are sufficient for investigating the effects of continuous and binary covariates. For categorical factors with several categories, such as the age group in the skin cancer data of Example 6.10, this process in PROC GENMOD would choose a baseline category and compare each other category versus the chosen baseline category. However, we are usually interested in the importance of the main effects, that is, one statistical test for each covariate, not each category of a covariate. This can be achieved using PROC GENMOD in two different ways: (i) treating the several category-specific effects as a group as seen in Example 6.11 (this would require two separate computer runs), or (ii) requesting the "type 3 analysis" option as shown in the following example.

■ **Example 6.12** Refer to the skin cancer data of Example 6.10, the type 3 analysis shows the following:

Source	df	LR Chi-Squared	p-Value
Age Group	7	2561.57	<0.0001
City	1	258.72	<0.0001

The result for "Age Group" main effect is identical to that of Example 6.11.

Note: A SAS program would include the following instruction:

```
MODEL CASES = AGEGROUP CITY/ DIST = POISSON
          LINK = LOG OFFSET = LN TYPE3;
```

Specific and Sequential Adjustments
In the type 3 analyses, or any other multiple regression analysis, we test the significance of the effect of each factor added to the model containing all other factors; that is to investigate the additional contribution of the factor to the explanation of the dependent variable. Sometimes, however, we may be interested in a hierarchical or sequential adjustment. For example, we have Poisson-distributed response Y and three covariates X_1, X_2, and X_3; we want to investigate the effect of X_1 on Y (unadjusted), the effect of X_2 added to the model containing X_1, and the effect of X_3 added to the model containing X_1 and X_2. This can be achieved using PROC GENMOD by requesting the type 1 analysis option.

■ **Example 6.13** Refer to the data set on emergency service of Example 6.5.

(i) Type 3 analysis shows the following:

Source	df	LR Chi-Squared	p-Value
Residency	1	1.09	0.2959
Gender	1	0.82	0.3641
Revenue	1	0.31	0.5781
Hours	1	4.18	0.0409

(ii) Type 1 analysis shows the following:

Source	df	LR Chi-Squared	p-Value
Residency	1	3.20	0.0741
Gender	1	0.84	0.3599
Revenue	1	0.71	0.3997
Hours	1	4.18	0.0409

The results for physician hours are identical because it is adjusted for the other three covariates in both types of analysis. However, the results for other covariates are very different. The effect of residency is marginally significant in type 1 analysis ($p = 0.0741$, unadjusted) and is not significant in type 3 analysis after adjusting for the other three covariates. Similarly, the results for revenue are also different: in type 1 analysis it is adjusted only for residency and gender ($p = 0.3997$; the ordering of variables is specified in the `INPUT` statement of the computer program) whereas in type 3 analysis it is adjusted for the three other covariates ($p = 0.5781$).

Note: A SAS program would include the following instruction:

```
MODEL CASES = RESIDENCY GENDER REVENUE HOURS/
      DIST = POISSON LINK = LOG OFFSET = LN TYPE1 TYPE3;
```

6.3.3 Overdispersion

The Poisson is a very special distribution; its mean μ and its variance σ^2 are equal. If we use the variance–mean ratio as a dispersion parameter, then it is 1 in a standard Poisson model, less than 1 in an underdispersed model, and greater than 1 in an overdispersed model. Overdispersion is a common phenomenon in practice and it causes concern because the implication is serious; the analysis that assumes the Poisson model often underestimates standard error(s) and thus wrongly inflates the level of significance.

Measuring and Monitoring Dispersion

After a Poisson regression model is fitted, dispersion is measured by the scaled deviance or scaled Pearson chi-squared statistic; it is the deviance or Pearson chi-squared statistic divided by the degrees of freedom (number of observations minus number of parameters). The deviance is defined as twice the difference between the maximum achievable log-likelihood and the log-likelihood at the maximum likelihood estimates of the regression parameters. The contribution to the Pearson chi-squared statistic from the ith observation is (Frome, 1983; Frome and Checkoway, 1985)

$$\frac{(y-\hat{\mu}_i)^2}{\hat{\mu}_i}$$

■ **Example 6.14** Refer to the data set on emergency service of Example 6.5 with all four covariates. We have the following:

Criterion	df	Value	Scaled Value
Deviance	39	54.52	1.3980
Pearson chi-squared statistic	39	54.42	1.3700

Both indices are greater than 1, indicating an overdispersion. In this example, we have a sample size of 44 but 5 degrees of freedom lost due to the estimation of the five regression parameters, including the intercept.

Fitting an Overdispersed Poisson Model

PROC GENMOD allows the specification of a scale parameter to fit overdispersed Poisson regression models. The GENMOD procedure does not use the Poisson density function; it fits generalized linear models of which the Poisson model is a special case. Instead of a variance equal to the mean,

$$\text{Var}(Y) = \mu$$

it allows the variance function to have a multiplicative overdispersion factor φ:

$$\text{Var}(Y) = \varphi\mu$$

The model is fitted in the usual way, and the point estimates of the regression coefficient are not affected. The covariance matrix, however, is multiplied by φ. There are two options available for fitting overdispersed models: users can control either the scaled deviance (by specifying DSCALE in the model statement) or the scaled Pearson chi-squared statistic (by specifying PSCALE in the model statement). The value of the controlled index becomes 1; the value of the other is close to but may not be equal to 1.

■ **Example 6.15** Refer to the data set on emergency service of Example 6.5 with all four covariates. By fitting an overdispersed model controlling the scaled deviance, we have the following:

Variable	Coefficient	Standard Error	z-Statistic	p-Value
Intercept	−8.1338	1.0901	−7.462	<0.0001
No Residency	0.2090	0.2378	0.879	0.3795
Female	−0.1954	0.2679	−0.758	0.4486
Revenue	0.0016	0.0033	0.485	0.6375
Hours	0.0007	0.0004	1.694	0.0903

Compared to the results in Example 6.8, the point estimates remain the same but the standard errors are larger; the effect of work load (hours) is no longer significant at the 5% level.

Note: A SAS program would include the following instruction:

```
MODEL CASES = GENDER RESIDENCY REVENUE HOURS/
        DIST = POISSON LINK = LOG OFFSET = LN {\bf DSCALE};
```

The measures of dispersion become:

Criterion	df	Value	Scaled Value
Deviance	39	39.00	1.000
Pearson chi-squared statistic	39	38.22	0.980

We would obtain similar results by controlling the scaled Pearson is chi-squared statistic.

6.3.4 Stepwise Regression

In many applications, our major interest is to identify important risk factors. In other words, we wish to identify from many available factors a small subset of factors that relate significantly to the outcome (e.g., the disease under investigation). In that identification process, of course, we wish to avoid a large Type I (false-positive) error. In a regression analysis, a Type I error corresponds to including a predictor that has no real relationship to the outcome; such an inclusion can greatly confuse the interpretation of the regression results. In a standard multiple regression analysis, this goal can be achieved by using a strategy that adds into or removes from a regression model one factor at a time according to a certain order of relative importance. Therefore the two important steps are:

1. Specify a criterion or criteria for selecting a model.
2. Specify a strategy for applying the chosen criterion or criteria.

Strategies This is concerned with specifying the strategy for selecting variables. Traditionally, such a strategy is concerned with which and whether a particular variable should be added to a model or whether any variable should be deleted from a model at a particular stage of the process. As computers became more accessible and more powerful, these practices became more popular.

FORWARD SELECTION PROCEDURE In the forward selection procedure, we proceed as follows:

Step 1: Fit a simple logistic linear regression model to each factor, one at a time.

Step 2: Select the most important factor according to a certain predetermined criterion.

Step 3: Test for the significance of the factor selected in Step 2 and determine, according to a certain predetermined criterion, whether or not to add this factor to the model.

Step 4: Repeat Steps 2 and 3 for those variables not yet in the model. At any subsequent step, if none meets the criterion in Step 3, no more variables are included in the model and the process is terminated.

BACKWARD ELIMINATION PROCEDURE In the backward elimination procedure, we proceed as follows:

Step 1: Fit the multiple logistic regression model containing all available independent variables.

Step 2: Select the least important factor according to a certain predetermined criterion; this is done by considering one factor at a time and treating it as though it were the last variable to enter.

Step 3: Test for the significance of the factor selected in Step 2 and determine, according to a certain predetermined criterion, whether or not to delete this factor from the model.

Step 4: Repeat Steps 2 and 3 for those variables still in the model. At any subsequent step, if none meets the criterion in Step 3, no more variables are removed in the model and the process is terminated.

STEPWISE REGRESSION PROCEDURE Stepwise regression is a modified version of forward regression that permits reexamination, at every step, of the variables incorporated in the model in previous steps. A variable entered at an early stage may become superfluous at a later stage because of its relationship with other variables now in the model; the information it provides becomes redundant. That variable may be removed, if meeting the elimination criterion, and the model is refitted with the remaining variables, and the forward process goes on. The whole process, one step forward followed by one step backward, continues until no more variables can be added or removed. Without an automatic computer algorithm, this comprehensive strategy may be too tedious to implement.

Criteria For the first step of the forward selection procedure, decisions are based on individual score test results (chi-squared, 1 df). In subsequent steps, both forward and backward, the ordering of levels of importance (Step 2) and the selection (test in Step 3) are based on the likelihood ratio chi-squared statistic:

$$X^2 = 2\{\ln L(\text{All factors included}) - \ln L(\text{One factor under investigation deleted})\}$$

In the case of Poisson regression, packaged computer programs, such as SAS PROC GENMOD, do not have the automatic stepwise option. Therefore the implementation is much more tedious and time consuming. In selecting the first variable (Step 1), we have to fit simple regression models to each and every factor separately, then decide, based on the computer output, on the first selection before coming back for computer runs in Step 2. At subsequent steps, we can take advantage of "type 1 analysis" results.

■ **Example 6.16** Refer to the data set on emergency service of Example 6.5 with all four covariates: work load (hours), residency, gender, and revenue. This time we perform a regression analysis using forward selection in which we specify that a variable has to be significant at the 0.10 level before it can enter into the model. In addition, we fit all overdispersed models using the `DSCALE` option in PROC GENMOD.

The results of the four simple regression analyses are as follows:

Source	LR Chi-Squared	*p*-Value
Hours	4.136	0.0420
Residency	2.116	0.1411
Gender	0.845	0.3581
Revenue	0.071	0.7897

Work load (Hours) meets the entrance criterion and is selected. In the next step, we fit three models each with two covariates: Hours and Residency, Hours and Gender, and Hours and Revenue. The following table shows the significance of each added variable to the model containing Hours using type 1 analyses.

Source	LR Chi-Squared	*p*-Value
Residency	0.817	0.3662
Gender	1.273	0.2593
Revenue	1.155	0.6938

None of these three variables meets the 0.10 level for entry into the model.

EXERCISE

6.1 Inflammation of the middle ear, or otitis media (OM), is one of the most common childhood illnesses and accounts for one-third of the practice of pediatrics

TABLE 6.4. Otitis Media Data

UIR	SIBHX	D-CARE	CIGS	CTNINE	FALL	NBER
1	0	0	0	0.00	0	2
1	0	0	0	27.52	0	3
1	0	1	0	0.00	0	0
1	0	0	0	0.00	0	0
0	1	1	0	0.00	0	0
1	0	0	0	0.00	0	0
1	0	1	0	0.00	0	0
0	0	1	0	0.00	0	0
0	1	0	8	83.33	0	0
1	0	1	0	89.29	0	0
0	0	1	0	0.00	0	1
0	1	1	0	32.05	0	0
1	0	0	0	471.40	0	0
1	0	0	0	0.00	0	1
1	0	1	0	12.10	0	0
0	1	0	5	26.64	0	0
0	0	1	0	40.00	0	0
1	0	1	0	512.05	0	0
0	0	1	0	77.59	0	0
1	0	1	0	0.00	0	0
1	0	1	0	0.00	0	0
0	0	1	0	0.00	0	0
1	0	1	0	0.00	0	3
1	0	0	0	0.00	0	1
0	0	1	0	21.13	0	0
1	0	0	0	15.96	0	1
1	0	0	0	0.00	0	1
0	0	0	0	0.00	0	0
1	0	0	0	9.26	0	0
1	0	1	0	0.00	0	0
1	0	0	0	0.00	0	1
0	0	0	0	0.00	0	0
0	1	0	0	0.00	0	0
0	0	1	0	0.00	0	0
1	0	1	0	525.00	0	2
0	0	0	0	0.00	0	0
1	0	1	0	0.00	0	0
1	0	1	0	0.00	0	1
1	0	0	0	0.00	0	0
0	0	0	0	57.14	0	0
1	1	1	0	125.00	0	0
1	0	0	0	0.00	0	0
0	1	0	0	0.00	0	0
1	0	1	0	0.00	0	3
1	0	1	0	0	0	0

TABLE 6.4. (*Continued*)

UIR	SIBHX	D-CARE	CIGS	CTNINE	FALL	NBER
0	0	0	0	0	0	1
1	0	1	0	0	0	1
1	0	1	0	80.25	0	2
0	0	0	0	0	0	0
1	0	1	0	219.51	0	1
1	0	0	0	0	0	0
1	0	0	0	0	0	1
0	0	0	0	0	0	0
1	0	0	0	0	0	0
1	0	1	0	0	0	2
1	0	1	0	8.33	0	2
1	0	0	0	12.02	0	5
0	1	1	40	297.98	0	0
1	0	1	0	13.33	0	0
1	0	0	0	0	0	3
1	1	1	25	110.31	0	3
1	0	1	0	0	0	1
1	0	1	0	0	0	1
1	0	0	0	0	0	1
0	0	0	0	0	0	0
0	0	0	0	0	0	0
1	0	1	0	0	0	1
1	0	1	0	0	0	1
1	0	0	0	285.28	0	1
1	0	0	0	0	0	1
1	0	0	0	15	0	2
1	0	0	0	0	0	0
1	0	0	0	13.4	0	1
1	0	1	0	0	0	2
1	0	1	0	46.3	0	0
1	0	1	0	0	0	0
1	0	1	0	0	0	1
0	0	0	0	0	1	0
1	0	1	0	0	1	0
1	0	1	0	0	1	2
1	0	1	0	0	1	0
1	0	1	0	0	1	1
1	0	0	0	0	1	2
1	1	0	0	0	1	6
1	0	0	0	53.46	1	0
1	0	1	0	0	1	0
1	0	0	0	0	1	0
1	0	1	0	3.46	1	3
1	0	1	0	125.00	1	1
0	0	1	0	0.00	1	0

(*continued*)

TABLE 6.4. (*Continued*)

UIR	SIBHX	D-CARE	CIGS	CTNINE	FALL	NBER
1	0	1	0	0.00	1	2
1	0	1	0	0.00	1	3
1	0	1	0	0.00	1	0
1	0	0	0	0.00	1	2
1	0	0	0	0.00	1	0
1	1	0	0	0.00	1	1
0	0	0	0	0.00	1	0
0	0	1	0	0.00	1	0
1	0	0	0	0.00	1	0
1	0	0	0	2.80	1	0
0	0	1	0	1950.00	1	0
1	0	1	0	69.44	1	2
1	0	1	0	0.00	1	3
1	0	0	0	0.00	1	0
1	0	0	0	0.00	1	2
1	1	0	0	0.00	1	0
1	1	1	0	0.00	1	4
1	0	0	0	0.00	0	0
1	0	1	0	0.00	0	0
1	0	1	0	0.00	0	1
1	0	0	0	0.00	0	1
1	0	1	0	0.00	0	1
0	0	1	0	31.53	0	0
0	0	1	0	0.00	0	0
1	0	1	0	11.40	0	3
0	0	0	0	0.00	0	0
0	1	0	0	750.00	0	0
1	0	0	0	0.00	0	0
0	0	0	0	0.00	0	1
1	1	0	0	0.00	0	0
1	1	0	0	0.00	0	0
0	0	0	0	0.00	1	0
1	1	0	0	0.00	1	2
0	0	0	0	0.00	1	1
1	0	1	0	0.00	1	1
0	1	1	22	424.22	1	1
1	0	0	0	0.00	1	0
0	0	0	0	0.00	1	2
1	1	0	0	0.00	1	1
1	1	1	25	384.98	1	2
1	0	1	0	0.00	1	2
0	0	0	0	0.00	1	0
1	0	1	0	29.41	1	0
1	0	0	0	0.00	1	0
1	0	0	0	0.00	1	0

TABLE 6.4. (*Continued*)

UIR	SIBHX	D-CARE	CIGS	CTNINE	FALL	NBER
0	0	1	0	0.00	1	0
0	0	0	0	35.59	1	0
0	0	0	0	0.00	1	0
1	0	1	0	0.00	1	3
1	0	1	0	0.00	1	4
1	0	0	0	0.00	1	1
1	0	0	0	0.00	1	0
1	0	0	0	0.00	1	1
1	1	1	35	390.80	1	2
1	0	1	0	0.00	1	0
1	0	1	0	0.00	1	2
0	0	1	0	0.00	1	0
0	0	1	0	0.00	1	0
1	1	1	0	0.00	1	3
1	1	0	22	1101.45	1	3
1	0	0	0	0.00	1	2
0	0	1	0	0.00	1	0
1	0	1	0	57.14	1	0
1	1	1	40	306.23	1	2
1	0	1	0	300.00	1	6
1	0	1	0	0.00	1	2
0	1	1	0	0.00	1	0
0	0	0	0	43.86	1	0
0	0	0	0	0.00	1	3
1	1	0	0	0.00	1	2
1	1	0	0	0.00	1	3
0	0	0	0	0.00	1	0
0	0	0	0	0.00	1	0
1	0	0	0	0.00	1	2
1	0	1	0	0.00	1	0
1	1	1	0	0.00	1	2
0	1	1	10	1000.00	1	1
0	1	0	10	0.00	1	0
0	1	1	1	0.00	0	0
1	0	0	0	0.00	0	1
1	0	0	0	0.00	0	3
1	0	0	0	0.00	0	0
0	0	0	0	0.00	0	0
0	0	1	0	0.00	0	0
1	0	0	0	0.00	0	1
0	0	0	0	0.00	0	3
1	0	0	0	0.00	0	1
0	1	1	23	400.00	0	1
1	1	1	0	0.00	0	1
0	1	0	10	0.00	0	0

(*continued*)

TABLE 6.4. (*Continued*)

UIR	SIBHX	D-CARE	CIGS	CTNINE	FALL	NBER
1	0	1	0	0.00	0	3
0	0	1	0	0.00	0	1
1	0	1	0	0.00	0	3
0	0	1	0	0.00	0	1
1	1	1	0	0.00	0	0
0	0	0	0	0.00	0	0
0	0	0	0	0.00	0	1
0	0	1	10	1067.57	0	1
1	1	1	3	1492.31	0	0
0	0	1	0	0.00	0	2
1	0	0	0	0.00	0	0
1	0	0	0	0.00	0	0
1	0	1	0	9.41	0	1
1	0	0	0	0.00	0	0
1	0	1	0	9.84	0	2
1	0	1	10	723.58	0	2
1	0	0	0	0.00	0	2
0	0	0	0	15.63	0	0
1	0	0	0	0.00	0	0
1	1	1	30	106.60	0	0
0	0	0	0	0.00	0	0
0	0	1	0	0.00	0	0
1	0	1	0	0.00	0	1
1	0	0	0	0.00	0	0
1	0	1	0	0.00	0	0
0	0	1	0	0.00	0	0
0	0	1	0	0.00	0	0
1	1	0	0	0.00	0	0
1	0	1	0	0.00	0	1
1	0	1	0	0.00	0	1
1	0	0	0	0.00	0	0
1	0	1	0	0.00	0	2
0	1	1	30	15375.00	0	0
0	1	0	75	11000.00	0	0
0	1	1	0	0.00	0	0
1	0	1	0	0.00	0	1
1	0	0	0	0.00	0	1
0	0	0	0	0.00	0	0
0	0	1	0	17.39	0	0
0	0	0	0	0.00	0	0
0	1	1	0	0.00	0	0
0	0	0	0	0.00	0	0
1	0	0	0	0.00	0	0
0	0	1	0	0.00	0	0
0	1	1	6	44.19	0	0

TABLE 6.4. (*Continued*)

UIR	SIBHX	D-CARE	CIGS	CTNINE	FALL	NBER
1	1	0	1	0.00	0	0
0	0	1	0	0.00	0	1
1	1	1	30	447.15	0	5
0	0	0	0	0.00	0	0
0	1	0	20	230.43	0	1
1	1	1	0	0.00	0	1
0	0	1	0	0.00	0	0
0	0	1	0	0.00	0	0
0	0	1	0	217.82	0	0
0	0	1	0	0.00	0	0
1	0	0	0	0.00	0	0
1	0	1	0	32.41	0	0
1	1	0	0	0.00	0	0
1	1	1	8	43.22	0	0
1	1	1	28	664.77	0	2
1	0	1	0	0.00	0	0

during the first five years of life. Understanding the natural history of otitis media is of considerable importance due to morbidity for the child as well as concern about long-term effects on behavior, speech, and language development. In an attempt to understand that natural history, large groups of pregnant women were enrolled and their newborns were followed from birth. The response variable is the number of episodes of otitis media in the first six months (NBER) and potential factors under investigation are upper respiratory infection (URI), sibling history of otitis media (SIBHX; 1 for yes), daycare, number of cigarettes consumed a day by the parents (CIGS), cotinine level (CTNINE) measured from the urine of the baby (a marker for exposure to cigarette smoke), and whether the baby was born in the Fall season (FALL). Table 6.4 provides about half of our data set.

(a) Taken collectively, do the covariates contribute significantly to the prediction of the number of otitis media cases in the first six months?

(b) Fit the multiple regression model to obtain estimates of individual regression coefficients and their standard errors. Draw your conclusion concerning the conditional contribution of each factor.

(c) Is there any indication of overdispersion ? If so, fit an appropriate overdispersed model and compare the results to those in (b).

(d) Refit the model (b) to implement this sequential adjustment:

$$\text{SIBHX} \rightarrow \text{D-CARE} \rightarrow \text{CIGS} \rightarrow \text{CNIN} \rightarrow \text{FALL}$$

(e) Within the context of the multiple regression model in (b), does daycare alter the effect of sibling history?

(f) Within the context of the multiple regression model in (b), is the effect of cotinine level linear?

(g) Focus on sibling history of otitis media (SIBHX) as the primary factor. Taken collectively, was this main effect altered by any other covariates?

7

CATEGORICAL DATA AND TRANSLATIONAL RESEARCH

Translational research consists of efforts to bring discoveries or works in the laboratories (basic sciences, such as biology and biochemistry) to the bedside (clinical practices)—referred to as T1; or to community interventions (public health)—referred to as T2. This chapter is focused on the first type, transition of basic sciences to clinical care and practices. Why does this topic of translation research belong in this book? Outcomes of work in laboratories could be on any measurement scales but we will focus on early-phase clinical trials where the main endpoints, such as *toxicity*, are often represented by a binary random variable with two possible outcomes, presence or absence of dose-limiting toxicity—that is, resulting data are categorical data.

Statistics is more than just a collection of long columns of numbers and sets of formulas. Statistics is a way of thinking, thinking about ways to gather and analyze data. By the nature of the topic—translational research—we emphasize the first part: study designs. The gathering part comes before the analyzing part; the first thing a statistician or a learner of statistics does when faced with data is to find out how the data were collected. How we analyze the data depends on how data were collected; formulas and techniques may be misused by a well-intentioned researcher simply because data were not collected properly. In other cases, many studies can be inconclusive because they were poorly planned and not enough data were collected to accomplish the goals and to support the hypotheses. In the case of translational

Applied Categorical Data Analysis and Translational Research, Second Edition, By Chap T. Le

research, the study design becomes even more important because, most of the time, the analyses are more simple.

7.1 TYPES OF CLINICAL STUDIES

In addition to surveys that are cross-sectional, as seen in many examples in previous chapters, study data may be collected many different ways. For example, investigators are more and more frequently faced with the problem of determining whether some specific factor or exposure is related to a certain aspect of health. Does air pollution cause lung cancer? Do birth control pills cause thromboembolic death? There are reasons for believing that the answer to each of these and other questions is yes, but all are controversial; otherwise, no studies are needed. Generally, biomedical research data may come from different sources; the two fundamental designs being *retrospective* and *prospective*. But strategies could often be further divided into four groups:

(i) Retrospective study (of past events)
(ii) Prospective study (of past events)
(iii) Cohort or prospective study (of ongoing or future events)
(iv) Clinical trials

Retrospective studies of past events gather past data from selected cases, persons are identified who have experienced the event in question, and controls, persons who have not experienced the event in question, are used to determine differences, if any, in the exposure to a suspected risk factor under investigation. They are commonly referred to as *case–control studies*; each case–control study is focused on a particular disease. In a typical case–control study, cases of a specific disease are ascertained as they arise from population-based registers or lists of hospital admissions, and controls are sampled either as disease-free individuals from the population at risk, or as hospitalized patients having a diagnosis other than the one under study. An example is the study of thromboembolic death and birth control drugs. Thromboembolic deaths were identified from death certificates, and the exposure to the pill was traced by interview with the woman's physician and a check of her various medical records. Control women were women in the same age range, under the care of the same physician.

Prospective studies of past events are less popular because they depend on the existence of records of a high quality. In these, samples of exposed subjects and unexposed subjects are identified in the records. Then the records of the selected persons are traced to determine if they have ever experienced the event to the present time. Events in question are past events but the method is called prospective because it proceeds from exposure forward to the event.

Prospective studies, also called cohort studies, are designs in which one enrolls a group of persons and follows them over certain periods of time; examples include occupational mortality studies and clinical trials. The cohort study design focuses on a particular exposure rather than a particular disease, as in case–control studies. There have been several major cohort studies with significant contribution to our understanding of important public health issues; but this form of study design is not very popular because cohort studies are time consuming and expensive.

In biomedical research, the sample survey is not a common form of study, and prospective studies of past events and cohort studies are not very often conducted. Therefore more emphasis is given to the designs of clinical trials, which are important because they are experiments on human beings. However, clinical trials are just a small subset of clinical studies.

Clinical studies form the class of all scientific approaches to evaluate medical disease prevention, diagnostic techniques, and treatments. Among this class, trials—often called clinical trials—form a subset of those clinical studies that evaluate investigational drugs.

Trials, especially cancer trials, are classified into five phases. The area of translational research consists of the first three; the last phase is referred to as *clinical research*:

(i) Preclinical studies are done in two different forms: *in vitro* and *in vivo*. *In vitro* studies are studies outside the living body and in an artificial environment. *In vivo* studies are studies in the living body of a plant or an animal. *In vivo* experiments or studies are designed and analyzed similar to those for clinical trials (for human subjects) whereas *in vitro* experiments are done in laboratories—with very different concerns.

(ii) Phase I trials focus on the safety of a new investigational medicine. These are the first human trials after successful animal trials (*in vivo* studies).

(iii) Phase II trials are small trials to evaluate efficacy and focus more on a safety profile.

(iv) Phase III trials are well-controlled trials, the most rigorous demonstration of a drug's efficacy prior to federal regulatory approval.

(v) Phase IV trials are often conducted after a medicine is marketed to provide additional details about the medicine's efficacy and more complete safety profile.

Phase I trials apply to patients who failed to get results from standard treatments and are at high risk of death in the short term. As for the new medicine or drug to be tested, there is no efficacy at low doses; at high doses, there will be unavoidable toxicity, which may be severe or even fatal. There is little known about the dose range, and animal studies may not be helpful enough. The goal in a Phase I trial is to identify a *maximum tolerated dose* (MTD)—a dose that has reasonable efficacy, that is, toxic enough to kill cancer cells, but not toxic enough to kill the patient.

Phase II trials, the next step, are often the simplest ones: the drug, at the optimal dose (MTD) found from a previous Phase I trial, is given to a small group of patients who meet predetermined inclusion criteria. The most common form are single-arm studies, where investigators are seeking to establish the antitumor activity of a drug usually measured by a response rate. A patient responds when his/her cancer condition improves (e.g., the tumor disappears or shrinks substantially). The response rate is the proportion or percentage of patients who respond. A Phase II trial may be conducted in two stages, as will be seen briefly in Section 12.6, when investigators are concerned about severe side effects. A second type of Phase II trials consists of small comparative trials where we want to establish efficacy of a new drug against a control or standard regimen. In these Phase II trials, with or without randomization, investigators are often concerned about their validity and pay careful attention to inclusion and exclusion criteria. Inclusion criteria focus on the definition of patient characteristics required for entry into a clinical trial. These describe the population of patients for whom the drug is intended. There are also exclusion criteria as well to keep out the patients for whom the drug is not intended.

Phase III trials and Phase IV trials are similarly designed. Phase III trials are conducted before regulatory approval and Phase IV trials, which are often optional, are conducted after regulatory approval. These are larger, controlled trials. The control is achieved by a randomization. Patients enter the study sequentially and, upon his/her enrollment, each is randomized to receive either the investigational drug or a placebo (or standard therapy). As medication, the placebo is "blank," that is, without any active medicine. Placebo, which has similar size and shape as the drug, is used to control psychological and emotional effects—possible prejudices on the part of the patient and/or investigator. Randomization is a technique to ensure that the two groups, the one receiving the real drug and the one receiving the placebo, are more comparable, more similar with respect to known as well as unknown factors (so that the conclusion is more valid). For example, the new patient is assigned to receive the drug or the placebo by a process similar to that of flipping a coin. Trials in Phases III and IV are often conducted as double blind, that is, blinding of the patient (he/she does not know if a real drug is given so as to prevent psychological effects; of course, the patient's consent is required) and blinding of the investigator (to prevent bias in measuring/evaluating outcomes). Some member of the investigational team, often designated a priori, keeps the "code" (the list of which patients received drug and which patients received placebo), which is only broken at the time of study completion and data analysis. The term *triple blind* may be used in some trials to indicate the blinding of regulatory officers.

A Phase III or Phase IV trial usually consists of two periods: an enrollment period, when patients enter the study and are randomized, and a follow-up period. The later is very desirable if a long-term outcome is needed. As an example, a study may consist of 3 years of enrollment and 2 years of follow-up; no patients are enrolled during the last 2 years.

7.2 FROM BIOASSAYS TO TRANSLATIONAL RESEARCH

The last steps in laboratory studies are perhaps biological assays—or bioassays—which are methods for estimating the potency or strength of an "agent," a potential future drug. Then the process moves into the preclinical stage. Preclinical studies can have two different forms: *in vitro* and *in vivo*. *In vitro* studies are studies outside the living body and in an artificial environment. *In vivo* studies are studies in the living body of a plant or an animal. *In vivo* experiments or studies are designed and analyzed similar to those for clinical trials (for human subjects) whereas *in vitro* experiments are done in laboratories—with very different concerns. The term *in vitro* has different meanings in different areas of science; for example, to a cell biologist, it refers to an experiment using tissue culture cells. We introduce here only two simple types of experiments at the beginning of the process eventually leading to clinical trials of investigational drugs: one for single agents and one for a combination of agents.

7.2.1 Analysis of *In Vitro* Experiments

Estimation of drug potencies or toxicity potencies is a regular practice in medical research, environmental health, and the food industry. Potency refers to the amount or dose of drug needed to produce a prespecified level of effect (Tallarida and Murray, 1981). The following experiment in cancer research is illustrative of an *in vitro* experiment. Cells from a tumor-derived cell line were deposited as fixed volumes in 12-well cell culture plates containing 2 mL of complete growth medium in each cell. After waiting for 72 hours (for further growth), the wells were then exposed to different concentrations of a drug; cells in wells not exposed to drug served as controls. At the end of the experiment, the numbers of surviving cells were counted on a hemocytometer for all wells—including the control wells. Data from a typical protocol are presented in Table 7.1, where each drug concentration was used in triplicate. The experiment giving rise to these data was designed to investigate the ability of a drug called vincristine to kill cancer cells. The parameter used to measure the drug potency is ED_{50}, the median effective dose. Given two different drugs or, (in Table 7.1) the same drug used under different conditions, the ratio of the two median effective doses represents the relative potency. If the relative potency, of treating recurrent tumor versus treating original tumor, is greater than 1, it is a good indication of drug resistance.

This *in vitro* experiment (actually two experiments, one for the original tumor and one for the recurrent tumor) fits the description of a *quantal assay* of Chapter 4 with one notable exception: the number of cells at the beginning of the experiment in each well, prior to drug exposure, may be large but always unknown. A quick and conventional approach is to assume that all wells, prior to drug exposure, contained the same number of cells; this is controlled by depositing the same volumes. The investigator would use the cell count from the "control well," the well exposed to dose zero: $n_1 = n_2 = \cdots = n_k = n_0$. The next step in the preliminary data analysis process is to divide the number of surviving cells at each dose level, r_i, by the original cell count (taken from $n_i = n_0$) to obtain the proportions p_i the surviving fractions, where

$$p_i = \frac{r_i}{n_0}$$

TABLE 7.1. Cell Counts from an In Vitro Experiment of All Cells

Original Tumor		Recurrent Tumor	
Dose (μg/mL)	Cell Count ($\times 10^4$)	Dose (μg/mL)	Cell Count ($\times 10^4$)
0.000	233.8	0.000	70.8
	212.6		69.3
	221.5		66.6
0.001	202.0	0.001	58.2
	205.3		54.9
	197.9		58.9
0.010	183.3	0.010	52.9
	186.1		54.4
	187.2		57.0
0.050	104.4	0.050	39.1
	100.9		38.4
	106.6		38.2
0.100	86.8	0.100	30.2
	88.5		26.7
	91.3		29.0
1.000	21.1	1.000	28.1
	31.0		28.0
	23.2		27.8
5.000	24.0	5.000	27.5
	26.8		27.0
	22.4		27.9
10.000	18.9	10.000	24.7
	17.0		26.1
	17.5		25.5
50.000	14.5	50.000	22.2
	17.5		23.8
	12.4		22.7
100.000	5.9	100.000	21.4
	6.9		21.1
	7.2		21.4

As in the case of quantal bioassays, we observe occurrences of an event—here the death of the tumor cell. With this binary dependent variable, one would proceed to apply the logistic regression model:

$$P = \frac{e^{\alpha+\beta x}}{1+e^{\alpha+\beta x}}$$

$$\log\left(\frac{P}{1-P}\right) = \alpha+\beta x$$

The common parameter LD_{50} (for median lethal dose—more appropriate in this case than ED_{50}) is the dose of the drug that kills 50% of the tumor cells. It is a measure of the drug's potency, which could be used to form relative potency (recurrent tumor versus original tumor—to study drug resistance). To obtain "log potency" (log of LD_{50}), M, one would set

$$\begin{aligned} P &= \frac{e^{\alpha+\beta x}}{1+e^{\alpha+\beta x}} \\ &= \tfrac{1}{2} \\ M &= -\frac{\alpha}{\beta} \end{aligned}$$

We estimate the parameters α and β by the maximum likelihood method, as in Chapter 4, to obtain MLEs a and b, then estimate LD_{50} or ED_{50} by exponentiation:

$$m = -\frac{a}{b}$$

■ **Example 7.1** With data from Table 7.1, we can average the three cell counts at each dose, form the surviving fraction p, and then graph the "log odds" or $\ln[p/(1-p)]$ versus log(dose). Figure 7.1 shows two fairly good straight lines; the slope associated with the recurrent tumor is smaller. The logistic model fits very well. On the graph, $\log(ED_{50})$—on the horizontal axis—is at the point where $p = \frac{1}{2}$ or $\log[p/(1-p)] = 0$ on the vertical axis.

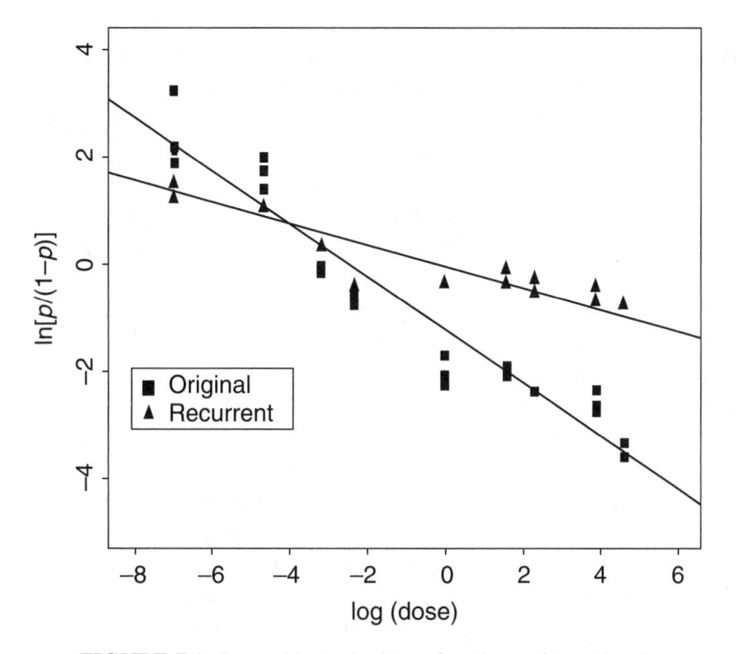

FIGURE 7.1. Log odds (or logit) as functions of log (dose).

One can argue with the accuracies of the results:

(1) The number of cells at the beginning of the experiment in the wells, prior to drug exposure, may be large but always unknown; they were aimed to be equal but are most likely not the same even in a well-controlled experiment.
(2) The number of surviving cells may not be counted precisely, counting error is unavoidable.

However, as the case of most experiments in basic sciences, these are minor inaccuracies. We can see that the triplicate counts at each dose are very similar. There are ways to improve the results but most of the time, the quick and easy solution of using a control well works rather well most of the times.

In the last few years, many laboratories have stopped counting cells because it can be too time consuming. Instead, some substrate is incubated in the plate wells—mixed with an indicator dye—which binds into the cells (similar to capturing proteins in ELISA). The extent of the color change read in a spectrophotometer results in an *optical density* (or *absorbance*), which relates linearly to the number of cells present in the well. After subtracting for the *background absorbance*, the ratio of numbers of cells—such as r_i/n_0—is equal to the ratio of absorbances or optical densities.

7.2.2 Design and Analysis of Experiments for Combination Therapy

The use of a combination of two therapeutic agents, either a radiation–drug combination or drug–drug combination (combination chemotherapy), has becomes more and more popular. There are a number of reasons: for example, better spatial coordination, improved quality of life, or less drug resistance. But the most important and compelling rationale is still the enhancement of tumor response; the drug combination (sometimes called a drug cocktail) is more lethal.

The basic desire, when both of the agents in a combination are active, is to produce positive tumor response, and frequently we wish to compare the therapeutic result of the combination with the results achieved by the component agents. For example, for simplicity, when we use a dose equal to half of the ED_{50} of drug 1 and a dose equal to half of the ED_{50} of drug 2, is the result equivalent to using either drug alone at its ED_{50}? Or putting it in a different way: Is the effect of the combination greater than the sum of the individual effects?

When the addition of one agent apparently increases the effect of the other, so that the effect of a combination appears to be greater than would be expected, the term *synergism* is used to describe these situations with enhancement of tumor response. On the other hand, the term *antagonism* is used to describe negative interaction. The benefits or effects of combination therapies are thoroughly tested before they are allowed to proceed to clinical trials. Experiments are done where these effects are validated from tumor response: ability of drugs and combination of drugs to kill cancer cells; to see if the drug combination is more lethal. These early, preclinical experiments are *in vitro* with some special form of data.

The Isobologram

When the addition of one agent increases the effect of the other, the drugs or agents act synergistically, and tumor response is enhanced. When the addition of one agent increases the effect of the other, the drugs or agents act antagonistically. How do we illustrate synergism or antagonism?

An isobologram is a graphical device used to evaluate two agents used in combination (Steel and Peckham, 1979). It is formed as follows:

(i) The effective doses (EDs) of two drugs are put on the axes.
(ii) A straight line joining two points having the same effect level, say, from the ED_{50} of drug 1 on the x-axis to the ED_{50} of drug 2 on the y-axis, is called an *isoeffect line*: 50% iso effect line, 80% isoeffect line, and so on.

It should be noted that, if we consider an isoeffect line, a combined dose (x,y) satisfying the equation of the line will produce the effect level (say, π) of that line if there is no interaction. That equation is

$$\frac{x}{ED_{X,\pi}} + \frac{y}{ED_{Y,\pi}} = 1$$

For example, if there is no interaction between drugs, 20% of ED_{50} of drug X mixed with 80% of ED_{50} of drug Y should produce 50% effects; if there are more than 50% effects, this indicates synergism. The experiments leading to an isobologram would proceed as follows:

Step 1: We fix any dose x (point A on Figure 7.2) of drug #1 and increase the dose of the other drug so as to reach—by trial and error—some preselected response level, say, 50%. If we need an amount of drug #2 smaller than y^* calculated from

$$\frac{x}{ED_{1,50\%}} + \frac{y^*}{ED_{2,50\%}} = 1$$

(i.e., to stop below the line, point B on Figure 7.2), then the effect of the combination is greater than if there were no interaction. That is an indication of synergism. By varying the chosen level x, the endpoints trace a concave curve; this concave curve lies in the *synergism region.*

Step 2: We fix any dose x (point A on Figure 7.2) of drug #1 and increase the dose of the other drug so as to reach—by trial and error—some preselected response level, say, 50%. If we need an amount of drug #2 larger than y^* calculated from

$$\frac{x}{ED_{X,50\%}} + \frac{y^*}{ED_{Y,50\%}} = 1$$

(i.e., to stop above the line, point D on Figure 7.2), then the effect of the combination is smaller than if there were no interaction. That is an indication

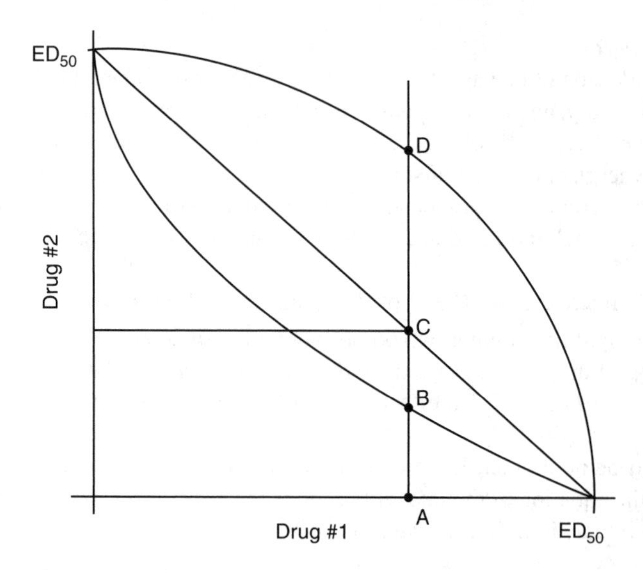

FIGURE 7.2. A typical isobologram.

of antagonism. By varying the chosen level x, the endpoints trace a convex curve; this convex curve lies in the *antagonism region.*

Step 3: We fix any dose x (point A on Figure 7.2) of drug #1 and increase the dose of the other drug so as to reach—by trial and error—some preselected response level, say, 50%. If we stop right on the is 0 effect line (of 50%), there is no interaction.

The very basic characterization of the isobologram approach is to use a *variable dose* to reach a *preset effect*: fixing one drug dose and the response level, then comparing required dose versus expected dose when the drugs are noninteractive. The design is labor intensive because it could be a long trial-and-error process of reaching precisely a preset response level. The implication is the use of very small samples (very often, size one or two) and, more important, no use of statistical inference. Tallarida (2000) raised, in a recent book on pharmacology, the need for statistics. Tallarida and Raffa (1996) have called for replacing the isobologram and have proposed a return to traditional probit analysis and logit analysis. However, the isobologram is popular with practitioners because they can "see" the result—even in one single drug combination. The concept is "common sense": more/less drug for the same response; probit and logistic analyses are more abstract. Le and Grambsch (2005) proposed to strengthen the method by adding a statistical component: a statistical framework, an experimental design, and a statistical test. The method is also based on the popular logistic regression model and is briefly summarized as follows.

The Experiment

If the isobologram experiment is characterized as *fixed effect–variable dose*, then the new experiments are *fixed dose–variable effect*. The experiment design consists of two steps:

Step 1: Choose several isoeffect lines, say, 50% and 80% or more; these are the expected response levels π.

Step 2: Choose several points on each line; doses (of drug X and of drug Y) at each point satisfy

$$\frac{x}{ED_{X,\pi}} + \frac{y}{ED_{Y,\pi}} = 1$$

The aim of the experiment is to compare the (variable) observed response p versus the (preset) expected response π at several levels of π.

■ **Example 7.2** A simple experiment for the combination of two drugs could be designed as follows:

(i) Determine the two ED_{50} values of drugs X and Y; use two samples, one for each drug at a number of doses from low to high and ED_{50} is determined using logistic regression as explained in Section 7.2.1.

(ii) Mix $x\%$ of $ED_{X,50\%}$ to $(100 - x)\%$ of $ED_{Y,50\%}$ with $x = 10$, 20, 30, 40, 50, 60,70, 80, and 90 ($n = 9$).

The Data and The Model

Let p be the observed proportion of response, the "surviving fraction," at a particular mixture of doses, where drug X is at dose $= x$ and drug Y is at dose $= y$:

$$p = r/n$$

where r is the number of surviving cells. The observed proportion p is the estimate of the true probability π. We then apply the logistic regression model:

$$\ln\left(\frac{\pi}{1-\pi}\right) = \alpha + \beta_1 \ln x + \beta_2 \ln y + \gamma \ln x \ln y$$

The last term in the model is the *interaction term.*

The Statistical Test

From the likelihood function,

$$\begin{aligned} L(\gamma) &= \prod (p)^r (1-p)^{n-r} \\ p &= \frac{1}{1+\exp(-\alpha-\beta_1 \ln x-\beta_2 \ln y-\gamma \ln x \ln y)} \end{aligned}$$

A score test statistic can easily be derived:

$$z = \frac{\sum (\ln x)(\ln y)(r-n\pi)}{\sqrt{\sum (\ln x)^2 (\ln y)^2 n\pi(1-\pi)}}$$

This can also be seen as a weighted average of results from the n binomial samples; in each sample, we compare the observed versus the expected numbers of cell survivors.

7.3 PHASE I CLINICAL TRIALS

We often have to deal with Phase I and Phase II clinical trials—more often than most statisticians are aware. Out of about ten Phase I trials, these may be only two or three drugs that are found suitable (i.e., safe enough) to proceed to Phase II trials. Similarly, out of about ten Phase II trials, there may be only one or two drugs that are found suitable (i.e., effective enough) for Phase III trials.

Different from clinical trials in subsequent phases, Phase I clinical trials—especially Phase I clinical trials in cancer—have several main features. First, the efficacy of chemotherapy, or any cancer treatment, is indeed frequently associated with a nonnegligible risk of severe toxic effect, often fatal, so that ethically, the initial administration of such drugs cannot be investigated in healthy volunteers, but only in cancer patients. Usually only a small number of patients are available to be entered into the Phase I cancer trials. Second, these patients are at very high risk of death in the short term under all standard therapies, some of which may have already failed. At low doses little or no efficacy is expected from the new therapy, and a slow intrapatient dose escalation is not possible. Third, there is not enough information about the drug's activity profile. In addition, clinicians often want to proceed as rapidly as possible to Phase II trials where there is more emphasis on efficacy. The lack of information about the relationship between dose and probability of toxicity causes a fundamental dilemma inherent in Phase I clinical cancer trials: the conflict between scientific and ethical intent. We need to reconcile the risks of toxicity to patients with the potential benefit to these same patients and make an efficient design to use no more patients than necessary. Thus the Phase I clinical cancer trial may be viewed as a problem in optimization: maximizing the dose–toxicity evaluation, while minimizing the number of patients treated.

Phase I trials focus on safety of a new investigational medicine. These are the first human trials after successful animal trials (*in vivo* studies); the goal in a Phase I trial is to identify a maximum tolerated dose (MTD), a dose that has reasonable efficacy (i.e., toxic enough, say, to kill cancer cells) but with tolerable toxicity (i.e., not toxic enough to kill the patient). Although recently their ad hoc nature and imprecise determination of maximum tolerated dose (MTD) have been called into question, a cohort-escalation trial design has been widely employed for years, called *standard design*. In the last several years a competing design, called *fast track* is getting more popular. These two cohort-escalation trial designs can be described as follows.

7.3.1 Standard Design

The starting dose selection of a Phase I trial depends heavily on pharmacology and toxicology from preclinical studies (Section 7.2). Although the translation from animal to human is not always a perfect correlation, the first dose is often chosen at about one-tenth of the mouse LD_{10} (at which 10% of mice die). Generally, toxicity as a function of body weight is assumed roughly constant across species. It is also often implicitly assumed that mouse LD_{10} is about the same as human MTD. However, use of a second species is necessary because in approximately 90 reviewed drugs, mouse data alone was insufficient to safely predict the human MTD (Arbuck, 1996). Some proposed to use the smaller of (i) one-tenth of the mouse LD_{10} and (ii) one-third of the beagle dog LD_{10}.

Toxicology studies further offer an estimation range of the drug's dose–toxicity profile and the organ sites that are most likely to be affected in humans. Since a slow intrapatient dose escalation is either not possible or not practical, investigators often use 3–8 doses selected from "safe enough" to "effective enough," most with 5 doses.

Once the starting dose is selected and the number of doses is decided, a reasonable dose-escalation scheme needs to be defined. There is no one optimal or universally efficient escalation scheme for all drugs. Generally, the dose levels are selected in order that the percentage increments between successive doses diminish as the dose is increased; for example, (i) equally spaced on the log scale or (ii) a modified Fibonacci sequence is often employed (increases of 100%, 67%, 50%, 40%, then 33% for subsequent doses if more than 5 are planned); this follows a diminishing pattern, with modest increases. Note that no intrapatient escalation is used to evaluate the doses to avoid the confounding effect of carryover from one dose to the next.

The standard design uses three-patient cohorts and begins with one cohort at the lowest possible dose level. It observes the number of patients in the cohort who experience toxicity seriously enough to be considered a dose-limiting toxicity (DLT). The trial escalates through the sequence of doses until enough patients experience the DLT. In the standard design, if no patients in a cohort experience a DLT, the trial continues with a new cohort at the next higher dose; if two or three patients experience a DLT, the trial is stopped; if exactly one patient experiences a DLT, a new cohort of three patients is employed at the same dose. In this second cohort evaluated at the same dose, if no severe toxicity is observed, then the dose is escalated to the next highest level, otherwise the trial is terminated. We can refer to the standard design as a *3-and-3 design* because, at each new dose, it enrolls a cohort of three patients with the option of enrolling an additional three patients evaluated at the same dose. Some slight variation of the standard design can also be found in various trials. If dose escalation not possible (i.e. it is stopped at a given dose), this dose is considered as above the MTD and one would have to go "down" to find the MTD:

(i) When a dose is judged as above the MTD, the next lower dose is (often) declared the MTD in the designs without dose de-escalation (it's not completely universal; some investigators only step down half a dose).

(ii) In designs with dose de-escalation, the next lower dose is declared the MTD if six patients have already been treated at that dose. Otherwise, three more patients are treated and this next lower dose is taken as the MTD if none or one experiences DLT; if two or three experience a DLT, there is further de-escalation according to the same rule. The reason is that one may not know enough if only three patients are treated at a dose—even if all three experience no DLT.

In summary, in designs without dose de-escalation, the MTD is often the highest dose at which either (i) a total of three patients are treated with none showing a DLT, or (ii) a total of six patients are treated with no more than one experiencing a DLT. In designs with de-escalation, the MTD is often the highest dose at which a total of six patients are treated with no more than one experiencing a DLT. Dose de-escalation is up to investigators.

Standard design is popular, used in 70–80% of trials; but the performance of this design has only recently been investigated and is still not fully understood by all medical investigators concerning its many aspects. The following are some *statistical operating characteristics* of this popular design. The rule of governing dose escalation from one level to the next relies on *no* assumptions concerning the shape of the dose–toxicity curve/relationship but based solely on toxicity result of the current dose.

(i) If r is the toxicity rate of the current dose, the probability of escalating after only three patients (none with toxicity) is

$$\mathrm{UP}_3 = (1-r)^3$$

(ii) If r is the toxicity rate of the current dose, the probability of stopping after only three patients (two or three with toxicity) is

$$\mathrm{STOP}_3 = 3r^2(1-r) + r^3$$

(iii) If r is the toxicity rate of the current dose, the probability of stopping (with three or six patients) is

$$\mathrm{STOP}_{3/6} = [3r^2(1-r) + r_3] + [3r(1-r)^2][1-(1-r)^3]$$

(iv) The probability that a second cohort of three patients be enrolled/needed is

$$\mathrm{NEED}_6 = 1-(\mathrm{UP}_3 + \mathrm{STOP}_3)$$

■ **Example 7.3** Suppose $r = 0.4$ or 40%. We have

$$\begin{aligned}
\mathrm{UP}_3 &= 0.6^3 = 0.216 \\
\mathrm{STOP}_3 &= (3)(0.4)^2(0.6) + 0.4^3 = 0.504 \\
\mathrm{STOP}_{3/6} &= [(3)(0.4)^2(0.6) + 0.4^3] + [(3)(0.4)(0.6)^2][1-(0.6)^3] = 0.843 \\
\mathrm{NEED}_6 &= 1-(0.216)-(0.504) = 0.280
\end{aligned}$$

That means the trial is less likely to go to the next step after three patients (21.6%), less likely to require a second cohort (28%), and more likely to stop (84.3%).

The following table gives some other similar results:

	Rate, r							
Probability	0.05	0.10	0.20	0.30	0.40	0.50	0.60	0.70
$STOP_{3/6}$	0.03	0.09	0.29	0.51	0.69	0.83	0.92	0.97
UP_3	0.86	0.73	0.51	0.34	0.22	0.13	0.06	0.03
$STOP_3$	0.01	0.03	0.10	0.22	0.35	0.50	0.65	0.78
$NEED_6$	0.13	0.24	0.39	0.44	0.43	0.37	0.29	0.19

From these typical results, for example, we can observe and conclude the following:

(i) For more extreme DLT probabilities (5% and 70%), the cohort is only expanded to six patients with probability of less than 20%.
(ii) When the rate is 30% or higher, the probability to stop is 51%.
(iii) The cohort is limited to three patients each with 56% probability.
(iv) If two or three patients experience a DLT ($STOP_3$), we are 90% sure that $r = 20\%$.
(v) If no patients experience a DLT (UP_3), we are 90% sure that $r = 55\%$.
(vi) If $r = 10\%$, there is a 91% chance to escalate; if $r = 60\%$, there is a 92% chance of stopping.

In other words, a trial with standard design is rather safe (e.g., when the rate is 30% or higher, the probability to stop is 51%) and economical (e.g., the cohort is limited to three patients each with more than 56% probability); the total number of patients needed is usually small and manageable. The results show why the standard design is popular.

More Research Questions

In order to provide a more comprehensive picture about the standard design, we can further investigate the following three basic research questions:

1. What is the actual expected toxicity rate of the MTD selected by the standard design?
2. Is the expected toxicity rate of the selected MTD by the standard design robust?
3. Can improvement be made to the standard design so that we can get to the next phase quicker but still safe enough?

In order to answer the last question, we can investigate two important characteristics: (i) expected trial size and (ii) expected suboptimal trial sizes, the number of patients treated with rather high toxicity rates (overtreated; say, with toxicity rates over 40%) or with rather low toxicity rates (i.e., not enough efficacy; say, toxicity rates below 30%).

Because the probability of a toxic reaction is set for each dose in the scenario and each design has a fixed pattern of decision, escalation, and stopping rules, we can calculate the characteristics for each designs, including the expected value of the toxic rate for the selected dose, the expected trial size, the expected overtreated trial size, and the expected undertreated trial size. No simulations are needed; we can calculate the above parameters using binomial probabilities. Recall that with the standard design, for example, if r_i is the toxicity rate of dose i, the probability of stopping at dose i, denoted by $p[i]$, after three or six patients is

$$\begin{aligned} p[i] &= \text{STOP}_{3/6} \\ &= [3r^2(1-r)+r^3]+[3r(1-r)^2][1-(1-r)^3] \end{aligned}$$

The probability of escalating to dose i, denoted by $p(i)$, is calculated as follows:

$$\begin{aligned} &p(1) = 1 \\ &p(2) = 1 - p[1] \\ &p(i) = p(i-1)[1-p[i-1]] \quad \text{for } i = 2, 3, \ldots \end{aligned}$$

Then we have the following:

(i) Expected toxicity rate of the MTD:

$$r(\text{MTD}) = \sum_i r_i\, p_{[i+1]}$$

(ii) Expected trial size:

$$E(n) = \sum_i p_{(i)}[(3)(\text{UP}_3 + \text{STOP}_3) + (6)(\text{NEED}_6)]$$

(iii) Expected overtreated trial size:

$$E(n_{\text{over}}) = \sum_{r \geq .40} p_{(i)}[(3)(\text{UP}_3 + \text{STOP}_3) + (6)(\text{NEED}_6)]$$

(iv) Expected undertreated trial size:

$$E(n_{\text{under}}) = \sum_{r \leq .30} p_{(i)}[(3)(\text{UP}_3 + \text{STOP}_3) + (6)(\text{NEED}_6)]$$

Variations of Standard Design

When the effects of dose-limiting toxicity are less severe (nonfatal and treatable), one can consider other variations of the standard design, which can also accelerate the process a little by reducing the size of the cohorts to fewer than three patients. In other words, instead of a 3-and-3 design (standard), some use a 3-and-2, 2-and-3, or 2-and-2 design. For example, a 2-and-2 design uses two-patient cohorts only and starts at the lowest dose level. If no patients experience a DLT, trial continues with a new two-patient cohort at the next highest dose level; if both patients experience a DLT, trial is stopped; if one patient experiences a DLT, a new cohort of two patients is employed at the same dose. In the second two-patient cohort, if no patients experience a DLT, the trial continues with a new two-patient cohort at the next higher dose; if one or two patients experience a DLT, the trial is stopped. Various probabilities are calculated similar to, but more simple than, those for the standard design:

(i) If r is the toxicity rate of the current dose, the probability of escalating after only two patients (none with toxicity) is

$$\text{UP}_2 = (1-r)^2$$

(ii) If r is the toxicity rate of the current dose, the probability of stopping after only two patients (both with toxicity) is

$$\text{STOP}_2 = r^2$$

(iii) If r is the toxicity rate of the current dose, the probability of stopping (with two or four patients) is

$$\text{STOP}_{2/4} = r^2 + [2r(1-r)][1-(1-r)^2]$$

(iv) The probability that a second cohort of two patients be enrolled/needed is

$$\text{NEED}_4 = 1-(\text{UP}_2 + \text{STOP}_2)$$

The results for other variations, 2-and 3 and 3-and-2 designs, are also similar.

7.3.2 Fast Track Design

The fast track design is another variation of the standard design. It was also created by modifying the standard design to move through low toxic rate doses using fewer patients. The design uses a predefined set of doses and cohorts of one or three patients and escalates through the sequence of doses using a one-patient cohort until the first DLT is observed. After that, only three-patient cohorts are used. When no DLT is observed, the trial continues at the next highest dose with a cohort of one new patient. When a DLT is observed in a one-patient evaluation of a dose, the same dose is evaluated a second time with a cohort of three new patients. If no patient in this cohort experiences a DLT, the design moves to the next highest dose with a new cohort of three patients. From this point, the design progresses as a standard design; when one or more patients in a three-patient cohort experiences a DLT, the current dose is considered the MTD. If a one-patient cohort is used at each dose level throughout, then six patients are often tested at the very last dose. Similar to the case of the standard design, no intrapatient escalation is allowed in the fast track design.

■ **Example 7.4** For a simple illustration, let us consider seven scenarios in which each scenario is assumed to be composed of a sequence of seven doses with toxic rates increasing by 5% in an effort to evaluate the performances of various Phase I designs. The scenarios differ by the toxic rate of the initial dose; each is 5% lower than the previous scenario. The sequences of toxic rates for the scenarios are as follows:

Scenario 1: 40%, 45%, 50%, 55%, 60%, 65%, and 70%.
Scenario 2: 5%, 40%, 45%, 50%, 55%, 60%, and 65%.
Scenario 3: 30%, 35%, 40%, 45%, 50%, 55%, and 60%.
Scenario 4: 25%, 30%, 35%, 40%, 45%, 50%, and 55%.
Scenario 5: 20%, 25%, 30%, 35%, 40%, 45%, and 50%.
Scenario 6: 15%, 20%, 25%, 30%, 35%, 40%, and 45%.
Scenario 7: 10%, 15%, 20%, 25%, 30%, 35%, and 40%.

Note that with our assumption or desire to have toxicity rates between 30% and 40%, the first and the last scenario are truly undesirable: the first starts too high and the last ends at a rate too low. We have the following results.

Toxicity Rates for MTD from the Standard Design

Scenario	Toxicity Rates	Expected MTD Rate
1	40–70%	42%
2	35–65%	38%
3	30–60%	34%
4	25–55%	30%
5	20–50%	27%
6	15–45%	25%
7	10–40%	23%

Because 30% and 40% toxicity rates are used as undertreated rate and overtreated "borders," the acceptable targeted level should be bounded by these two dose levels. The expected values of the toxicity rate range from 23% to 42%; most of them are at or below the lower end of the 30–40% range. The standard design is not robust in this aspect. Expected rate of the selected MTD is strongly influenced by the doses used; these doses are often selected arbitrarily by investigators and receive very little—if any—attention from reviewers or DSMB.

Expected Trial Sizes from the Standard Design

Scenario	Toxicity Rates	Trial Size	Oversize	Undersize
1	40–70%	6.0	1.7	0.0
2	35–65%	6.7	0.7	0.0
3	30–60%	7.6	0.3	0.0
4	25–55%	8.8	0.2	4.3
5	20–50%	10.4	0.1	7.2
6	15–45%	12.4	0.1	9.8
7	10–40%	14.8	0.0	12.6

The expected trial size has a range of 6.0–14.8. The expected numbers of patients being overtreated are low, less than two in all scenarios, which is the desired outcome with regard to medical ethics—patient's safety. The expected undertreated trial sizes are a large proportion of the expected trial size when the sequences include many doses with low toxicity. This indicated an unsatisfactory use of patient resources. Given these results, the standard design can be regarded as a conservative design; maybe it's *too* conservative.

Toxicity Rates for MTD from the Fast Track Design

Scenario	Toxicity Rates	Expected Rate
1	40–70%	47%
2	35–65%	43%
3	30–60%	40%
4	25–55%	37%
5	20–50%	34%
6	15–45%	32%
7	10–40%	30%

The expected values for the toxicity rate range from 30% to 47%; the range is narrower than that of the standard design. However, the expected rate of the selected MTD is still not quite robust; in addition, this design has higher than expected toxicity rates in two high-dose-level scenarios (i.e., 43% and 47%).

Expected Trial Sizes from the Fast Track Design

Scenario	Toxicity Rates	Trial Size	Oversize	Undersize
1	40–70%	6.1	3.9	0.0
2	35–65%	6.6	2.7	0.0
3	30–60%	7.2	2	0.0
4	25–55%	8.0	1.5	1.8
5	20–50%	8.9	1.2	3.4
6	15–45%	9.9	1	5
7	10–40%	10.9	0.0	6.5

The expected trial size ranged from 6.1 to 10.9, more stable than that of the standard design. This design decreases the number of patients needed, especially in the low-dose-level scenarios. The values for expected overtreated trial size are greater than those seen in the standard design. The values for the expected undertreated trial size are smaller than those of the standard design. This shows an improved use of patient resources but with a moderate compromise of patient safety; safety could be a problem with inexperienced investigators who might select high doses to start.

The challenge of Phase I clinical trials in cancer is to reach the maximum tolerated dose efficiently and quickly, while offering patients safe and appropriate treatment. Thus an optimal Phase I trial design must fully evaluate drug toxicity but must also minimize the likelihood of both subtherapeutic (dose too low) and severely toxic (dose too high) treatment. Meeting both criteria may be difficult when both the maximum tolerated and minimally effective doses are expected after a six-dose escalation.

(i) *About Toxicity Rates.* Among the five designs, the 2-and-2 design has an expected toxicity range closer to the desired level, the standard design has more expected toxicity rates lower than the desired level, while the fast track design has more expected toxicity rates higher than the desired level.

(ii) *About Trial Sizes.* The 3-and-2 design, the 2-and-3 design, the 2-and-2 design, and the fast track design all decrease the number of patients needed in the study compared to the standard design. The fast track design has the smallest expected trial sizes for most scenarios. The 2-and-2 design has the next smallest and similar expected trial sizes for most scenarios. As for the expected overtreated trial sizes and the expected undertreated trial sizes, more patients in the fast track design are likely to encounter drug toxicity than in the other four designs. Fewer patients in the fast track design are expected to receive subtherapeutic doses compared to those in the other four designs. The situation of the standard design is just the opposite of that of the fast track design: few patients are likely to encounter the drug toxicity, and more patients are expected to receive subtherapeutic doses compared to the other four designs.

Is there a compromise? We can see that there seems to be no perfect solution. The standard design is more popular and more conservative (i.e., safer); very few patients are likely to be over treated by doses with undesirable levels of toxicity. However, in the standard design, many patients enter early in the trial and are likely to be treated suboptimally and perhaps only a few patients are left after an MTD is reached, especially if there are many doses below the MTD. Generally, the use of a fast track design seems very attractive because some clinicians want to proceed to a Phase II trial as fast as they can, to have a first look at efficacy. The fast track design quickly escalates through early doses that have a low expected risk of dose-limiting toxicity, thereby reducing the number of patients treated at the lowest-selected toxicity in single-patient cohorts. On the other hand, the fast track design may allow a higher percentage of patients to be treated at very high toxic doses; it uses a single-patient cohort until the first DLT is observed and this may seem too risky for some investigators. With more experienced investigators, the fast track design presents an improved use of patient resources with a moderate compromise of patient safety; but safety could be a problem with inexperienced investigators who might select high doses to start with. The common problem for both designs is the lack of robustness: the expected rate of the selected MTD is strongly influenced by the doses used and these doses may be selected arbitrarily by investigators, which makes their experience a crucial factor.

7.3.3 Continual Reassessment Method

When designing cancer clinical trials for development and evaluation of therapeutic interventions, two special aspects must be taken into consideration: (i) the target population and (ii) the fact that all anticancer drugs under investigation are cytotoxic agents. It is true that newer drugs are safer and more efficient, but toxicity is still a very serious concern—especially in cancer drug testing. That is why the dose-escalation rule in the design of Phase I clinical trials remains an important issue in cancer research. At this stage, the investigator is facing a classic, fundamental dilemma: a conflict of scientific versus ethical intent. We put some patients at risk for their own benefit and the benefit of others; it's an unavoidable conflict because these patients have failed to improve under all standard therapies.

Standard design is safe; that is, few patients are exposed to lethal toxicities or die because of toxicities. However, "safe" does not necessarily mean "good"; if not given enough medication, the patient would be killed by cancer, the disease. According to the (unstated) principle of "good medicine," each patient should be treated optimally. That is, each should be treated with the best treatment that the doctor has to offer. According to this principle, each patient in a Phase I trial should be given a dose equal to the MTD—if the doctor knows what it is. In most cases, doctors may not know what the MTD is, but they all know that, according to the standard design, the first few doses are likely below the MTD. Despite all of its weaknesses, the standard design is still widely used in practice because of its simplicity in logistics for the clinical teams to carry out (most of the time responsibility falls on nurses, not doctors). When an agent being tested is relatively safe, the fast track design represents some improvement but it is not a satisfactory solution. However, let's get back to the very basic aim of reaching the maximum dose with an acceptable and manageable safety profile. What is considered "acceptable"? How high is high? How do you know if you have no target? This is even more serious because the result is not robust: actually few investigators ever know what the toxicity rate of their resulting MTD is regardless of the design they use, standard or fast track. The common critiques for both are the following:

(i) They do not target a particular toxicity rate associated with the MTD.
(ii) They do not make use of all available toxicity data; the escalation rule depends solely on the toxicity outcomes of the current dose.

Let us examine the objective of Phase I clinical trials in a more systematic and more statistical way.

Statistical Formulation

The MTD could be statistically interpreted as some *percentile* of a tolerance distribution or dose–response curve in terms of the presence or absence of the DLT. In other words, the MTD is the dose at which a specified proportion of patients, say, θ, experience a DLT. Storer (1989) indicated that the value of θ is usually in the range from 0.1 to 0.4. In the previous section, we used 0.3–0.4 or 30–40%. There is no magic number; it depends on the severity of the side effects and if they are treatable. It's kind of strange but in employing the standard design, investigators set no goal for θ, the proportion of patients experiencing a DLT (that's why we studied to see what it was in Example 7.2). The result is implicit that with the stopping of "2-out-of-6" in standard design or 3-and-3 design; it's usually 30–40% but, again, the design is not robust. The following could serve as a plausible statistical model.

Let Y be the binary response such that $Y = 1$ denotes the occurrence of a predefined dose-limiting toxicity (DLT) and $\{d_i; i = 1, 2, \ldots, n\}$ be a set of fixed dose levels used in a Phase I clinical trial. Let

$$p(x) = \Pr[Y = 1|x]$$

and

$$\text{logit}\,[p(x)] = \log\,\{p(x)/[1 - p(x)\}$$

The relationship between Y and dose level d could be described by the logistic model:

$$\text{logit}\,[p(x)] = \alpha + \beta x$$

where x is usually the log of the dose d; or x could be the dose d itself. In general, one can consider $p(x) = F(+ \beta x)$, where $F(.)$ is some cumulative distribution function (cdf). For example, in addition to the logistic, we may consider the probit model. Some even consider the one-parameter family—such as the hyperbolic tangent or exponential. However, the logistic regression model has had extensive empirical supports, such as the *median effects principle* in pharmacology. Let

$$l_\theta = \text{logit}(\theta)$$

then the MTD is defined by

$$x_\theta = \log(\text{MTD}) = (l_\theta - \alpha)/\beta$$

(it's likely not to be one of the doses used in a Phase I trial under standard design). It should be noted that if we have some estimates of the toxicity rates associated with the dose levels used in a Phase I standard design, we could estimate the toxicity rate of the resulting MTD and compare it to the θ of choice. Of course, the exact rates are not known, but clinicians should have some estimates; otherwise, it would be difficult for the investigator to justify the selected dose levels.

The underlying process of Phase I clinical trials could be considered as consisting of these steps:

Step 1: Choose the maximum tolerated level, θ.

Step 2: Choose a design and calculate the expected toxicity rate r_0 of the resulting MTD.

Step 3: Compare the calculated expected rate r_0 to the selected level θ.

Besides many elements of arbitrariness (choosing the level θ for the problem and estimating the rates r_i for the planned doses), the basic problem is, according to Storer (1989), the standard dose-escalation design "frequently failed to provide a convergent estimate of MTD" (so, even if we know what we want, i.e., θ, the standard design might not get us there). The alternative is a newer design—the *continual reassessment method.*

The Continual Reassessment Process

The primary objective of Phase I trials is to find the maximum tolerated dose (MTD) with an acceptable and manageable safety profile for use in subsequent Phase II trials. But that's the investigator's objective, not the patient's. Patients in Phase I trials are mostly terminal cancer patients for whom the new antitumor agent being tested may be the last hope. Designs, such as the standard design, do not serve them—at least not ethically. According to the principle of "good medicine," the patient should be treated with the best treatment the doctor knows. Patients entered early in a Phase I trial with standard design are likely treated suboptimally; they receive a treatment level that the attending physician knows to be inferior. Some of these patients would likely die before any other therapy can be attempted. The newer design, the continual reassessment method (CRM), is an attempt to correct that by giving each patient a better chance of a favorable response. In addition to the attempt to treat each patient more ethically, the CRM also updates the information of the dose–response relationship as observations on DLT become available and then uses this information to concentrate the next step of the trial around the dose that might correspond to the anticipated target toxicity level. It does so using a Bayesian framework, even though it has been argued that the CRM could be explained by the likelihood approach. The CRM is very attractive and has fostered heated debate that has lasted for more than a decade. There are many variations of the CRM. We'll describe here a scheme based on a specific prior; the principle and the process are the same if another model is selected. This scheme consists of seven steps.

Step 1: Choose the maximum tolerated level (θ, the toxicity rate at the recommended dose level or MTD (say, conventionally 0.33 or a different level fitting what the investigator knows about the agent being investigated); this is a basic difference compared to standard design (SD). The standard design does not have this step.

Step 2: Choose a fixed number of patients to be enrolled; usually $n = 19$–24; this is another difference compared to the SD, where the number of patients needed is variable and is unknown to start.

Step 3: The CRM uses binary response (DLT or not). Let Y be the binary response such that $Y = 1$ denotes the occurrence of a predefined DLT. Let

$$p(x) = \Pr[Y = 1|x] \quad \text{and} \quad \text{logit}\,[p(x)] = \log\{p(x)/[1 - p(x)]\}$$

The next step is to choose a statistical model representing the relationship between Y and dose level; for example, it could be described by the logistic (or probit) model:

$$\text{logit}\,[p(x)] = \alpha + \beta x$$

where x is the log of the dose d; or x is dose d.

Step 4: Use the baseline toxicity or adverse-effect rate (dose $= 0$) to calculate and then fix the "intercept" α of the logistic model.

Step 5: Under the Bayesian framework, choose a prior distribution for the "slope" β; for example, "unit exponential"—one with probability density function $g(\beta) = \exp(-\beta)$.

Step 6: From the model, logit $[p(x); \beta] = \alpha + \beta x$, with β placed at *the prior mean* and setting $p(x)$ equal to the target rate θ, solve for dose x. This is the dose for the first patient, a dose determined to reflect the current belief of the investigator/doctor as the dose level that produces the probability of DLT closest to the target rate θ the maximum dose with an acceptable and manageable safety. This step fits the principle of good medicine: the patient is treated at the MTD.

Step 7: After the first patient's toxicity/adverse-effect result becomes available, the *posterior distribution* of β is calculated and the posterior mean of β is substituted in logit $[p(x); \beta] = \alpha + \beta x$.

The next patient is treated at the dose level x whose probability $p(x)$ is the target rate θ (with calculated posterior mean of β). This step is repeated in subsequent patients every time toxicity or adverse-effect result becomes available and the posterior distribution of β is recalculated.

There is more than one way to calculate the posterior mean, one of which can proceed as follows—without going through the posterior distribution. From the model logit $[p(x_i; \beta)] = \alpha + \beta x_i$, the (Bernoulli) likelihood at dose i is

$$L(\beta) = p(x_i; \beta)^{\delta_i}[1 - p(x_i; \beta)]^{1-\delta_i}$$

and the mean β is calculated from

$$E(\beta) = \frac{\int \beta L(\beta)g(\beta)d\beta}{\int L(\beta)g(\beta)d\beta}$$

Finally, the MTD is estimated as the dose level for the hypothetical $(n + 1)$st patient; n has been predetermined, usually 19–24.

The original CRM, proposed by O'Quigley et al. (1990), drew some critiques and/or some strong opposition. Korn et al. (1994), Goodman et al. (1995), and Ahn (1998) pointed out the following. Corrections have been proposed by those authors and also by Thall et al. (1999), Zohar and Chevret (2001), Storer (2001), among others. First, the CRM might start the trial with an initial dose far above the "customary" lowest dose that is often one-tenth the LD_{10} in mice. This possibility makes many clinicians and regulatory agencies (e.g. FDA) reluctant to implement the CRM. After all, this is the first trial in humans, and little is known about the dose range—except results from animal studies. Some might go higher at the first dose, but not more than one-third the LD_{10} in mice. Some proposed that the trial always starts with the lowest dose as the dose for the first patient; CRM would start with the second dose for the first patient. Second, there is a possibility that dose could be escalated for more than one dose level at a time (traditionally, as in standard design, doses are equally spaced on the log scale or following a modified Fibonacci sequence with increases of 100%, 67%, 50%, 40%, and 33% for fifth and subsequent doses). Moller (1995) gave an example showing that the first dose could be escalated to the top level when the first patient has no LTD. It should be noted that the dose has been "escalated" or updated after obtaining a sample of size one.

To overcome the problems, some proposed that one could predetermine a set of doses to be used in the trial just as under the standard design. From the model logit $[p(x); \beta] = \alpha + \beta x$, we would use the prior mean of β to solve for doses with (prior) probabilities 5%, 10%, 15%,.... We would start the first patient at the lowest dose (as previously proposed) and the magnitude of dose escalation would be limited or imputed to one dose level only between two consecutive patients. If doses with (prior) probabilities 5%, 10%, 15%,... are predetermined, then each patient is treated at a dose that is closest to the calculated dose from the Bayesian CRM process. As previously mentioned, some investigators are uncomfortable with another feature of the CRM, where a cohort of only one patient is used for the dose adjustment, just as with the fast track design. Of course, it can easily be changed by increasing the size of the cohort to two or three patients. The problem is that, if all of those modifications are implemented, the resulting design would be similar to the standard design (Korn et al., 1994). There seems to be no perfect solution to the very fundamental dilemma, the conflict between scientific efficiency and ethical intent—even with this new and improved continual reassessment method.

The strengths of the CRM method are still its three unique properties:

(i) It has a well-defined goal of estimating a percentile of the dose–toxicity relationship.

(ii) It should converge to this percentile with increasing sample sizes.
(iii) The accrual is predetermined.

The standard design does not have these characteristics. In addition, there seems to be no way to overcome the problem that, under CRM and cohorts of size one, the dose for the next patient can be determined only after the result on the DLT for the current patient becomes available. This goes against the desire by most investigators to complete Phase I trials as rapidly as possible— not only with the minimum number of patients but also in a minimal amount of time. This is mostly due to the urgent need to identify new active drugs and Phase II efficacy trials cannot begin until completion of the Phase I trial. In other words, Korn et al. (1994) pointed out two severe deficiencies of the CRM. First, trials will take too long to complete—using cohorts of size one—especially when there is no shortage of patients (if one did not have concerns with this, one could accrue one patient at a time to the SD rather than three at a time—where one enrolls the second and third patient without waiting for toxicity results from the first). Second, some patients, especially the first few, may be treated at a dose level that is higher than the intended MTD. A typical problem with the Bayesian method is that, at early times when very little data are available, results or decisions are dominated by the choice of the prior. With a poorly chosen prior, some early patients may be treated at doses higher than the MTD, which is defined as the highest possible dose with acceptable toxicity. And the dose–severity relationship is not linear; toxicity seen at doses higher than the MTD will likely be more serious than toxicity seen at the MTD. Of course, as mentioned, an obvious remedy to the first deficiency is to have CRM accrue more than one patient—say, three at a time, to a dose level. This, however, worsens the second deficiency of the CRM—its tendency to treat patients at high doses. Regulatory agencies, protocol review panels, and Institutional Review Boards (IRB) may be too rigid; they are more concerned about side effects:

(i) Patients may die from side effects—some are fatal. But patients may also die (and more likely so) from the disease if not treated with enough medication (as mentioned, these are mostly terminal patients; some would likely die before any other therapy can be attempted).
(ii) Some side effects are not serious or are treatable or reversible; in such situations, methods such as CRM should seriously be considered.

The problem for the time being is only the ease of application, or lack of it. There are some software in the public domain but only for the original algorithm proposed by O'Quigley et al. (1990). It would be more efficient if available in SAS with some options for more flexibility. The next problem is logistic arrangement: Can we afford to have a statistician calculating the next dose for a CRM design when we need it? (About 100–200 active early-phase trials are open at any time in any large medical center.) How do we make it simple enough to make its implementation practical? In addition, we need immediate access to toxicity results, which might not be possible without advances in information technology.

7.4 PHASE II CLINICAL TRIALS

Phase II clinical trials, the next step, are often the simplest ones: the drug, at the optimal dose (MTD) found from a previous Phase I trial, is given to a small group of patients who meet predetermined inclusion criteria. The most common form of Phase II clinical trials is single-arm studies, where investigators are seeking to establish the antitumor activity of a drug usually measured by a response rate. A patient responds when his/her cancer condition improves; for example, the tumor disappears or shrinks substantially. The response rate is the proportion or percentage of patients who respond.

There are three basic objectives in conducting Phase II clinical trials:

Objective 1: Benefit the patients.

Objective 2: Screen the agent/drug for antitumor activity.

Objective 3: Extend knowledge of the toxicology and pharmacology of the agent or drug.

In addition, safety is always a major concern; however, in Phase II clinical trials, efficacy is the outcome of interest whereas safety is embedded to serve as the stopping rule.

The first objective is to benefit the patients enrolled in the trial; it must be a primary objective of any therapeutic intervention. It is always the primary "motivation"; however, it's not often stated in research protocols as an "objective." The concerns for patients should always be taken very seriously—by everyone—in the design stage; benefiting patients is the first major objective but not a "research objective" simply because we would not be able to evaluate it. The second objective is to screen the experimental agent for antitumor activity in a given type of cancer; agents found to have substantial antitumor activity and an appropriate spectrum of toxicity are generally incorporated into combinations to be evaluated for patient benefit in controlled Phase III clinical trials. For many investigators, this process of screening for antitumor activity is considered as "the" activity of Phase II trials—as far as research is concerned. We should clearly distinguish, from the research point of view, Objective 1 (benefit the patients) from Objective 2 (screen agent used in the trial for antitumor activity); the second objective only benefits investigators. Primary outcome used in Phase II trials is often the *response*, which is defined as having a 50% decrease in tumor size, for example, lasting for 4 weeks. The analysis of the resulting binary data is simply based on the *response rate*. But response rate is only an appropriate endpoint for evaluating the second objective (screen agent used in the trial for antitumor activity)—not the first objective (benefit the patients).

Generally, we cannot adequately evaluate the extent to which the first objective (benefit the patients) is achieved in one-arm Phase II trials; and that is why it is not often stated. First, response is only meaningful to patients and benefits the patients if causing tumor shrinkage means extending survival or, at least, improving quality of life. This may or may not be the case. Logically, we believe so but it has not been proved. Second, when an untreated control group is not available, we generally cannot properly evaluate whether the new agent influences survival so as to benefit the patients. One could compare, in terms of survival, responders versus nonresponders; however, this not a valid way of demonstrating that there has been an impact of treatment on survival. Such comparisons are biased by the fact that responders must live long enough for a response to be documented. In addition, responders may have more favorable prognostic factors than nonresponders, leading to a difference in survival which then is wrongly credited to the treatment. Response is still used as a surrogate for the more relevant, more important endpoint of survival, even though no one can prove that they are equivalent (we can even say that everyone knows that they are not equivalent). Response is still used and is popular because (i) it can be observed on all (or almost all) patients, and (ii) it can typically be determined rather quickly.

The third basic objective of a Phase II clinical trial is to extend our knowledge of toxicology and pharmacology of a drug/agent. Ironically, this objective is often listed as "secondary" and therefore is overlooked by statisticians. Most of the time, details, such as the data analysis plan, are missing. Most biostatistics students do not often see pharmacology data, which are mostly nonlinear regression. However, this is an important area; pharmacology could even be used to guide the dose-escalation plan in Phase I clinical trials because how much medication is given may not be as important as how much and how long it stays in a patient's body.

The second objective of a Phase II clinical trial is to screen the agent used in the trial for antitumor activity. There is frequently great variability in the response rates reported from different Phase II trials of the same agent. There are a number of factors that contribute to this variability. For example, *response criteria* and *response assessment* are often subjective without universal guidelines, in addition to a number of factors related to the conduct of the trials— dosage, protocol compliance, reporting procedure ("evaluable" vs. "nonevaluable" patients), and so on. The most important factor leading to variability (of reported response rate) comes from the patient selection process dictated by *inclusion criteria* and *exclusion criteria*—some sections that few statisticians read. Patients in Phase I and Phase II clinical trials are mostly terminal cancer patients who have failed to respond to all standard therapies and for whom the new antitumor agent being tested may be the last hope. Response rates generally decrease as the extent of prior therapy increases. Patients who have failed to respond to several prior regimens are more likely to have tumors composed of large numbers of resistant cells, and such patients are also less likely to be able to tolerate full doses of the investigational drug. Probably the most frequent problem with Phase II trials is that some selected patients are so debilitated by disease and prior therapy that an adequate evaluation of antitumor activity is impossible. Such patients are more likely to die or withdraw early in the course of treatment; and some investigators consider these patients "nonevaluable." The variable proportion of such patients—from study to study—contributes to the variability in reported response rates. Combining results—based on response rates—in a method such as *meta-analysis* is therefore even judged as invalid or questionable; it's not a matter of sample size. To overcome this problem to a certain extent, it is recommended that the practice of "intent-to-treat analysis" in Phase III trials also be used in the analysis of Phase II trials.

7.4.1 Sample Size Determination for Phase II Clinical Trials

The most common form of Phase II clinical trials is single-arm studies, where investigators are seeking to establish the antitumor activity of a drug usually measured by a response rate. In addition to one-arm trials, however, there are randomized Phase II clinical trials; some with a control arm, some without a control arm. In any case, the determination of the size of a trial is a crucial element in the design of a clinical trial. Methods presented here are similar to those in Section 2.5 but framed in the context of Phase II clinical trials so it would make it easier to compare to the designs for selection in Section 7.4.2.

One-Arm Phase II Clinical Trials

In designing any study, one of the first questions that must be answered is: How large must the sample be to accomplish the goals of the study? Depending on the study goals, the planning of sample size can be approached accordingly. One-arm Phase II clinical trials are the simplest ones; the drug, at the optimal dose (MTD) found from a previous Phase I trial, is given to a small group of patients who meet predetermined inclusion criteria. The focus is often on the response rate. Because of this focus, the planning of sample size can be approached in terms of controlling the width of a desired confidence interval for the parameter of interest—the response rate.

Suppose the goal of another study is to estimate an unknown response rate π. For the confidence interval to be useful, it must be short enough to pinpoint the value of the parameter reasonably well with a high degree of confidence. If a study is unplanned or poorly planned, there is a real possibility that the resulting confidence interval will be too long to be of any use to the researcher. In this case, we may determine to have an error of the estimate not exceeding d, an upper bound for the margin of error. The 95% confidence interval for the response rate π, a population proportion, is

$$p \pm 1.96\sqrt{\frac{p(1-p)}{n}}$$

where p is the sample proportion. Therefore our goal is expressed as

$$(1.96)\sqrt{\frac{p(1-p)}{n}} \leq d$$

leading to the required minimum sample size:

$$n = \frac{(1.96)^2 p(1-p)}{d^2}$$

(rounded up to the next integer). This *required sample size* is affected by three factors:

(i) The degree of confidence (i.e., 95%, which yields the coefficient 1.96.
(ii) The upper bound for the margin of error, d, determined by the investigator.
(iii) The proportion p itself.

This third factor is unsettling. In order to find n so as to obtain an accurate value of the proportion π, we need the estimate of the proportion itself. There is no perfect, exact solution for this. Usually, we can use information from similar studies, past studies, or studies on similar populations. If no good prior knowledge about the proportion is available, we can replace the term $p(1-p)$ by its maximum value (i.e. 0.25) and use a "conservative" sample size estimate:

$$n_{\max} = \frac{(0.25)(1.96)^2}{d^2}$$

because $n_{\max} \geq n$ regardless of the value of the true response rate π. Most Phase II trials are small; investigators often set the *maximum tolerated error* or upper bound for the margin of error, d, at 10% (or 0.10) or 15%, and some even set it at 20%.

■ **Example 7.5** If we set the maximum tolerated error, d, at 10% (or 0.10), then the required minimum sample size is

$$n_{\max} = \frac{(0.25)(1.96)^2}{(0.10)^2}$$

or 97 patients, which is usually too high for a small Phase II trial, especially in the field of cancer research where very few patients are available to participate.

If we set the maximum tolerated error, d, at 15% (or 0.15), then the required minimum sample size is

$$n_{\max} = \frac{(0.25)(1.96)^2}{(0.15)^2}$$

or 43 patients, which is more manageable. About 24 patients are needed if d is set at 0.20 or 20%.

Two-Arm Phase II Clinical Trials

In addition to one-arm trials, where we are seeking the antitumor activities of a drug measured by response rate, there are a variety of other Phase II trials—including randomized Phase II trials, some with control arms and some without a control arm. However, these are not very popular because Phase II sample sizes are often small. In addition, large controlled Phase III trials involving real-life treatment regimens are often involved combinations, not single agents.

1. Some Phase II trials are randomized comparative studies. These are most likely to be cases where we have established the activity for a given drug (from a previous one-arm, nonrandomized trial) and wish to add another drug to that regimen. In these randomized Phase II trials, which we refer to as *designs for selection*, the goal is to select the better treatment and the sample sizes for these Phase II trials are covered in the next subsection.
2. There are Phase II designs involving randomization between an investigational agent and an active standard treatment; those are randomized trials with a *control arm*. The purpose, however, is not to determine if the new agent is better or worse than the active control. The major objective of the randomization is to help in the interpretation of a poor response rate of the investigational agent. If the new agent has a high observed response rate, say, 25%, then the drug would be identified as having antitumor activity regardless of the magnitude of the response rate to the control treatment. Suppose, however, that the new agent has a low observed response rate (say, less than 10%); the conclusion would depend on the result from the control treatment: (i) if the control response rate is high, we would conclude that the new agent is "inactive," and (ii) if both response rates are poor, then the trial is indeterminate, especially if the sample size is not large. The "borders," 10% and 25%, are just conventional numbers; there is no magic in forming and using these numbers. This type of randomized design is not very popular because (i) it's only potentially useful where an adequate response rate on the active control is not known or not assured (most of the time, we know more about the controls), and (ii) if we do not know enough about the control—with the usual Phase II small sample sizes, it may be difficult to reliably determine whether the patients are sufficiently responsive to the control treatment. Plus there is the underlying ethical concern of having some patients serve as controls because for them the new antitumor agent being tested may be the last hope. In any case, if such a study is needed—a randomized trial with a control—the focus is exclusively on the treated arm and the sample size is determined as in the case of one-arm Phase II clinical trials.
3. Two or more treatment arms are possible in a Phase II clinical trial and the arms may be all experimental; those are randomized trials without a control arm. Investigators might be puzzled at the rationale for conducting a large randomized Phase III trial to compare the two arms, either one of which may have no activity (i.e., efficacy) in the disease. Phase II trials may provide needed early stopping rules because toxicity profiles are still not known. The major advantages are (i) randomization helps to ensure that patients are centrally registered before treatment starts; (ii) central registration is essential for checking a patient's eligibility, terminating accrual when the target sample size is reached, and establishing reliable records; and (iii) there will be some limited form of comparison—in addition to response rate—such as the degree of antitumor activity (extent of tumor shrinkage) and the durability of responses.

There are some Phase II trials that deal with assessing the activity of a biologic agent where tumor response is not the main endpoint of interest. We may be attempting to determine the effect of a new agent, for example, on the prevention of toxicity, and the primary endpoint may be measured on a continuous scale. In other trials, a pharmacologic- or biologic-to-outcome correlative objective may be the target. Whatever the rationale or reason, this type of two-arm randomized design is only used for certain limited forms of comparison; it does not involve formal statistical tests of significance. However, the sample sizes of the two arms in the trial are determined as in typical designs to compare two means or two proportions; the approach is focused on the planning of sample size in terms of controlling the risk of making Type II errors. We briefly present here only the latter case, the design of a Phase II trial to compare two response rates.

Suppose patients or subjects are to be randomized into two groups of equal size: a control (group 1) and an experimental agent group (group 2). The null hypothesis to be tested is

$$H_0\colon\ \pi_1 = \pi_2 \quad \text{versus} \quad H_A\colon\ \pi_1 < \pi_2$$

How large a total sample should be used to conduct this trial?

Suppose that it is important to detect an improvement in the response rate of the level

$$d = \pi_2 - \pi_1$$

And suppose we decide to preset the size of the study at $\alpha = 0.05$ and want the power $(1-\beta)$ to detect the above (alternative) difference d. Then the required (total) sample size (for both arms) is given by the following complicated formula:

$$\begin{aligned} N &= 2n \\ &= 4(z_{1-\alpha} + z_{1-\beta})^2 \frac{p(1-p)}{d^2} \end{aligned}$$

In this formula the quantities $z_{1-\alpha}$ and $z_{1-\beta}$ are percentiles of the standard normal distribution associated with the choice of Type I errors α and the choice of the statistical power $(1-\beta)$, and π is the average true response rate:

$$\pi = \frac{\pi_1 + \pi_2}{2}$$

It is obvious that the problem of planning sample size is more difficult and a good solution requires a deeper knowledge of the scientific problem: some good idea of the magnitude of the response rates π_1 and π_2 themselves.

■ **Example 7.6** Suppose we wish to conduct a clinical trial to test a new agent, where the rate of success in the control group (standard treatment) was known to be about 5%.

Furthermore, we would consider the new agent to be superior—cost, risks, and other factors considered—if its rate of success is about 15%. Suppose also that we decide to preset $\alpha = 0.05$ and want the power to be about 90% (i.e., $1 - \beta = 0.10$). In other words, we have

$$\begin{aligned} z_{1-\alpha} &= 1.96 \\ z_{1-\beta} &= 1.28 \\ N &= 2n \\ &= 4(1.96 + 1.28)^2 \frac{(0.10)(0.90)}{(0.15-0.05)^2} \\ &= 378 \text{ patients or } 189 \text{ patients per arm} \end{aligned}$$

This result is obviously impractical; we might need to consider a Type I error level of 20% and statistical power of 80%—or even as low as 70%.

7.4.2 Phase II Clinical Trial Designs for Selection

Perhaps the most prevalent form of randomized trials without a control consists of what we usually call designs for selection or screening trials.

The needs for selection can be seen as follows. The process starts with a dose-finding Phase I trial leading to the MTD. In the next phase, Phase II, a small one-arm clinical trial is conducted to study antitumor activity through response rate; if the results from this Phase II clinical trial are promising—that is, the agent is found safe enough and effective enough—the agent then becomes a candidate for the next phase. The problem is that there may be too many candidates for Phase III trials to compare—in some combination form—the agent's efficacy to a standard treatment or placebo; sometimes differences between candidates are small. For example, in the study of thymidylate synthase inhibitor 5, fluorouracil (5-FU), for metastatic colorectal cancer, we might have two choices: (i) continuous infusion (which requires placement and maintenance of a central intravenous access devise and use of a pump) versus (ii) injection over a short-time interval, through a temporary venous catheter, which is easier to apply but could be somewhat less efficacious. How should we decide?

Specific Aim

At this stage, the aim is not to make a definite conclusion about the superiority of one treatment (or one mode of administration) as compared to the other. Randomized Phase II clinical trials of this type do not fit the framework of tests of significance. In performing statistical tests or tests of significance, we have the option to declare "not significant" when data from a study do not provide enough support for a treatment difference; that is not enough evidence, say, to support the conclusion of superiority of the new agent as compared to the standard treatment. In those cases, we decide not to pursue the new treatment, and we do not choose the new treatment because the new treatment does not prove any better than the placebo effect, and we have a standard therapy on which to rely. The problem is, for some cancers, we may not have a standard therapy; or if we do, some subgroup of patients may have failed to respond to the standard therapy. Suppose further that we have established activity for a given drug from a previous one-arm, nonrandomized trial and the only remaining question is scheduling; for example, between continuous infusion of 5-FU versus injection over a short time interval or, for some other medication, between daily versus one every other day schedules. We do not have the option to declare "not significant" in these trials because (i) one of the treatments or schedules has to be chosen because patients have to be treated, and (ii) it is inconsequential to choose one of the two treatments or schedules even if they are equally efficacious. In other words, we face Type II errors; Type I errors become less important. The aim of these randomized trials is to choose the *better* treatment. If correct ordering is the goal, a properly powered Phase III trial would be required; we can't afford a Phase III trial before a Phase II trial. Therefore the goal is just to ensure that if one treatment is clearly inferior, it is less likely carried forward to the Phase III trial (to compare versus standard or placebo arm).

In summary, the basic characterizations of a design for selection are the following:

(i) There is no standard or placebo arm; both treatments (or modes of treatment) A and B are experimental. Treatments or agents A and B were identified, likely separately from Phase II trials. So this screening trial could be referred to as Phase II and a half; but just Phase II for simplicity.

(ii) A decision (i.e., selection) has to be made, but the setting does not fit the framework of a statistical test of significance, where "not statistically significant" is a possibility. We cannot afford it!

Other characteristics are:

(iii) Because the goal is not superiority, Type I errors are less relevant; emphasis is on the probability of a correct selection—called designs for selection or screening trials.

(iv) The trial is randomized.

(v) In addition to efficacy, other criteria may also be considered (toxicity, cost, ease of administration, or quality of life); investigators want that flexibility.

Criteria

When the primary outcome of a trial is measured on the binary scale, the focus is on a proportion—the response rate. At the end of the study, we select the treatment or schedule with the larger sample proportion. But first, we have to define what we mean by "better treatment." The selection criteria could be as follows:

(i) If the observed outcome (e.g., response rate, but could be sample mean in some trials) of one arm is d units greater than the other, the arm with better observed outcome (larger proportion) is selected for use in the next Phase III trial.

(ii) If the difference is smaller than d units (d may or may not be 0), selection may be based on other factors.

Correct Outcome

Suppose that the outcome variable is response rate and treatment A is assumed to be better:

$$\pi_{\text{A}} - \pi_{\text{B}} = \delta$$

The probability of correct outcome is

$$P_{\text{Corr}} = \Pr[p_{\text{A}} - p_{\text{B}} > d | \pi_{\text{A}}, \pi_{\text{B}}]$$

Correct Selection

If the observed outcome is ambiguous, that is, the difference is less than d, treatment A could still be chosen (by factors other than efficacy) with, say, probability ρ; then the probability of correct selection is

$$\begin{aligned} P_{\text{Corr}} &= \Pr[p_{\text{A}} - p_{\text{B}} > d | \pi_{\text{A}}, \pi_{\text{B}}] \\ P_{\text{Amb}} &= \Pr[p_{\text{B}} - d \leq p_{\text{A}} \leq p_{\text{B}} + d] \\ \lambda &= P_{\text{Corr Sel}} \quad \text{(probability of correct selection)} \\ &= P_{\text{Corr}} + \rho P_{\text{Amb}} \end{aligned}$$

For simplicity, we could set $d = 0$; in that case the decision rule requires that at the end of the trial, whichever arm is ahead by any margin be carried forward to the Phase III trial. However, this may be less desirable because the rule does not allow the inclusion of factors other than efficacy to be included in the decision process. If the observed outcome is ambiguous, that is, the difference is less than d, treatment A could still be chosen (by factors other than efficacy) with, say, probability ρ; conservatively (and likely), $\rho = 0$ but we could have $\rho = 0.5$.

The Role of Statistics

Population parameters (such as π_{A} and π_{B}, or π_{A} and d) are used in "alternative hypotheses"—ideas from separate Phase II trials. The size of d is a clinical decision. At the end of the trial the analysis is simple: compute $(p_{\text{A}} - p_{\text{B}})$ and compare to d. However, the role of statistics is clear and crucial; the statistician is responsible for the design, to find sample size n (per arm) to ensure that the probability of correct selection exceeds a certain threshold, say, $\lambda \geq 0.90$ or 90% (similar to power). The details are as follows.

The difference $(p_A - p_B)$ is distributed normally with mean μ and variance s^2:

$$\begin{aligned}\mu &= (p_A - p_B) = \delta \\ \sigma^2 &= \frac{1}{n}[\pi_A(1-\pi_A) + \pi_B(1-\pi_B)]\end{aligned}$$

These imply that

$$\begin{aligned}P_{\text{Corr}} &= \Pr[p_A - p_B > d | \pi_A, \pi_B] \\ &= 1 - \Phi\left(\frac{d-\delta}{\sigma}\right) \\ P_{\text{Amb}} &= \Pr[-d \le p_A - p_B \le d] \\ &= \Phi\left(\frac{d-\delta}{\sigma}\right) - \Phi\left(\frac{-d-\delta}{\sigma}\right) \\ \lambda &= P_{\text{Corr Sel}} = P_{\text{Corr}} + \rho P_{\text{Amb}}\end{aligned}$$

Example 7.7 Let us assume that $\pi_A = 0.35$ and $\pi_B = 0.25$ (or $d = 0.10$) and $d = 0.05$, $n = 50$ ($\sigma = 0.09$). We have

$$\begin{aligned}P_{\text{Corr}} &= 1 - \Phi\left(\frac{d-\delta}{\sigma}\right) \\ &= 1 - \Phi(-0.55) = 0.71 \\ P_{\text{Amb}} &= \Phi\left(\frac{d-\delta}{\sigma}\right) - \Phi\left(\frac{-d-\delta}{\sigma}\right) \\ &= \Phi(-0.55) - \Phi(-1.65) = 0.24 \\ \lambda &= P_{\text{Corr Sel}} \\ &= P_{\text{Corr}} + \rho P_{\text{Amb}} \\ &= \begin{cases} 0.71 \text{ for } \rho = 0 \\ 0.83 \text{ for } \rho = 0.5 \end{cases}\end{aligned}$$

Required Sample Size

Changing n will change σ, and therefore the probability of correct selection λ. That's key to sample size determination—even by trial and error. In the simple case with $d = 0$, we can impose a minimum probability of selection and easily derive the required sample size:

$$P_{\text{Corr}} = \Pr[p_A - p_B > 0 | \pi_A - \pi_B = \delta] = \lambda$$

For example, if we want to be 99% sure that the better treatment will be the one with larger sample proportion, we can preset $\lambda = 0.99$. In general, the total sample size must be at least

$$N = 4z_{\lambda}^{2}\frac{\pi(1-\pi)}{(\pi_{\mathrm{A}}-\pi_{\mathrm{B}})^{2}}$$

assuming that we conduct a balanced study with each group consisting of $n = N/2$ subjects. In this formula, π is the average proportion:

$$\pi = \frac{\pi_{\mathrm{A}}+\pi_{\mathrm{B}}}{2}$$

It is obvious that the problem of planning sample size is more difficult and a good solution requires a deeper knowledge of the scientific problem—some good idea of the magnitude of the proportions π_{A} and π_{B} themselves. In many cases, that may be impractical at this stage. If a selection (of one treatment over the other) is made when two treatments are equally effective, it's Type I error. But in our context, it's fine because whatever treatment is selected patients are equally well served. A design for selection has no concern for Type I errors, so required sample size is much smaller: the probability of correct selection is like the counterpart of statistical power. If it was a "test of significance," at the conclusion of the trial, one would compute the sample proportions and reject the null hypothesis if

$$\frac{|p_{\mathrm{A}}-p_{\mathrm{B}}|}{\sigma} \geq z_{\alpha/2}$$

The statistical power of this test, which is very different from the probability of correct selection, would be

$$1-\beta = 1-\Phi\left(z_{\alpha/2}-\frac{\delta}{\sigma}\right)+\Phi\left(-z_{\alpha/2}-\frac{\delta}{\sigma}\right)$$

■ **Example 7.8** Let us assume that $\pi_{\mathrm{A}} = 0.35$, $\pi_{\mathrm{B}} = 0.45$ ($\delta = 0.10$), $d = 0.05$ and take $n = 100$. Then

$$\begin{aligned} P_{\text{corr}} &= 0.88 \\ P_{\text{corr}} &+ (0.5)P_{\text{Amb}} = 0.93 \end{aligned}$$

We have for the normal approximation to the binomial distribution

$$\begin{aligned} \sigma &= \sqrt{\frac{0.35(1-0.35)+0.45(1-0.45)}{100}} \\ &= 0.07 \end{aligned}$$

If it was a test of significance, the power would be very low compared to the probability of correct selection of 88% or 93%:

$$\begin{aligned}\text{Power} &= 1-\Phi\left(z_{\alpha/2}-\frac{\delta}{\sigma}\right)+\Phi\left(-z_{\alpha/2}-\frac{\delta}{\sigma}\right)\\ &= 1-\Phi\left(1.96-\frac{0.10}{0.07}\right)+\Phi\left(-1.96-\frac{0.10}{0.07}\right)\\ &= 1-\Phi(0.53)+\Phi(-3.39)\\ &= 1-0.7019+0.0003\\ &= 0.2984\end{aligned}$$

■ **Example 7.9** The following table gives a few examples, taken from Sargent and Goldberg (2001). We can see that the probabilities of correct selection are quite high even with a small sample, while it can be verified that the corresponding statistical powers would be much lower. The parameters used in the tables are:

(i) Odd rows: $\pi_A = 0.35$, $\pi_B = 0.45$ ($\delta = 0.10$), $d = 0.05$.
(ii) Even rows: $\pi_A = 0.35$, $\pi_B = 0.50$ ($\delta = 0.15$), $d = 0.05$.

n	d	P_{Corr}	P_{Amb}	$P_{\text{corr}} + (0.5)\text{P}_{\text{Amb}}$
50	0.10	0.71	0.24	0.83
50	0.15	0.87	0.12	0.93
75	0.10	0.82	0.15	0.89
75	0.15	0.95	0.05	0.97
100	0.10	0.88	0.10	0.93
100	0.15	0.98	0.02	0.99

7.4.3 Two-Stage Phase II Design

Phase I trials provide information about the MTD; it is important because most cancer treatments must be delivered at maximum dose for maximum effect. Patients may die from toxicity or side effects, but if not treated with "enough" therapy, they might die from the disease too. Phase I trials provide little or no information about efficacy; patients are diverse with regard to their cancer diagnosis and are treated at different doses—only three or six at a dose, even one at a dose in fast track design.

A Phase II trial is the next step in the study of an investigational drug. A Phase II trial of a cancer treatment is an uncontrolled trial (most trials of Phase II are one-arm, open-label) to obtain an estimate of the *degree* of antitumor effect. The proportion of patients whose tumors shrink by at least 50% and which lasts for at least 4 weeks is often the primary endpoint. The aim is to see if the agent has sufficient activity against a specific type of tumor to warrant its further development (to combine with other drugs in a Phase III trial comparing survival results with a standard treatment). It is desirable to find out about the antitumor capacity of new agents and to determine if a treatment is sufficiently promising to warrant a major controlled evaluation. However, recall that there are three basic objectives in conducting Phase II clinical trials:

Objective 1: Benefit the patients.
Objective 2: Screen the agent/drug for antitumor activity.
Objective 3: Extend knowledge of the toxicology and pharmacology of the agent or drug.

The first is to benefit the patients. The problem is that, if the agent has no or low antitumor activity, patients in the Phase II trial might die from the disease. Therefore in order to indirectly achieve the first objective of benefiting the patients, we often wish to minimize the number of patients treated with an ineffective drug. Early acceptance of a highly effective drug is permitted but very rare in Phase II trials; however, it is ethically imperative to exercise early termination when the drug has no or low antitumor activity. In other words, it is desirable to use as few patients as possible in a Phase II trial if the regimen under investigation has low antitumor activity. When such ethical concerns are of high priority, investigators often choose a two-stage design. Investigators choose a two-stage design when they do not want to enroll a large group of patients (in conventional one-stage designs) and when they are not sure if the treatment is effective. There are more than one method but the ideas are similar. A small group of patients are enrolled in the first stage; and the enrollment of another group of patients in stage 2 is "conditional" on the outcome of the first group. Then the activation of the second stage depends on an "adequate" number of responses observed from the first stage.

Gehan's Two-Stage Design

The first and most commonly used design for many years was developed by Gehan (1961). This design was popular much more so in the 1970s than in the last decade. It has two stages: the primary aim of Gehan's design is to estimate the response rate and the two-stage feature is an option for screening agents worthy of further development. This early and simple design can be briefly described as follows.

1. The first stage enrolls 14 patients; if *no* responses are observed, trial is terminated.
2. If at least one response is observed among the first 14 patients of stage 1, the second stage of accrual is activated in order to obtain an estimate of the response probability having a prespecified standard error (SE).
3. Patients from both stages are used in the estimation of the response rate.

Since the probability of observing no responses among 14 patients is less than 0.05 if the response probability is greater than 20%,

$$(0.20)^0(0.80)^{14} = 0.044$$
$$\pi^0(1-\pi)^{14} \leq 0.044 \;\; \text{if } \pi \geq 0.20$$

that means, implicitly, that response rates over 20% are considered promising for further studies. That explains the rationale of Gehan whose primary goal is to estimate the response rate.

If no responses are observed in the first stage of 14 patients, trial is terminated. It is stopped because we can conclude that $\pi < 0.2$ or 20%, and thus the drug is not worthy of further investigation. The number of patients n_2 accrued in the second stage depends on the number of responses observed in the first stage (because patients from both stages are used in the estimation of the response rate) and the predetermined standard error:

$$SE(p) = \sqrt{\frac{\pi(1-\pi)}{14+n_2}}$$

Gehan's design is often used with a second stage of $n_2 = 11$ patients. This accrual provides for estimation with no more than 10% standard error ($\mathrm{SE} \leq 0.10$):

$$SE(p) = \sqrt{\frac{\pi(1-\pi)}{25}} \leq \sqrt{\frac{(0.50)(0.50)}{25}} = 0.10$$

The following are common critiques:

(i) The size of the first stage is fixed; it may not be optimal for the underlying aim of early termination if the drug has no or low antitumor activity.

(ii) It serves investigators and drug companies more—not the patients enrolled in the trial.

(iii) With a standard error of 10%, it corresponds to a very broad 95% confidence interval; reducing SE to, say, 5% would lead to a sample size too large for Phase II trials (but this is a universal problem for Phase II trials, not just Gehan's).

The more serious problem is in the first critique. Gehan's design provides an option for screening of agents worthy of further development—those with response rates of 20% or more. However, it does not help to achieve the aim of early termination when the drug has no or low antitumor activity, a very important ethical concern. For example, even for a poor drug with a true response probability of 5%, there is a 51% chance of obtaining at least one response in the first 14 patients and therefore activating the second stage accrual.

$$\Pr(\text{at least 1 response}) = 1-(0.05)^0(0.95)^{14} = 0.51$$

That is, there might be high probabilities to enroll more patients (in the second stage) to be treated by an inefficient drug. Gehan's two-stage design was popular in the 1970s and is still being used in some trials because of its simplicity.

Simon's Two-Stage Design

In the conduct of Phase II trials, the ethical imperative for early termination occurs when the drug has low antitumor activity. The Simon's two-stage design is currently a very popular tool to achieve that. The trial is conducted in two stages with the option to stop the trial after the first or after the second stage (and not recommending the agent for further development). The basic approach is to minimize expected sample size when the true response is low—say, less than some predetermined uninterested level. This design uses a computer search to meet certain optimal requirements; it does require some special programs but most research organizations and health centers have this software.

The Statistical Setup for Simon's Design

The two-stage design can be formally statistically framed as follows:

Endpoint: (Binary) tumor response: yes/no.

Null Hypothesis: H_0: $\pi = \pi_0$; π is the true response (say, proportion of patients whose tumors shrink by at least 50%) and π_0 is a predetermined uninterested or undesirable level.

Alternative Hypothesis: H_A: $\pi = \pi_A$; π_A is some desirable level that warrants further development.

Type I and Type II Errors: α and β.

Basis for Decision: Minimize the number of patients treated in the trial if H_0 is true.

The response rate under the null hypothesis, π_0, could be considered as "dangerous" because patients might die from the disease. Remember that patients in early-phase trials are mostly terminal cancer patients for whom the new antitumor agent being tested may be the last hope.

Design Summary

The increasingly popular Simon's design could be summarized as follows:

(i) Enroll n_1 patients in stage 1. The trial is stopped if r_1 or fewer responses are observed; it goes on to the second stage otherwise.

(ii) Enroll n_2 patients in stage 2. The trial is not recommended for further development if a total of r (of course $r > r_1$) or fewer responses are observed in both stages.

The main practical consideration is that evaluation of a patient's response is usually not instantaneous and may require observations for weeks or months. Consequently, patient accrual at the end of stage 1 may have to be suspended until it is determined whether the criterion for continuing is satisfied. Such suspension of accrual is awkward for physicians who are contributing patients to the study and is the main reason for not considering more than two stages.

Operating Characteristics

(i) *Probability of Early Termination.* The probability of early termination is the probability of having r_1 responses or fewer in the first stage,

$$PET(\pi) = B(r_1; n_1, \pi)$$

where $B(.)$ denotes the cumulative binomial probability, and π is the true response.

(ii) *Expected Sample Size.*

$$EN(\pi) = n_1 + [1 - PET(\pi)] * n_2$$

It can be seen that both $PET(\pi)$, the probability of early termination, and $EN(\pi)$, expected sample size, are functions of the response rate π.

Decision Not to Recommend

The drug may not be recommended after either one or two stages; the probability is $PNC(\pi)$.

(i) The trial is terminated at the end of the first stage and the drug is not recommended if r_1 or fewer responses are observed; or

(ii) The drug is not recommended at the end of the second stage if r ($r = r_1 + r_2$) or fewer responses are observed; some of the responses may come from stage 1 (r_1), some from stage 2.

The probability of not recommending is

$$PNC(\pi) = B(r_1; n_1, \pi) + \sum_{x=r_1+1}^{\min[n_1, r]} b(x; n_1, \pi) B(r - x; n_2, \pi)$$

The probability of not recommending, $PCN(\pi)$, is also a function of the response rate π. The decision of not recommending leads to two types of errors with the following probabilities:

$$\text{Type I errors:} \quad \alpha = 1 - PNC(\pi_0)$$
$$\text{Type II errors:} \quad \beta = PNC(\pi_A)$$

Simon's Approach

The design approach considered by Simon is to specify the parameters π_0, π_1, α, and β; then determine the two-stage design that satisfies the errors probabilities α and β and minimizes the expected sample size EN when the response probability is π_0, that is, minimizing $EN(\pi_0)$. It's a search with the help of a computer program.

Step 1: For each integer n and n_1, in the range $(1, n-1)$, determine the integers r_1 and r that satisfy the error constraints and minimize $EN(\pi_0)$.

(i) This is found by searching over the range r_1 in $(0, n_1)$; for each value of r_1, determine the value of r satisfying the Type II error rate.

(ii) See whether the set of parameters (n, n_1, r_1, r) satisfy the Type I error rate; if it did, compare the expected sample size $EN(\pi_0)$ to the one achieved previously, and continue to search with a different r_1.

Step 2:

(i) Keeping n fixed, search over the range of n_1, $(1, n-1)$, repeating the process in Step 1.

(ii) Search over the range of n up to, say, 50, as commonly used in Phase II trials, repeating the process in Step 1.

Simon's approach leads to an *optimal design*, in the sense of achieving the objective of early termination when the drug has no or low antitumor activity (key step: check $EN(\pi_0)$). There is an option where, in Step 1, we only check against the constraints imposed by the two types of errors but skip checking for $EN(\pi_0)$; then keep the same Steps 2(i) and 2(ii). The first two-stage design found would satisfy the error constraints and has the smallest total maximum/potential sample size n. Simon called the result the *minimax design*. Usually, the minimax two-stage design has the same maximum sample size n as the smallest single-stage design that satisfies the error probabilities. However, because of the early-termination option (after the first stage), the minimax two-stage design has a smaller expected sample size under H_0.

■ **Example 7.10** Let us consider $\pi_0 = 0.10$ and $\pi_1 = 0.30$ and take $\alpha = 0.05$ and $\beta = 0.20$. If we consider a one-stage design, we would need $n = 32$ patients. The following are the results of Simon's approach.

Optimal Design

Stage 1: Reject the drug if response $\leq 1/11$ (enroll 11 and terminate if 0 or 1 response).

Stage 2: (Two or more responses from Stage 1.) Reject the drug if response $\leq 5/29$ (enroll 18 in the second stage; reject if total responses ≤ 5).

$$\text{Expected sample size } EN(\pi = 0.10) = 15.0$$
$$PET(\pi = 0.10) = 0.74$$

Minimax Design

Stage 1: Reject the drug if response $\leq 1/15$ (enroll 15 and terminate if 0 or 1 response).

Stage 2: (Two or more responses from stage 1.) Reject the drug if response $\leq$5/25 (enroll 10 in the second stage; reject if total responses $\leq$5).

$$\text{Expected sample size } EN(\pi = 0.10) = 19.5$$
$$PET(\pi = 0.10) = 0.55$$

Both of Simon's designs require fewer patients than the conventional one-stage design; both would stop earlier if the drug is not effective. Minimax design enrolls no more than 25 patients (versus 29 for optimal); however, it (always) enrolls more subjects in the first stage (15 versus 11), (always) has a lower probability for early termination if the drug performs poorly (here, 0.55 versus 0.74), and therefore it has larger than expected sample size (than optimal design, 19.5 versus 15). In some cases, the minimax design may be more attractive than the optimal design. This is the case where the difference in expected sample sizes $EN(\pi_0)$ is small and the patient accrual rate is low/slow. We would always need to check out both.

■ **Example 7.11** Let us consider $\pi_0 = 0.10$ and $\pi_1 = 0.30$ and take $\alpha = 0.10$ and $\beta = 0.10$. If we consider a one-stage design, we would need $n = 35$ patients. The following are results of Simon's approach:

Optimal Design

Stage 1: Reject the drug if response $\leq$1/12 (enroll 12 and terminate if 0 or 1 response).

Stage 2: (Two or more responses from stage 1.) Reject the drug if response $\leq$5/35 (enroll 23 in the second stage; reject if total responses $\leq$5).

$$\text{Expected sample size } EN(\pi = 0.10) = 19.8$$
$$PET(\pi = 0.10) = 0.65$$

Minimax Design

Stage 1: Reject the drug if response $\leq$1/16 (enroll 16 and terminate if 0 or 1 response).

Stage 2: (Two or more responses from stage 1.) Reject the drug if response $\leq$4/25 (enroll 9 in the second stage; reject if total responses $\leq$4).

$$\text{Expected sample size } EN(\pi = 0.10) = 20.4$$
$$PET(\pi = 0.10) = 0.51$$

It is still true that minimax design enrolls no more than 25 patients (versus 35), has more subjects in the first stage (16 versus 12), has lower probability for early termination if the drug performs poorly (0.51 versus 0.65), and has a larger than expected sample size (20.4 versus 19.6). But the reduction in expected sample size is negligible; on the other hand, if the accrual is slow—say, 10 per year—it could take a year longer to complete the optimal design than the minimax design.

7.4.4 Toxicity Monitoring in Phase II Trials

AE is the common abbreviation for adverse effects, also referred to as side effects; many investigators use the terms interchangingly for these unwanted events.

Phase I and Phase II clinical trials present special difficulties because they involve use of agents whose spectrum of toxicity and likelihood of benefits are poorly understood or defined. Recall that there are three basic objectives in conducting Phase II clinical trials:

Objective 1: Benefit the patients.

Objective 2: Screen the agent or drug for antitumor activity.

Objective 3: Extend the knowledge of toxicology and pharmacology of the agent or drug.

In Phase II trials, *efficacy* is the outcome of interest, whereas *safety* is embedded to serve as a stopping rule. In planning a clinical trial of a new treatment, we should always be aware that severe, even fatal, side effects are a real possibility. If the accrual or treatment occurs over an extended period of time, we must anticipate the need for a decision to stop the trial—at any time—if there is an excess of these unwanted events.

We focus on Phase II clinical trials. In Phase I trials, toxicity may be considered the outcome variable and a dose-escalation plan serves as the stopping rule. In Phase II trials, we start to focus on efficacy, which requires conventional analysis at the end. Response becomes the outcome variable; however, toxicity (or other adverse effects) may still turn out to be a problem during the trial. The monitoring of side-effect events is a separate activity that may require special consideration. Two-stage designs stop trials for efficacy reasons; however, here we want rules to stop trials for safety reason—both reasons, not treated enough or excessive adverse effects, put patients at risk. Two-stage designs are optional (decision by investigators) but stopping rules for safety reasons are required by regulatory affairs agencies or entities. The most common method for monitoring toxicity or adverse effects is to design a formal sequential *stopping rule* based on the limit of acceptable side-effect rates; the sequential nature of the rule allows investigators to stop the trial as soon as the event's rate becomes excessive. In multisite trials, a data safety and monitoring board (DSMB) is required; in local Phase II trials, it's the statistician's responsibility to form the rule and the Clinical Trial Office's staff is responsible for its implementation. For practical use, the rule has to be simple. At larger institutions, statisticians usually have to monitor these events on a daily basis.

A simple example of needs can be seen in a field of treatment called *bone marrow transplant* (BMT). Bone marrow is a spongy tissue found inside the bones; it contains *stem cells* that produce the body's blood cells, including white blood cells that fight infection. In patients with leukemia (and a few other diseases), the stem cells malfunction, producing excessive defective cells that interfere with the production of normal white and red blood cells; the defective cells also accumulate in the bloodstream and invade other tissues and organs. Bad bone marrow needs to be replaced. In bone marrow transplant, the patient's diseased marrow is destroyed (usually by radiation), and healthy marrow is then infused into the patient's bloodstream. In successful BMT, the new bone marrow migrates to the cavities of the bones (i.e., engrafts) and begins producing normal blood cells. If the marrow from a donor is used, the transplant is called *allogeneic BMT* or *syngeneic BMT* if the donor is an identical twin. If the donor cells used are from the patient (after being treated), the transplant is called *autologous BMT*; this has a lower success rate. Bone marrow transplantation is a complex procedure that exposes patients to the high risk of a variety of complications, many of them associated with death; these risks are in exchange for even higher risks associated with the leukemia or other disease for which the patient is being treated. Since the patient's immune system has been weakened or destroyed, a complication that usually develops in half or more of BMT patients is *graft-versus-host disease* (GVHD). One way to prevent GVHD is to treat the donor's marrow prior to transplantation; unfortunately, such a treatment may cause some patients to have engraftment problems (either delayed or failed). Since the patient's own marrow was destroyed in preparation for BMT, if the donor's marrow does not engraft, the patient does not have the capacity to produce blood cells—and the transplant fails. A sequential monitoring for nonengraftment is desirable so as not to have more failed transplants.

The most common method for monitoring toxicity or adverse effects is to design a formal sequential stopping rule. A sequential stopping rule could be formed in two different ways:

(i) A Bayesian approach to evaluating the proportion of patients with side effects
(ii) A hypothesis testing approach, using the sequential probability ratio test (SPRT) to see if the normal, acceptable side-effect rate has been exceeded.

Hypothesis Testing Approach

Let us start with the hypothesis testing approach, because it's more conventional (with statisticians). Let π be the proportion of patients with adverse side effects as defined specifically for the trial; for example, toxicity grade III or IV. As with any other statistical test of significance, the decision is concerned with testing a null hypothesis:

$$H_0: \; \pi - \pi_0$$

against an alternative hypothesis,

$$H_A: \; \pi = \pi_A$$

where π_0 is the investigator's hypothesized value for the incidence rate π or the normal baseline side-effect rate (say, 5%) and π_A is the maximum tolerated rate (say, 20%)—anything over that is considered excessive. Baseline rate is determined or estimated from historical data but the setting of a ceiling rate is subjective. The other figure, π_A, is the maximum tolerated level for the incidence rate π of severe side effects. The trial has to be stopped if the incidence rate π of severe side effects exceeds π_A. In addition to the null and alternative parameters, π_0 and π_A, a stopping rule also depends on the chosen level of significance α (usually 0.05) and the desired statistical power $(1-\beta)$; β is the size of Type II error associated with the alternative H_A: $\pi = \pi_A$. Power is usually preset at 80% or 90%.

Statistical Model We can assume that the number of adverse events e follows the usual binomial distribution $B(n, \pi)$, where n is the total number of patients. This leads to the log-likelihood function:

$$L(\pi;\ e) = \text{constant} + e\ \ln\pi + (n-e)\ln(1-\pi)$$

Sequential Probability Ratio Test When e adverse events are observed out of n evaluable patients, the test for null hypothesis H_0: $\pi = \pi_0$ against the alternative hypothesis H_A: $\pi = \pi_A$ can be based on the log-likelihood ratio statistic, LR_n:

$$LR_n = e(\ln\pi_A - \ln\pi_0) + (n-e)[\ln(1-\pi_A) - \ln(1-\pi_0)]$$

In conventional sequential testing, the statistic is calculated as each patient's evaluation becomes available and plotted against n; the trial is stopped if the plot goes outside predefined boundaries, which depend on preset Type I and Type II errors. In testing the null hypothesis H_0: $\pi = \pi_0$ against the alternative H_A: $\pi = \pi_A$, the decision is:

(i) To stop the trial and reject H_0 if $LR_n \geq \ln(1-\beta) - \ln\ \alpha$
(ii) To stop the trial and accept H_0 if $LR_n \leq \ln\ \beta - \ln(1-\alpha)$
(iii) To continue the study otherwise

In (i) there are too many events and in (ii) there are too few events—not enough to make a decision. In side effects monitoring, we do not stop the trial because there are too few events; we only stop the trial early for an excess of side effects, that is, when

$$e(\ln\pi_A - \ln\pi_0) + (n-e)[\ln(1-\pi_A) - \ln(1-\pi_0)] \geq \ln(1-\beta) - \ln\alpha$$

The lower boundary is ignored and trial continues; solving the equation for e yields an upper boundary. We can also solve the same equation for n. The resulting sequential rule is to stop the trial as soon as n, as a function of e, satisfies the following equation—where $n(e)$ is the number of evaluable patients for having e of them with adverse effects:

$$n(e) = \frac{\ln(1-\beta) - \ln\alpha + e[\ln(1-\pi_A) - \ln(1-\pi_0) - \ln\pi_A + \ln\pi_0]}{\ln(1-\pi_A) - \ln(1-\pi_0)}$$

With this stopping rule, we monitor for the side effects by sequentially counting the number of events e (i.e., number of patients with severe side effects) and the number of evaluable patients $n(e)$ at which the eth event is observed. The trial is stopped when this condition is first met; in other words, the above formula gives us the maximum number of evaluable patients $n(e)$ at which the trial has to be stopped if e events have been observed. Actually, we have to stop the trial when we have e adverse effects before reaching a total of $n(e)$ patients.

Some Phase II trials may be randomized; however, even in these randomized trials, toxicity monitoring should be done separately for each study arm. That is, if the side effect can reasonably occur in only one of the arms of the study—most likely the arm treated by the new therapy—the incidence in that group alone is considered. Otherwise, the sensitivity of the process to stop the trial would be diluted by the inclusion of the other group. Sometimes the goal is to compare two treatments according to some composite hypothesis that the new treatment is equally effective but has less toxicity. In those cases, both efficacy and toxicity are endpoints, and the analysis should be planned accordingly, but the situations are not that of monitoring in order to stop the trial as intended in the above rule.

■ **Example 7.12** Suppose in planning for a Phase II trial, an investigator or clinician in the study committee decideds that $\pi_0 = 3\%$, or 0.03, based on some prior knowlege and that $\pi_A = 15\%$, or 0.15, should be the upper limit that can be tolerated (as related to the risks of the disease itself). For this illustrative example, we find $n(1) = -7.8$, $n(2) = 5.4$, $n(3) = 18.6$, $n(4) = 31.8, \ldots$, rounding off ("down") to $\{-, 5, 18, 31, \ldots\}$, when we preset the level of significance at 0.05 and statistical power at 80%. In other words, we stop the trial if there are 2 events among the first 5 evaluable patients, 3 events among the first 18 patients, 4 events among the first 31 patients, and soon. Here, we use only the positive solutions and the integer proportion of each solution from the above equation. The negative solution $n(1) = -7$ indicates that the first event will not result in stopping the trial (because it is judged as not excessive yet). The stopping rule would be more stringent if we want higher (statistical) power or if incidence rates are higher. For example, consider the following:

(i) With $\pi_0 = 3\%$, or 0.03 and $\pi_A = 15\%$, or 0.15, as previously set, if we preset the level of significance at 0.05 and statistical power at 90%, the results become $\{-, 4, 17, 30, \ldots\}$. That is, we stop the trial if there are 2 events among the first 4 evaluable patients, 3 events among the first 17 patients, 4 events among the first 30 patients, and so on.

(ii) On the other hand, if we keep the level of significance at 0.05 and statistical power at 80%, and we decide on $\pi_0 = 5\%$, or 0.05 and $\pi_A = 20\%$, or 0.20, the results become $\{-, 2, 11, 20, \ldots\}$. That is, we stop the trial if there are 2 events among the first 2 evaluable patients, 3 events among the first 11 patients, 4 events among the first 20 patients, and so on.

It can be seen that the rule accelerates faster with higher rates than with higher power.

The hypothesis testing-based approach has two problems or weaknesses:

(i) At times, the result might appear to be overaggressive; the trial is stopped when the observed rate of adverse events (i.e., $p = e/n$) is below the ceiling rate π_A.
(ii) The statistical power falls short of the preset level because we apply the rejection rule for a two-sided test to a one-sided alternative.

Is it really overaggressive? Take the example where we know that the baseline rate is $\pi_0 = 3\%$ and investigator sets a ceiling rate of $\pi_A = 15\%$; the stopping rule is $\{-, 5, 18, 31, \ldots\}$. But, at the fourth event, the observed rate is 4/31 or 12.9%, still below the ceiling set at 15%. In the context of the statistical test, at that point, even though the observed rate is only 12.9% it is enough to reject H_0 (3%) and accept H_A(15%), the rate at which the trial should be stopped. Of course, it is kind of unsettling to a clinician to stop a trial when the observed rate is still not yet considered unsafe (to him/her). Actually, the rule $\{-8, 5, 18, 31, \ldots\}$ is not very aggressive. In addition, the problem only appears so when the clinician is too aggressive to go on by setting the ceiling rate way over the baseline rate (15% vs. 3%). It would not appear as a problem when the gap is set smaller; for example, if the baseline rate is $\pi_0 = 3\%$ and the investigator sets a ceiling rate of $\pi_\text{A} = 10\%$, the stopping rule would be $\{-, -, 10, 23\}$. Here, we did not stop before the ceiling rate.

What about the second possible weakness concerning the test's statistical power? The problem with statistical power is that it falls short of the preset level because we apply the rejection rule for a two-sided test to a one-sided alternative. We can compute the actual/achieved power and compare to the preset power. For example, we decide to enroll a total of N patients and apply the rule LR_N; the true power is $1 - \Pr(N; \pi_A, LR_N)$, where $\Pr(N; \pi_A, LR_N)$ is the probability of reaching N patients without having stopped the trial.

Example 7.13 Suppose the rule is $LR_N = \{-, n(2), n(3), N\}$ and let u, v, and w be the numbers of adverse events that occur in each of the three segments of the trial $[0, n(2)]$, $[n(2), n(3)]$, and $[n(3), N]$. The probabilities for the three segments are $b[u; n(2), \pi_A]$, $b[v; n(3) - n(2), \pi_A]$, and $b[w; N - n(3), \pi_A]$, where $b[i; n, \pi_A]$ is the binomial probability of having exactly i events in n trials when the true rate is π_A. Reaching N patients without stopping the trial means that $u < 2$, $v < 3 - u$, and $w < 4 - (u + v)$. The true power is

$$1 - \Pr(N; \pi_A, LR_N) = 1 - \sum_{u=0}^{1} \sum_{v=0}^{2-u} \sum_{w=0}^{3-u-v} b[u;\ n(2), \pi_A] b[v;\ n(3) - n(2), \pi_A] \times b[w; N - n(3),\ \pi_A]$$

By a similar calculation, but replacing π_A by π_0, we can calculate and check for the "size" of the test (Type I error rate). For example,

$$1 - \Pr(N; \pi_0, LR_N) = 1 - \sum_{u=0}^{1} \sum_{v=0}^{2-u} \sum_{w=0}^{3-u-v} b[u; n(2), \pi_0] b[v; n(3) - n(2), \pi_0] \times b[w; N - n(3), \pi_0]$$

The problem of being underpowered is correctable; since the power falls short, the boundary needs to be pulled downward to retain the preset level. For example, with $\pi_0 = 3\%$ and $\pi_A = 15\%$, the stopping rule found for 80% power was $\{-, 5, 18, 31, \ldots\}$. The true power is only 74%; we need to stop, say, for the fourth event before $n(4) = 31$. But when or how much earlier?

Goldman (1987) described an algorithm for computing exact power (and Type I error rate). Goldman and Hannan (2001) proposed to repeatedly use that algorithm to search for a stopping rule that almost achieves the preset levels of Type I error rate and statistical power; they also provided a FORTRAN program allowing users to set their own size and power (and design parameters). The Goldman–Hannan algorithm works, but choosing one among many rules is sometimes not an easy job; several found could be "odd." The gain may be small; it is true that the power falls short without a correction, but it's only a few percentage points. It does not solve the perceived problem that the observed rate may be below the preset ceiling rate. Perhaps it would be more simple just to set the power higher, say, 85% when we want 80%.

The Bayesian Approach

Consider a binomial distribution $B(n, \pi)$. If we assume that the probability π has a prior distribution, say, Beta(α, β), then after e adverse events having been observed, the posterior distribution of π becomes Beta $(\alpha + e, \beta + n - e)$. From this, we have

$$\begin{aligned} P(\pi_*) &= \Pr(\pi > \pi_*) \\ &= 1 - \int_0^{\pi_*} \frac{\Gamma(n+\alpha+\beta)}{\Gamma(\alpha+e)\Gamma(\beta+n-e)} y^{e+\alpha-1}(1-y)^{n-e+\beta-1} dy \end{aligned}$$

Mehta and Cain's Rule By assuming a uniform prior (where $\alpha = \beta = 1$), Mehta and Cain (1984) provided a simple formula:

$$\begin{aligned} P(\pi_*) &= \Pr(\pi > \pi_*) \\ &= 1 - \int_0^{\pi_*} \frac{\Gamma(n+\alpha+\beta)}{\Gamma(\alpha+e)\Gamma(\beta+n-e)} y^{e+\alpha-1}(1-y)^{n-e+\beta-1} dy \\ &= \sum_{i=0}^{e} b[i; n+1, \pi_*] \end{aligned}$$

They then proposed a rule for which the trial is stopped when $P(\pi_0)$ is large, say, exceeding 97%, where π_0 is the baseline side-effect rate.

■ **Example 7.14** We have approximately

$$\begin{aligned} 0.97 &= \binom{n(1)+1}{0} \pi_0^0 (1-\pi_0)^{n(1)+1} + \binom{n(1)+1}{1} \pi_0^1 (1-\pi_0)^{n(1)} \\ 0.97 &= (1-\pi_0)^{n(1)+1} + \{n(1)+1\}\pi_0(1-\pi_0)^{n(1)} \\ n(1) &\cong 8 \text{ when } p_0 = 0.03 \end{aligned}$$

Similarly,

$$0.97 = \binom{n(2)+1}{0}\pi_0^0(1-\pi_0)^{n(2)+1} + \binom{n(2)+1}{1}\pi_0^1(1-\pi_0)^{n(2)}$$
$$+ \binom{n(2)}{2}\pi_0^2(1-\pi_0)^{n(2)-2}$$
$$0.97 = (1-\pi_0)^{n(2)+1} + \{n(2)+1\}\pi_0(1-\pi_0)^{n(2)}$$
$$+ \frac{[n(2)+1]n(2)}{2}\pi_0^2(1-\pi_0)^{n(2)-1}$$
$$n(2) \cong 21 \text{ when } p_0 = 0.03$$

By applying the Mehta–Cain Bayesian rule, we come up with pairs of numbers $[e, n(e)]$; it works just like the stopping rule obtained from the hypothesis testing-based approach. The major difference is that this Bayesian rule does not require the setting of a ceiling rate. At first it appears reasonable: if the usual normal rate is π_0, then the trial should be stopped when this rate is exceeded because the rate is no longer "normal."

■ **Example 7.15** With $\pi_0 = 0.03$ or 3%, the Mehta–Cain rule yields the stopping rule $\{8, 21, 38, \ldots\}$; that is, to stop at 1 event out of 8 patients, 2 events out of 21 patients, 3 events out of 38 patients, and so on. As a comparison, with $\pi_0 = 3\%$ and $\pi_A = 15\%$, the test-based stopping rule found for 80% power was $\{-, 5, 18, 31, \ldots\}$; that is, to stop at 2 events out of 5 patients, 3 events out of 18 patients, 4 events out of 38 patients, and so on.

Goldman (1987), after consulting her collaborators/clinicians, concluded that even though the Mehta—Cain Bayesian boundaries are philosophically very attractive but rather liberal, especially since it allows for the stopping of a trial after a single event. In fact, it seems too aggressive to trial simply because $\pi > \pi_0$; say, when $\pi_0 = 3\%$ and $\pi = 3.5\%$, because patients benefit from the treatment as well. To overcome having an overaggressive Bayesian rule, Goldman (1987) considered raising the cut-point 0.97 for the posterior probability or formulating a rule using $P(\pi_A)$ instead of $P(\pi_0)$, where π_A is the ceiling or maximum tolerated rate. For example, the trial is stop when $P(\pi_A)$ is large, say, exceeding 95% or 97%. However, she concluded that "various adjustments did not seem to remedy the problem." It is true that setting a stopping rule based on large values of $P(\pi_0)$, say, when $\pi_0 = 3\%$ and $\pi_A = 3.5\%$, may be too aggressive; the increase in the rate may not be large enough to be clinically significant (or to outweigh the benefits of the treatment). On the other hand, setting a stopping rule based on large values of $P(\pi_A)$ alone seems "unsettling" because it ignores the baseline rate and never reveals the impact of the treatment on having side effects. It is true that setting a ceiling rate is always subjective; but by seeing both π_0 and π_A, one would know how reasonable the parameters are. To have a fair comparison with the corresponding hypothesis-based stopping rule, perhaps we should stop the trial based on large values of $P(\pi_A)$, say, the trial is stop when $P(\pi_A)$ is large, say, exceeding 80% or 90%—whatever the number usually used as the preset value for statistical power—not 97%. But this would make the resulting Bayesian rule even more aggressive! The problem was the choice of the *prior*. With the uniform prior (where $\alpha = \beta = 1$), the mean is 0.5; we really need some prior distribution with an expected value more in line with the concept of "rare" side effects. Usually, in Bayesian analysis, the choice of the prior carries only moderate weight—sometimes not that important, a non-informative prior does the job. But here we conduct mostly very small trials, and using a sequential rule, it carries very heavy weight. For example, if we observe 3 events from 7 patients then (i) the posterior mean is still 0.5 (4/8) (leaning to stopping) with choice $\alpha = \beta = 1$, but (ii) the posterior mean is 0.1 (4/40) (leaning to nonstopping) with choice $\alpha = 1$ and $\beta = 32$.

A Revised Bayesian Rule There seems no perfect choice for a prior distribution. Uniform prior may be popular but it is biased toward stopping (its mean is 0.5), and the resulting rule may be too aggressive. Perhaps one should choose $(\alpha + \beta)$ small (e.g., take $\alpha = 1$), but it is not easy to set the mean $\alpha/(\alpha + \beta)$.

(i) Setting $\alpha/(\alpha + \beta) = \pi_A$ may also be somewhat biased toward stopping—unless we want to be more cautious.

(ii) Setting $\alpha/(\alpha + \beta) = \pi_0$ may be biased toward nonstopping; perhaps this is the right choice when an investigator believes that the treatment is safe.

Suppose we choose, for example, as prior, $\alpha = 1$ and $\beta = m$ (e.g., $m = 32$ so that the prior mean is $\alpha/(\alpha + \beta) = \pi_0 = 0.03$). A revised rule could be formed using

$$\begin{aligned} P(\pi_*) &= \Pr(\pi > \pi_*) \\ &= 1 - \int_0^{\pi_*} \frac{\Gamma(n+\alpha+\beta)}{\Gamma(\alpha+e)\Gamma(\beta+n-e)} y^{e+\alpha-1}(1-y)^{n-e+\beta-1} dy \\ &= \sum_{i=0}^{e} b[i; n+m, \pi_*] \end{aligned}$$

That is, the trial is stopped when $P(\pi_A)$ is large, say, exceeding 80% or 90%; π_A being the ceiling side-effect rate.

■ **Example 7.16** (i) If we choose $m = 32$ so that the prior mean is $\alpha/(\alpha + \beta) = \pi_0 = 0.03$, then

$$0.80 = \binom{n(1)+32}{0} \pi_A^0 (1-\pi_A)^{n(1)+32-0} + \binom{n(1)+32}{1} \pi_A^1 (1-\pi_A)^{n(1)+32-1}$$

$$0.80 = (1-\pi_A)^{n(1)+32} + [n(1)+32]\pi_A(1-\pi_A)^{n(1)+31}$$

$n(1)$ is negative, no stopping—just as in the test-based rule

This choice would result in a rule that is even more conservative the test-based one.

(ii) If we choose $m = 6$ so that the prior mean is $\alpha/(\alpha + \beta) = \pi_A = 0.15$, then

$$0.80 = \binom{n(1)+6}{0}\pi_A^0(1-\pi_A)^{n(1)+6-0} + \binom{n(1)+6}{1}\pi_A^1(1-\pi_A)^{n(1)+6-1}$$

$$0.80 = (1-\pi_A)^{n(1)+6} + [n(1)+6]\pi_A(1-\pi_A)^{n(1)+5}$$

$n(1)$ is still negative, no stopping

This choice would result in a rule that is closer to the hypothesis test-based one.

After a rule is formed, including the Bayesian rule, we can always calculate its Type I error rate and check to see if it is overaggressive:

$$1-\Pr(N; \pi_0, Rule)$$

7.4.5 Multiple Decisions

Phase II clinical trials are undertaken, in addition to patient care, to assess the antitumor therapeutic efficacy of a specific treatment regimen (investigational drug). They are the first human trials looking at *efficacy*; the *dose* used in a Phase II trial is the MTD found in previous Phase I trials. Primary outcome used in Phase II clinical trials is often the *response*, which is defined as having a 50% decrease in tumor size, for example, lasting for 4 weeks. The analysis of the resulting binary data is simply based on the *response rate*. The main analysis is often a simple *interval estimation* (95% confidence interval of response rate).

These trials are primarily designed to lead to a recommendation of whether or not the treatment regimen deserves further investigation (in future Phase III clinical trials—which are often in combination with other drugs). The agent or drug being tested is recommended if the response is "high" and not recommended if the response is "low." The definition of "high" and low" is always subjective but, generally, a response of 10% or less is considered "poor or low" and a response of 20% or higher is considered "promising or high." A response of 50% or more is considered very high, a great success—a rare finding.

Early Termination and Multiple Decision

Typically, a Phase II clinical trial is designed to have one sample, and a single stage in which n evaluable patients accrue, are treated, and are then observed for possible response. There are also some two-arm randomized Phase II trials. Recommendation is made at the end of the trial. However, for ethical reasons, the conduct of the trial sometimes should allow for early termination if initial/early results are extreme. If the early estimate of response is "low," say, 10% or less, the trial should be stopped so (next) patients could get better treatment (agents being tested are these patients' last hope); if the early estimate of response is "very high," say, 50–60%, the trial should be stopped so as to proceed to Phase III trial faster (more patients could benefit from this good/effective treatment).

The termination of a Phase II trial involving a poor agent, due to its low response rate, can be accomplished by a proper study design—for example, the popular Simon's two-phase design. If the early estimate of response is "very high," the trial should also be stopped so as to proceed to Phase III trial faster. But this practice of early termination cannot be achieved by a design. Therefore the current conventional approach is to achieve early termination for poor efficacy by study design and early termination for excellent efficacy by data analysis. One can also handle both types of early termination in the same trial by data analysis but it's more complicated and a bit confusing. In order to reach a decision "early", we would need to analyze data more than once; at least once before the (planned) end of the study. Each analysis leads to a decision by a statistical test of significance. For binary outcomes, such as response, the test is chi-squared; and one can apply the statistical test once or more than once. Of course, we are all aware of the multiple-decision problem with its possible inflation of Type I errors.

Concerns in Multiple Decisions

The central objective is to essentially preserve the *size*, the *power* (involved in the decision to recommend or not to recommend further investigation), and—as much as possible—the *simplicity* of a single-stage procedure. The most obvious/serious concern is the size. To perform many tests increases the probability that one or more of the comparisons will result in a Type I error (test is significant but null hypothesis is true); for example, suppose the null hypothesis is true and we perform 100 tests—each has a 0.05 probability of resulting in a Type I error. Then 5 of these 100 tests would be wrongly statistically significant simply as the result of Type I errors (false positives). We do not do any test 100 times, but every time we do it more than once, the probability of Type I errors exceeds 0.05. We should be prepared to preserve this size—to keep the probability of Type I errors at or below a preset value, say, 5%. One can think of a simple adjustment like Bonferroni, by dividing 0.05 by the number of tests. But that is not optimal and, in addition, (i) in a two-arm trial, the most important test is still the last one (when we have more data), and (ii) if you do the test often enough, you would not be able to prove anything.

Randomized Two-Arm Trials

There are more one-arm, open-label Phase II trials but it is easier to handle multiple decisions in two-arm trials. Let us start with those. The usual settings for randomized two-arm Phase II clinical trials are:

(i) Response is dichotomous and immediate.
(ii) They are single-phase trials, with sample sizes fixed in advance.

(iii) At the end of a trial, compare success rates (i.e., proportions) using a formal test of significance based on the usual Pearson, is chi-squared test.

The usual concern is ethical—the possibility of early termination of the study should early results indicate a marked superiority of one treatment over the other treatment. Therefore we want to build in a provision for multiple decisions. However, in common applications, results indicate a marked superiority of a new treatment over the placebo; the trial is often stopped so future patients can get this better treatment. Whatever the underlying rationale, the aim is to form a multiple testing procedure that provides investigators with an opportunity to conduct periodic reviews of the data as they accumulate and thereby offers the chance for early termination should one treatment prove superior to the other early on while continuing to use essentially the single-phase decision rule should early termination not occur. The following is a brief description of the O'Brien–Flemming procedure:

(i) Investigators plan to test k times, including the final comparison at the end of the trial.

(ii) Data are reviewed periodically, with m_1 subjects receiving treatment 1 and m_2 subjects receiving treatment 2, between successive tests; there are a total of $(k)(m_1 + m_2)$ subjects.

(iii) The constraint is to maintain an overall size α, say, $\alpha = 0.05$.

(iv) *Rule:* After the nth test, $1 \leq n \leq k$, the study is terminated and H_0 is rejected if

$$(n/k)X^2 \geq P(k, \alpha)$$

where X^2 is the usual Pearson's chi-squared statistic.

Using the theory of Brownian motion, O'Brien and Fleming (1979) obtained the values for $P(k,\alpha)$ but, more importantly, they concluded that they are approximately the $(1 - \alpha)$th percentile of the chi-squared distribution with 1 degree of freedom—almost independent of k.

■ **Example 7.17** Let us take $k = 2$ (one "interim" analysis and one "final" analysis) and $\alpha = 0.05$. Then $P(2,\ 0.05) = 3.928$ ("exact"). The O'Brien–Flemming rule is:

(i) Reject the null hypothesis after the interim analysis if $X^2 \geq (2)(3.928) = 7.86$, which is equivalent to having the p-value less than 0.005.

(ii) Reject the null hypothesis after the final analysis if $X^2 \geq (1)(3.928)$, which is equivalent to having the p-value less than 0.045.

The O'Brien–Flemming rule can be implemented as follows:

1. Calculate the cut-point for the p-value for the interim analyses (in application of this rule, one would assign 0.5% to the interim analysis) and subtract them

out of the planned size (say, 5%) to obtain the cut-point for the p-value for the final analysis.

2. Use an approximate value (the 95th percentile of the chi-squared distribution with 1 degree of freedom, i.e., 3.84) instead of $P(k,\alpha)$ (i.e., 3.928).
3. In the example, we can use the usual chi-squared tests at 0.5% and 4.5%, respectively.

■ **Example 7.18** Let us take $N = 3$ (two interim analyses and one final analysis) and $\alpha = 0.05$. The rule would be to reject the null hypothesis after the first interim analysis if $X^2 \geq (3)(3.84) = 11.52$, which is equivalent to having the p-value less than 0.001; and to reject the null hypothesis after the second analysis if $X_2 \geq (3/2)(3.84) = 5.76$, which is equivalent to having the p-value less than 0.014. In application of this rule, one would assign 0.1% and 1.4% to the interim analyses and 3.5% to the final analysis for an overall 5% size.

The overall observation is that if you want to look very often ($k \geq 3$), early inspections (say, first or second in a series of 4 or 5) are very unlikely to lead to rejection. For larger N, the threshold for the p-value will be reduced substantially. For example, if $N = 5$, we reject the null hypothesis after the first interim analysis if

$$X^2 \geq (5)(3.84) = 19.20$$

which is equivalent to having the p-value less than 0.00002—almost impossible with only 1/5th of the planned sample size. The problem and the rule were formulated assigning constant $(m_1 + m_2)$ subjects accrued between successive periodic tests. However, the O'Brien–Fleming simulations showed that their conclusions and results virtually remain valid in more general settings. In other words, the only major factors affecting the rule are the number of tests N and the overall size α. The number of tests k affects the rule but not $P(k, \alpha)$; in practice, most use $k = 2$ or 3.

One-Arm Nonrandomized Trials

Multiple decisions are more popular for two-arm randomized trials but it is possible to make multiple decisions in one-arm, open-label trials too. The idea sounds kind of "new" but the methods are not that new; we just form it as an extension of a one-sample test.

The usual settings for randomized two-arm Phase II clinical trials are as follows:

(i) Assume that a Phase I trial has been completed from which an MTD has been established.

(ii) The investigator is first asked to specify the largest response rate, π_0, which if true the treatment does not warrant further investigation.

(iii) The investigator is next asked to judge the smallest response rate, π_A, which if true would imply that the treatment has adequate therapeutic efficacy to warrant further investigation. It seems it would be easier to understand if we set $\pi_0 = \pi_A$; however, the "gap" would also allow for consideration of other factors: cost, ease of application safety, and so on. It is also similar to the case of monitoring for toxicity or adverse effects (where π_0 is the baseline rate but the study is only stopped if the toxicity rate exceeds π_A).

After acquiring the needed components, we state the (one-sided) hypotheses to be tested as

$$H_0: \ \pi = \pi_0 \quad \text{versus} \quad H_A: \ \pi = \pi_A$$

In addition, we should specify the size (α) and the power ($1 - \beta$) from which we determine the sample size n.

Single-Stage Decision The number of responses x is distributed as a binomial $\text{Bin}(n, \pi)$; but if $n\pi = 10$, say, the distribution of Y is approximately normal:

$$Y(\pi) = \frac{x - n\pi}{\sqrt{n\pi(1-\pi)}}$$

Using the normal approximation, the null hypothesis H_0 is rejected when

$$\begin{aligned} Y(\pi_0) &= \frac{x - n\pi_0}{\sqrt{n\pi_0(1-\pi_0)}} > z_{1-\alpha} \\ x > x_r &= n\pi_0 + z_{1-\alpha}\sqrt{n\pi_0(1-\pi_0)} \end{aligned}$$

The rejection rule (rejecting H_0 when $y(\pi_0) > z_{1-\alpha}$) preserves the size α but may lead to a power somewhat less than $(1 - \beta)$; we therefore need some "continuity correction." In order to maintain the power (may result from the normal approximation), the single-stage procedure should more appropriately reject H_0 whenever

$$x \geq x_r = \left[n\pi_0 + z_{1-\alpha}\sqrt{n\pi_0(1-\pi_0)}\right]^* + 1$$

where x^* denotes the nearest integer to x (round up and add 1). It should be noted that, in applying this rejection rule, we do not need to specify the power; however, if statistical power is specified, we can determine sample size:

$$n = \left\{ \frac{z_{1-\beta}\sqrt{\pi_A(1-\pi_A)} + z_{1-\alpha}\sqrt{\pi_0(1-\pi_0)}}{(\pi_A - \pi_0)} \right\}^2$$

It is relatively simple to specify the size α and π_0, but more difficult to specify π_A. For example, one can take as π_0 the response rate of the standard treatment; then, with a practical estimate of n, we can see what it can do.

Flemming's Result Instead of independent specification of power $(1-\beta)$ and alternative response rate π_A, Fleming (1982) proves that, for a given n, the usual single-stage procedure, where one rejects π_0 if $y(\pi_0) > z_{1-\alpha}$, has power $(1-\alpha)$ against the alternative rate:

$$\pi_A^+ = \frac{\{\sqrt{n\pi_0} + z_{1-\alpha}\sqrt{1-\pi_0}\}^2}{n + z_{1-\alpha}^2}$$

■ **Example 7.19** (i) Let us take $\pi_0 = 0.20$, $\alpha = 0.05$, and $n = 20$. Then the rejection rule, $y(\pi_0) > z_{1-\alpha}$, has power 95% against the alternative rate:

$$\begin{aligned}\pi_A^+ &= \frac{\{\sqrt{n\pi_0} + z_{1-\alpha}\sqrt{1-\pi_0}\}^2}{n + z_{1-\alpha}^2} \\ &= \frac{\{\sqrt{(20)(0.2)} + (1.65)\sqrt{0.8}\}^2}{20 + (1.65)^2} \\ &= 0.53\end{aligned}$$

(ii) Let us take $\pi_0 = 0.20$, $\pi_A = 0.40$, and $n = 25$. The null hypothesis is rejected if there are 9 responses or more:

$$\begin{aligned}x_r &= [n\pi_0 + z_{1-\alpha}\sqrt{n\pi_0(1-\pi_0)}]^* + 1 \\ &= [(25)(0.2) + (1.65)\sqrt{(25)(0.2)(1-0.2)}]^* + 1 = 9\end{aligned}$$

(iii) If we take $\pi_0 = 0.25$, $\pi_A = 0.50$, and $n = 25$, the null hypothesis is rejected if there are 11 responses or more.

In a single-stage plan, the decision is made at the end; if the null hypothesis is rejected (i.e., $x = x_r$), the agent/drug is recommended for further investigation.

Suppose one decides to perform N tests (usually $N = 2$ or 3) and to allow n_i patients to accrue between the $(I-1)$th and ith tests, so that $n = n_1 + n_2 + \ldots + n_k$. Let $x_1, x_2, \ldots, x_k$ represent the number of responses among the n_1, $n_2, \ldots, n_k$ evaluable patients (so that $x = x_1 + x_2 + \ldots + x_k$). In addition, denote the set of (cumulative) rejection points (of H_0) by $\{x_{r1}, x_{r2}, \ldots, x_{rk}\}$, where x_{ri} is determined after i tests, $(n_1 + n_2 + \ldots + n_i)$:

$$x_{ri} = \left[\left(\sum_{j=1}^{i} n_j\right)\pi_0 + z_{1-\alpha}\sqrt{\left(\sum_{j=1}^{i} n_j\right)\pi_0(1-\pi_0)}\right]^* + 1$$

Schultz's Rule After test #n, $1 \le n \le N$, stop and reject H_0 if

$$\sum_{i=1}^{j} x_i > x_{rj}$$

■ **Example 7.20** Let us take $\pi_0 = 0.20$, $\pi_A = 0.40$, and $n = 30$, and decide to use $k = 3$ (2 interim analyses).

(i) After the first 10 patients: $x_r = 4$; reject H_0 and stop if 4 responses or more.
(ii) After the first 20 patients: $x_r = 7$; reject H_0 and stop if a total of 7 responses or more.
(iii) After all 30 patients: $x_r = 10$; reject H_0 and stop if a total of 10 responses or more.

Flemming's Rule Fleming (1982) observed that the true significance level of a multiple testing procedure can be considerably higher than the nominal level. To preserve the nominal significance level in multiple testing, Fleming proposed to "inflate" the variance used in determining the rejection points from

$$x_{ri} = \left[\left(\sum_{j=1}^{i} n_j \right) \pi_0 + z_{1-\alpha} \sqrt{\left(\sum_{j=1}^{i} n_j \right) \pi_0 (1 - \pi_0)} \right]^* + 1$$

to

$$x_{ri} = \left[\left(\sum_{j=1}^{i} n_j \right) \pi_0 + z_{1-\alpha} \sqrt{n \pi_0 (1 - \pi_0)} \right]^* + 1$$

■ **Example 7.21** Let us take $\pi_0 = 0.20$, $\pi_A = 0.40$, and $n = 30$, and decide to use $k = 3$ (2 interim analyses).

(i) After the first 10 patients: $x_r = 4$ by Schultz's rule and $x_r = 6$ by Fleming's rule.
(ii) After the first 20 patients: $x_r = 7$ by Schultz's rule and $x_r = 8$ by Fleming's rule.
(iii) After all 30 patients: $x_r = 10$ unchanged.

EXERCISES

7.1 When a Phase I cancer trial following the standard design reaches a dose level with a toxicity rate of 40%, what is the probability that it would pass to the next higher dose?

7.2 Suppose we consider conducting a Phase I trial using the standard design with only three prespecified doses. Suppose further that the toxicity rates of these three doses are 10%, 20%, and 30%, respectively. What is the probability that the middle dose would be selected as the MTD?

7.3 Suppose we consider conducting a Phase I trial using the standard design with only three prespecified doses. Suppose further that the toxicity rates of these three doses are 40%, 50%, and 60%, respectively. What is the probability that the first dose would be selected as the MTD? What is the probability that the last dose would be selected as the MTD?

7.4 Consider a Phase I cancer trial with three doses having toxicity rates of 15%, 35%, and 55%. In each of the two designs, standard and fast track, calculate the expected toxicity rate and the expected trial size. How many subjects are expected to be treated at the last dose?

7.5 We can refer to the standard design as a 3-and-3 design because, at each new dose, it enrolls a cohort of three patients with the option of enrolling an additional three patients evaluated at the same dose. Describe the dose-escalation plan for a 3-and-2 design and describe its possible effects on the toxicity rate of the resulting maximum tolerated dose (MTD).

7.6 The status of the maxillary lymp node basin is the most powerful predictor of long-term survival in patients with breast cancer. The pathologic analysis of the maxillary nodes also provides essential information used to determine the administration of adjuvant therapies. Until recently, a maxillary lymp node dissection (ALND) was the standard surgical procedure to identify nodal metastases. However, ALND is associated with numerous side effects including arm numbness and pain, infection, and lymphedema. A new procedure, sentinal lymp node (SLN) biopsy, has been proposed as a substitute and it has been reported with a successful identification rate of about 90%. Suppose we want to conduct a study in order to estimate and confirm this rate to identify nodal metastases among breast cancer patients because previous estimates were all based on rather small samples. How many patients are needed to confirm this 80% success rate with a margin of error of 10%? Does the answer change if we do not trust the 90% figure?

7.7 Metastasic melanoma and renal cell carcinoma are incurable malignancies with a median survival time of less than a year. Although these malignancies are refractory to most chemotherapy drugs, their growth may be regulated by immune mechanisms and there are various strategies for development and administration of tumor vaccines. An investigator considers conducting a Phase II trial for such a vaccine for patients with stage IV melanoma. How many patients are needed to estimate the response rate with a margin of error of 15%?

7.8 Suppose we consider conducting a study to evaluate the efficacy of prolonged infusional paclitaxel (96-hour continuous infusion) in patients with recurrent or metastatic squamous carcinoma of the head and neck. How many patients are needed to estimate the response rate with a margin of error of 20%?

7.9 Suppose in planning for a Phase II trial, an investigator believes that the incidence of severe side effects is about 5% and the trial has to be stopped if the incidence of severe side effects exceeds 20%. Preset the level of significance at 0.05; design a stopping rule that has 80% power.

7.10 Among the ovarian cancer patients treated with cisplatin, it is anticipated that 20% will experience either partial or complete response. If adding paclitaxel to this regimen can increase the response by 15% (to 35%) without undue toxicity, then that would be considered as clinically significant. Calculate the total sample size needed for a randomized trial that would have a 80% chance of detecting this magnitude of a treatment difference while probability of Type I error for a two-sided test is preset at 0.05.

7.11 Metastatic breast cancer is a leading cause of cancer-related mortality with no major change in the mortality rate over the past few decades. Therapeutic options are available with active drugs such as paclitaxel. However, a promising response rate is also accompanied by a high incidence of toxicities—especially neurotoxicity. An investigator considers testing a new agent, which may provide significant prevention, reduction, or mitigation of drug-related toxicity. This new agent is to be tested against placebo in a double-blind, randomized trial among patients with mestastatic breast cancer who receive weekly paclitaxel. The rate of neurotoxicity over the period of the trial is estimated to be about 40% in the placebo group and the hypothesis is that this new agent lowers this toxicity rate by one half to 20%. Find the total sample size needed using a two-sided level of significance of 0.05 if the hypothesis would be detected with a power of 80%.

7.12 In the above study on metastatic breast cancer, the investigator also focuses on tumor response rate, hoping to show that this rate is comparable in the two treatment groups. The hypothesis is that the addition of the new agent to a weekly paclitaxel regimen would reduce the incidence of neurotoxicity without a compromise of its efficacy. At the present time, it is estimated that the tumor response rate for the placebo group, without the new agent added, is about 70%. Assuming the same response rate for the treated patients, find the margin of error of its estimate using the sample size obtained from Exercise 7.11.

7.13 For the case of "design for selection" with two arms of equal size (n per arm), derive the formula for n in the special case of $\rho = 0$.

7.14 From the result in Exercise 7.13, show that:

(a) For fixed n, the probability of correct selection increases as d increases.

(b) For fixed d, the sample size n is decreased as d decreases.

7.15 Suppose we are conducting a small Phase II trial with $N = 25$ patients. We wish to form a sequential stopping rule with these two parameters: $\pi_0 = 0.05$ and $\pi_A = 0.20$.

(a) Form a rule by applying the SPRT and calculate its power and its Type I error rate.

(b) Form a rule by applying the Mehta–Cain Bayesian rule and calculate Type I error rate.

(c) Form a Bayesian rule by choosing $\alpha = 1$ and $\beta = \text{m}$ so that $\alpha/(\alpha + \beta) = \pi_0 = 0.05$. Calculate the Type I error rate.

8

CATEGORICAL DATA AND DIAGNOSTIC MEDICINE

Diagnosis is defined as the act or process of identifying or determining the nature of a disease through examination, and *screening* is defined as the act or process of separating or sifting out by means of an appraisal or a selection. Are these different processes? Yes, screening is a population-based process (public health) whereas diagnosis is individually based (medicine). However, the only difference is not in the make-up of the processes but in their uses. For the purpose of learning biostatistics, we make no such distinction between the terms "diagnosis" and "screening"; differences, if any, are minor. From a statistical point of view, screening tests are diagnostic tests; they are procedures geared toward detecting a condition (e.g., disease). The context in which they are applied in practice sets them apart. Screening tests are applied on a large scale; therefore they must be noninvasive and inexpensive. A positive result is usually not followed by treatment, but with further, more definitive procedures; thus their accuracy can be somewhat less than perfect. As a generic terminology, we use the term "diagnosis" to aim at the act or process of predicting a not yet observable condition—such as a disease—using an observable characteristic or characteristics (clinical observations and/or laboratory test results), referred to as separators or predictors. The process leads to a quick, easy, and economical way to classify individuals as *diseased* (condition present) or *healthy* (condition absent). A few simple examples are the skin test for tuberculosis and urine tests for early pregnancy. Why does this topic of diagnostic medicine belong here in this book? The aim of the diagnosis, the unobservable condition, is represented by a binary random variable with two possible outcomes, presence or absence; resulting data are categorical data.

8.1 SOME EXAMPLES

In general, the following are criteria for a useful diagnostic test:

1. Disease should be serious or potentially so; if the disease has no serious consequences in terms of longevity or quality of life, there is no benefit to be gained from diagnosing it.
2. Disease should be relatively prevalent in the target population; otherwise, the benefit of the test, used as a screening too, is rather minimal.
3. Disease should be treatable.
4. Treatment should be available to those who test positive; sometimes treatment exists but is not accessible because of the cost or social reasons.
5. The test should not harm the individuals. Of course, all tests have negative impact above money and time; they may cause physical or emotional discomfort and sometimes are even dangerous to the well-being of the subject. The overriding principle is that these should be reasonable and not outweigh the potential benefits.
6. The test should accurately classify diseased and nondiseased individuals; false negatives leave diseased subjects untreated and false positives result in unnecessary treatments—both should be minimal.

The following are a few common examples in diagnostic medicine.

1. *Otitis Media.* Otitis media is considered the second most prevalent disease on Earth—only less prevalent than the common cold, affecting 90% of children by age 2. It cost $3.8 billion in direct costs (physician visits, tube placements, cost of antibiotics, etc.) in 1995 dollars. Otitis media can cause hearing loss, learning disabilities, and other middle-ear sequelae. As a children's disease, it is the most common diagnosis at physician visits ahead of well-child, URI, injury, and sore throat; it was responsible for 24.5 million physician visits in 1990. Otitis media is an inflammation of the middle-ear space, often referred to as ear infection; the inflammation is hidden behind the eardrum as seen in Figure 8.1.

Diagnosis of otitis media is usually made on clinical grounds—that is, using otoscopic characteristics of the tympanic membrane (color, position, appearance, and mobility)—or persistent earache or bubbles of air. Otoscopy is often supplemented by tympanometry, a method that measures the compliance of the tympanic membrane;

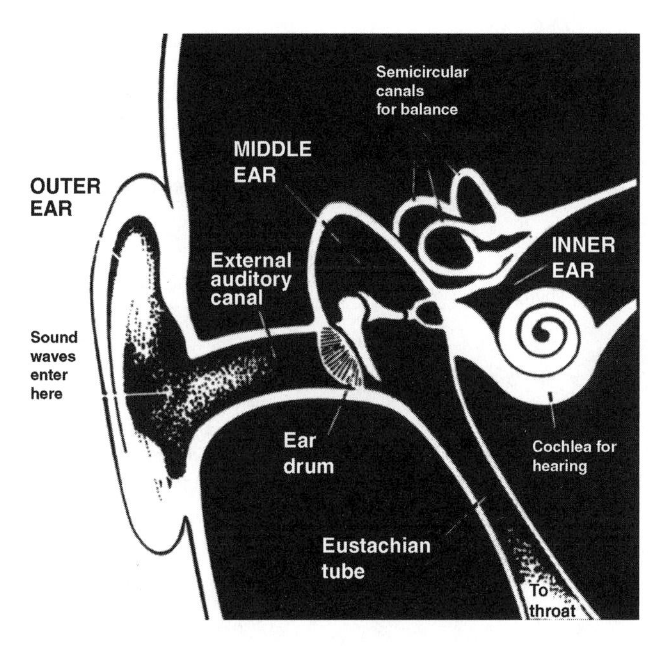

FIGURE 8.1. The middle-ear space.

two (continuous) characteristics have emerged as leading potential predictors of middle-ear fluid: the *static admittance* and the *tympanometric width*.

2. *Diabetes.* Diabetes is a disease in which the body is unable to properly use and store glucose (a form of sugar); glucose backs up into the bloodstream causing one's blood glucose to rise too high. There are two types of diabetes: type 1 (juvenile-onset or insulin-dependent—about 10% of all cases—in which the body stops producing insulin) and type 2 (adult-onset or non-insulin-dependent—about 90% of all cases—in which the body does not produce enough insulin or is unable to use insulin properly—called *insulin resistance*).

There may be symptoms (e.g., being very thirsty, blurry vision) or risk factors (e.g., family history, being overweight). Primary diagnosis, however, is based on plasma glucose often measured/tested early in the morning before eating meals (called *fasting plasma glucose*).

3. *Thyroid Diseases.* The thyroid is a small bowtie or butterfly-shaped gland, located in your neck, wrapped around the windpipe, behind and below the Adam's apple area. The thyroid, just below the voice box, has two parts, called lobes; the two lobes are separated by a thin section called the isthmus as seen in Figure 8.2.

The hypothalamus—part of the brain—releases thyroid-releasing hormone (TRH), which leads to the release of thyroid-stimulating hormone (TSH); THS tells the thyroid to make thyroid hormones and release them into the bloodstream as a feedback process. The thyroid produces several hormones, of which two are very important—one is tri-iodothyronine (called T3) and the other is thyroxin (called T4). These hormones help oxygen get into cells; the hormones then help cells to convert

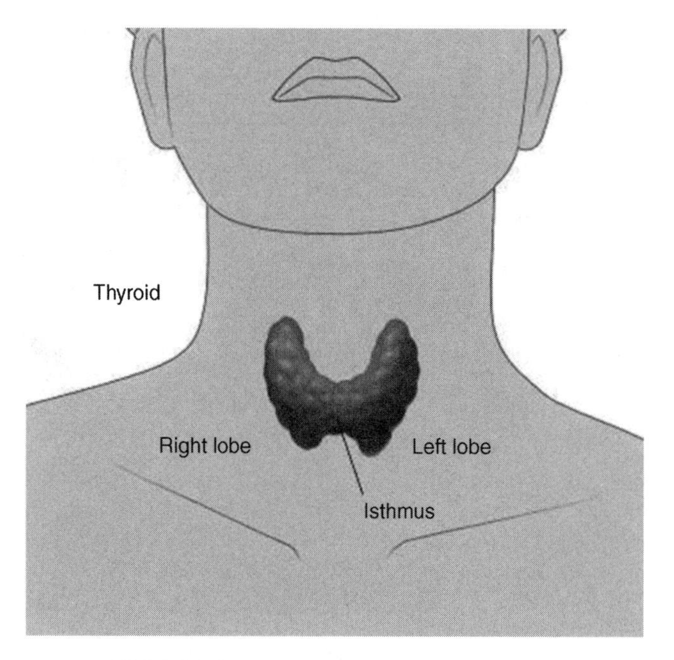

FIGURE 8.2. Location and shape of the thyroid.

oxygen and calories into energy, making the thyroid the "master gland of metabolism." There are hypothyroidism and hyperthyroidism; the latter is more consequential. When your thyroid starts producing too much thyroid hormones, your body goes into overdrive, causing an increased heart rate, increased blood pressure, and burning calories more quickly. These lead to weight loss, anxiety, muscle weakness, tremors, loss of concentration, and so on.

Like the case of diabetes, there are risk factors (family history, menopause—the disease is more prevalent among women, being exposed to radiation) and many more symptoms. Primary diagnosis, however, is based on a blood test to measure the levels of three major hormones: T3, T4, and TSH—all on a continuous scale.

4. *Prostate Cancer.* The prostate is part of a man's reproductive system. It is a gland surrounding the neck of the bladder and it contributes a secretion to the semen. A healthy prostate is about the size of a walnut and is shaped like a donut. The urethra (the tube through which urine flows) passes through the hole in the middle of that "donut" as seen in Figure 8.3.

If the prostate grows too large, it squeezes the urethra, causing a variety of urinary problems—with or without cancer. Cancer begins in cells, the building blocks of tissues. When the normal process goes wrong, new cells form unnecessarily and old cells do not die when they should. The result is an extra mass of cells called a tumor; and malignant tumors are cancer. No one knows the exact causes of prostate cancer . . . yet, but age is a significant factor. Most men with prostate cancer are over 65; if they live long enough a very large proportion of men would eventually have prostate cancer.

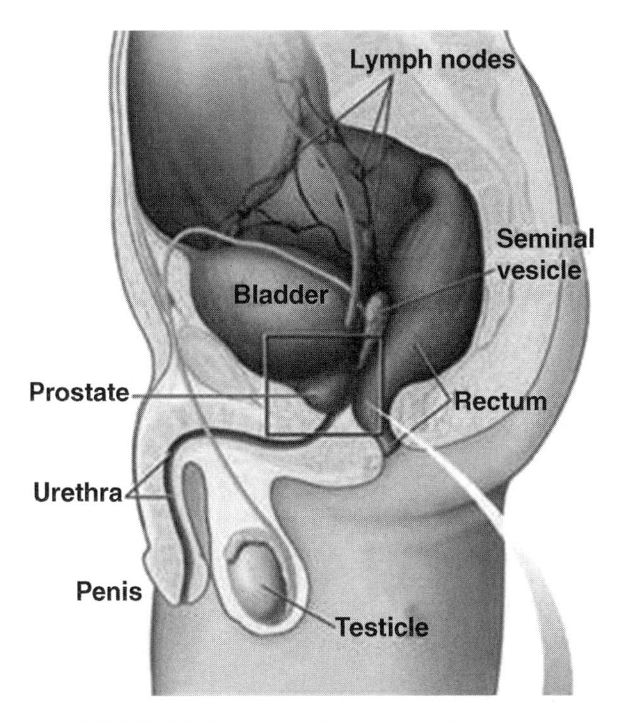

FIGURE 8.3. Prostate and the surrounding organs.

There are risk factors (e.g., age, family history) and symptoms (e.g., inability to urinate, frequent urination at night). Common screening is a blood test to measure prostate-specific antigen (PSA). However, a high level could be caused by benign prostatic hyperplasia (BPH—growth of benign cells); so the test is not specific. Another newer candidate is cathepsin B and related proteins.

5. *AIDS.* Acquired immunodeficiency syndrome (AIDS) is a severe manifestation of infection with the human immunodeficiency virus (HIV, identified in 1983). The virus destroys the immune system, leading to opportunistic infections of the lungs, brain, eyes, and other organs. Consequences include debilitating weight loss, diarrhea, and several forms of cancer. Currently, 40 million people are living with AIDS. In 2004, about 5 million were newly infected and 3 million died; the most affected region is Sub-Sahara Africa.

HIV infection is diagnosed by blood tests (e.g., using CD4 + T-cell count, or CD8 + T-cell count).

6. *Ulcers.* An ulcer is a break in the lining of the stomach or in the duodenum (first part of the small intestine). Gastric/peptic ulcers cost $3.2 billion in 1975 dollars. The two Australian physicians who discovered the responsible bacteria won the Nobel prize in 2005. Most ulcers are caused by *Helicobacter pylori* (identified in 1982), a bacterium living in the pylorus (the passage connecting the stomach and the duodenum) as seen in Figure 8.4

Ulcers are diagnosed by a blood test and even by a breath test.

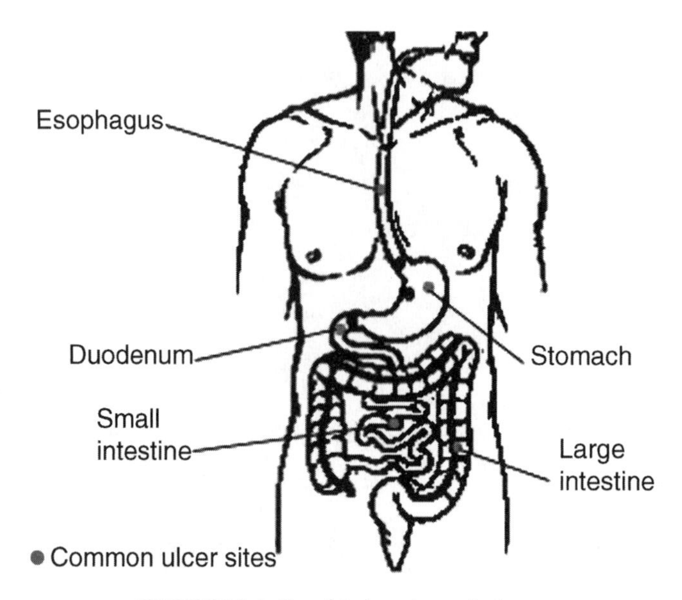

FIGURE 8.4. Possible locations of ulcers.

7. *Infections.* There are many bacterial and viral diseases such as ulcers and AIDS. Bacterial and viral diseases are often detected more easily and more accurately because (i) the diseases are better defined, and (ii) they are all implicated by a common marker: some form of agent-specific antibodies provided that we have the right assay.

8.2 THE DIAGNOSIS PROCESS

In general, a diagnosis starts with an "idea"; it could be accidental or the result of a long search. The idea then goes through a two-stage process

Stage 1: *Developmental Stage*. The question here is: Does the idea work?
Stage 2: *Applicational Stage*. The question here is: Does it work for me (i.e., the user)? or When does it work?

8.2.1 The Developmental Stage

In the developmental stage, the basic question is: Does the idea work? It's the investigator's (or producer's) burden to prove it does. The scientific approach is trying the test's idea on a "pilot population," where one compares the test results versus truth; the "true diagnosis" may be based on a more refined/accurate method or evidence that emerges after the passage of time. It should be noted that, at this stage and using this approach, we have data to "prove' what we want to prove.

Test Result

Some tests yield dichotomous results, such as the presence or absence of a bacterium or some specific DNA sequence; but many may involve assessments measured on an ordinal scale (e.g., 5-point scale for mammograms) or a continuous scale (e.g., blood glucose). Let us start with the simple binary case.

Parameters

Let D and T denote the true diagnosis and the test result, respectively, each with two possible outcomes, positive or presence (+) and negative or absence (−). The key parameters are two conditional probabilities:

$$\text{Sensitivity:} \quad S^{+} = \Pr(T = + | D = +)$$
$$\text{Specificity:} \quad S^{-} = \Pr(T = - | D = -)$$

Sensitivity is the probability of correctly identifying a diseased individual and specificity is the probability of correctly identifying a healthy individual.

If the "idea", in the developmental stage, was to classify people as "diseased" (condition present) or "healthy" (condition absent) based on a certain measurement on the continuous scale (from blood or urinary components), the basic question is "How high is high?" or "How low is low?" Consider the plausible model in Figure 8.5, where the measurement—also called the *separator X*—is assumed normally distributed with the same variance, but different means. It can be seen that no matter where you "cut," both errors would result: false positive and false negative (Figure 8.6).

The two types of errors, positive and negative, and the two basic parameters of the developmental stage, sensitivity and specificity, are related to each others as follows:

	Test = positive	Test = negative
Diseased	True positive	False negative
Healthy	False positive	True negative

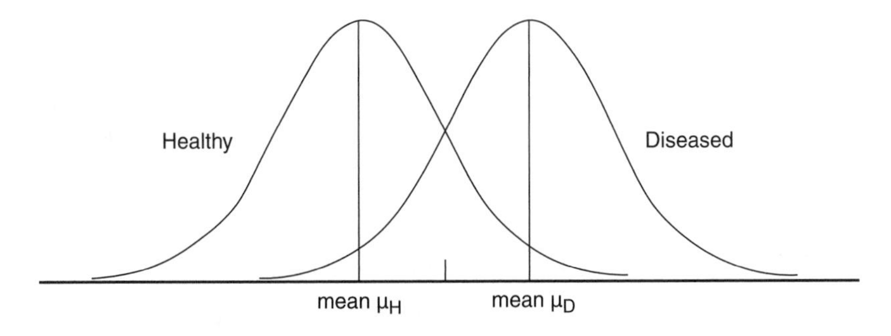

FIGURE 8.5. A plausible model for the distribution of separators.

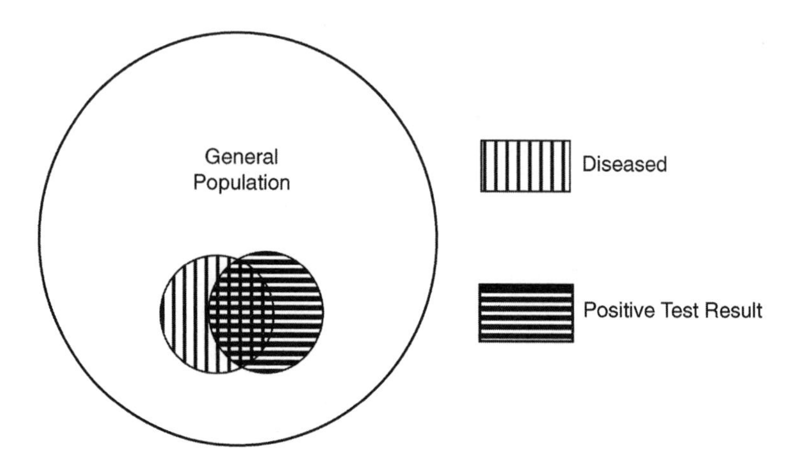

FIGURE 8.6. An illustration of misclassification.

- Sensitivity is (1 − false negativity); false negativity is the rate (or percentage, %) of false negatives.
- Specificity is (1 − false positivity); false positivity is the rate (or percentage %) of false positives.

Sensitivity and (1− Specificity) are also referred to as *true positive fraction/rate* and *false positive fraction/rate*, respectively. Clearly, it is desirable that a test or screening should be highly sensitive and highly specific; both error rates are small.

As compared to statistical tests of significance, the analogy can be expressed as follows:

- A false negative is a Type π error, $S^+ = 1 - \beta$.
- A false positive is a Type I error, $S^- = 1 - \alpha$.

Having a highly sensitive and highly specific diagnosis is equivalent to having both α and β—the most ideal hypothesis test.

Estimation of Parameters

Sensitivity and specificity can simply be estimated as proportions s^+ and s^- from the two samples: sensitivity is the proportion of diseased individuals detected as positive by the test; specificity is the proportion of healthy individuals detected as negative.

$$\text{Sensitivity} = \frac{\text{Number of diseased individuals who screen positive}}{\text{Total number of diseased individuals}}$$

$$\text{Sensitivity} = \frac{\text{Number of healthy individuals who screen negative}}{\text{Total number of healthy individuals}}$$

Standard errors and 95% confidence intervals, for example, are calculated accordingly.

■ **Example 8.1** The Pap test is highly specific (Specificity = 98.5%) but not very sensitive (Sensitivity = 40.6%). If a healthy person is tested, the result is almost sure negative; but if a woman with cancer is tested the chance is 59.4% that the disease is undetected.

	Test		
True	−	A +	Totals
−	23,362	362	23,724
+	225	154	379

$$\text{Sensitivity} = \frac{154}{379} = 0.406 \text{ or } 4.06\%$$

$$\text{Specificity} = \frac{23362}{23724} = 0.985 \text{ or } 98.5\%$$

The Pap test or Pap smear test (or cytology test) is an important part of women's health care. The smeared cells or cell suspension is placed on a glass slide, stained with a special dye (Pap stain), and viewed under a microscope. It can also be used to detect vaginal or uterine infections and has been substantially improved in recent years. It may not be sensitive due to cases in the early stage; there are a few abnormal cells among many thousands of cells—and it cannot detect precancerous lesions. The Pap test is rather ineffective in detecting cervical cancer, which can be diagnosed by a blood test to measure a protein called CA-125.

■ **Example 8.2** Thermography (thermal imaging) is a way of measuring and mapping the heat from the breast with the use of a special camera, based on the idea that changes in temperature may indicate disease. While thermography is approved by the FDA as safe, it is not approved as a standalone screening test for breast cancer. According to the American Cancer Society (ACS), "studies have not proven this to be an effective screening tool for early detection of breast cancer and it is not a replacement for mammograms. It is not considered a reliable detection test since it can miss some cancers and can give a high false positive rate." (A newer version of this test, known as computerized thermal imaging (CTI), is now being studied.) The following are some historical data.

In a study of thermography, used to detect breast cancer in the early stages, Moskowitz et al. (1976) had experienced thermographers read films from people with and without breast cancer.

	Thermography		
Cancer	Positive	Negative	Unreadable
Yes	181	239	42
No	233	302	70

Discarding unreadable films, we have

$$s^+ = \frac{181}{181+239} = 0.43$$
$$s^- = \frac{302}{302+233} = 0.56$$

Diagnostic Likelihood Ratios

Besides positive and negative predictive values, P^+ and P^-, likelihood ratios LR^+ and LR^- are sometimes used to describe the diagnostic value of a test. These parameters are the ratios of the test result in the diseased versus nondiseased populations:

$$LR^+ = \frac{\Pr(T = +|D = +)}{\Pr(T = +|D = -)} = \frac{S^+}{1-S^-}$$
$$LR^- = \frac{\Pr(T = -|D = +)}{\Pr(T = -|D = -)} = \frac{1-S^+}{S^-}$$

In a sense, the likelihood ratios are more like the sensitivity and specificity; they do not lead directly to prediction because they do not depend on prevalence. On the other hand, they can be considered as at a higher stage because they have *predictive quality* and they quantify the increase in knowledge about the presence of disease that is gained through the diagnostic test through the pretest and post-test odds.

8.2.2 The Applicational Stage

In the applicational stage, when the product is on the market, the basic question is: Does it work for "me" (i.e., the user)? or When does it work?). It's the user's or consumer's concern. The problem is that one can't resolve the concern, like comparing the test result versus the truth, because if one knows the truth one would not need the test; and waiting for evidence to emerge after some passage of time may be "too late."

Parameters

Again, let D and T denote the true diagnosis and the test result, respectively—each with two possible outcomes, positive or presence (+) and negative or absence (−); D is unknown at this stage. The key parameters are two other conditional probabilities: positive predictive value (denoted P^+) and negative predictive value (denoted P^-):

$$\begin{aligned} P^+ &= \Pr(D = + | T = +) \\ P^- &= \Pr(D = - | T = -) \end{aligned}$$

Positive predictive value, or positive predictivity, is the probability of having an accurate positive result and negative predictive value is the probability of having an accurate negative result. Perhaps users are more often concerned about P^+ than P^-.

Estimation of Parameters

Unlike sensitivity and specificity, predictive values P^+ and P^- cannot be estimated directly because there are no data. But at this stage, knowledge of the positive predictive value and the negative predictive value only benefits the user, that is, one person. Therefore we only need the point estimate; and these point estimates can be obtained indirectly using the Bayes' theorem or Bayes' rule. For any two events A and B, we have the Bayes' theorem:

$$\begin{aligned} \Pr(B|A) &= \Pr(B \text{ and } A)/\Pr(A) \\ &= \frac{\Pr(A|B)\Pr(B)}{\Pr(A \text{ and } B) + \Pr(A \text{ and Not } B)} \\ &= \frac{\Pr(A|B)\Pr(B)}{\Pr(A|B)\Pr(B) + \Pr(A|\text{Not})B)\Pr(\text{Not } B)} \end{aligned}$$

Let $A = (T = +)$ and $B = (D = +)$. Then (Not A) $= (T = -)$ and (Not B) $=$ $(D = -)$, respectively. We thus have

$$\Pr(D = + | T = +) = \frac{\Pr(T = + | D = +)\Pr(D = +)}{\Pr(T = + | D = +)\Pr(D = +) + \Pr(T = + | D = -)\Pr(D = -)}$$

$$\Pr(D = - | T = -) = \frac{\Pr(T = - | D = -)\Pr(D = -)}{\Pr(T = - | D = -)\Pr(D = -) + \Pr(T = - | D = +)\Pr(D = +)}$$

Those can be written as

$$P^{+} = \frac{S^{+}\pi}{S^{+}\pi + (1-S^{-})(1-\pi)}$$

$$P^{-} = \frac{S^{-}(1-\pi)}{S^{-}(1-\pi) + (1-S^{+})\pi}$$

In other words:

(i) Predictive values of a screening test depend not only on sensitivity and specificity but on disease prevalence too.
(ii) The higher the prevalence, the higher the positive predictive value.
(iii) However, the higher the prevalence, the lower the negative predictive value (much weaker influence here).

■ **Example 8.3**

Scenario	Sensitivity	Specificity	Prevalence	P^{+}	P^{-}
1A	0.9	0.9	0.002	0.0178	0.9998
1B	0.9	0.9	0.005	0.0433	0.9994
2A	0.9	0.8	0.002	0.0089	0.9997
2B	0.9	0.8	0.005		
3A	0.8	0.9	0.002	0.0158	0.9996
3B	0.8	0.9	0.005		

It can be seen from these different scenarios that, since all cancers are "rare" (low prevalence and incidence), specificity has its dominating effect on the positive predictive value but not much on the negative predictive value.

■ **Example 8.4** Let us assume that we have a good and reliable screening procedure that is 98% sensitive and 97% specific. Let us consider two examples: a low-risk subpopulation (prevalence is 0.1%) and a high-risk subpopulation (prevalence is 20%). From

$$P^{+} = \frac{S^{+}\pi}{S^{+}\pi + (1-S^{-})(1-\pi)}$$

we can show that:

(i) The positive predictive value for the low-risk subpopulation is very low—$P^{+} = 3.2\%$.

(ii) The positive predictive value for the high-risk subpopulation is in an acceptable range—$P^+ = 89.1\%$.

This example shows the following:

(i) Predictive values of a screening test depend not only on sensitivity and specificity but on disease prevalence as well. The higher the prevalence, the higher the positive predictive values.
(ii) We should only screen high-risk subpopulations; random screening does not do anyone any good—not users and not policy makers.

■ **Example 8.5** The following is a summary of what we know about breast cancer screening.

(i) About 5–10% of all breast cancer cases are associated with a patient's genetic predisposition to the disease.
(ii) A genetic predisposition accounts for at least 9100 new breast cancer cases in the United State per year.
(iii) *BRCA1* and *BRCA2*, the identified breast and ovarian cancer susceptibility genes, account for about 50% of the genetically induced breast cancer cases.
(iv) The lifetime risk of developing breast cancer accumulates to 80–90% for carriers of *BRCA1* mutations, and 60–80% for carriers of *BRCA2* mutations.
(v) Age at diagnosis is an important feature of familial breast cancer; by the age of 50 years, more than 50% of the *BRCA1* or *BRCA2* mutation carriers have already developed the disease.
(vi) Existing screening methods are self breast exam, mammography, ultrasound, and magnetic resonance imaging (MRI).

Mammography. Mammography is the process of using low-dose X-rays to examine the human breast. Like all X-rays, mammograms use doses of ionizing radiation to create an image. It is used to detect and diagnose breast disease, both in women who have no breast complaints or symptoms and in women who have breast symptoms. Modern mammography has only existed since 1969, when the first X-ray machines used just for breast imaging became available. Since then, the technology has advanced a great deal, so that today's mammogram is very different even from those of the mid-1980s. For a mammogram, the breast is squeezed between two plastic plates attached to the mammogram machine unit in order to spread the tissue apart. This squeezing or compression ensures that there will be very little movement, so that the image is sharper.

There are several issues with mammography:

(i) The need is not the issue; it decreases breast cancer mortality by 32% (Tabar et al., 2000). The test characteristics may not be the major issue: sensitivity is low (Kuhl et al., 2000) but the specificity ranges from 93% to 99% in high-risk women (Warner et al., 2001).

(ii) But is forty—even fifty—old enough to be at a high enough risk for screening?

(iii) There are guidelines, by federal panels and/or the ACS, but is there any justification? Why start at 40? Why start at 50? Or why not start at 21 or 35?

The following are some post-hoc overall data by the American Cancer Society (ACS): about 10% are recalled for more tests (because the first mammogram is positive); 8–10% of those need biopsy, and 20% of those with biopsy have cancer. That puts the positive predictive value at 1.6–2%.

Ultrasonography. In physics, the term ultrasound applies to all acoustic energy with a frequency above human hearing. Medical ultrasonography is an ultrasound-based diagnostic imaging technique used to visualize muscles and internal organs, their size, structures, and possible pathologies or lesions.

Magnetic Resonance Imaging. Magnetic resonance imaging (MRI) is a noninvasive method used to render images of the inside of an object. It uses radio waves and a strong magnetic field rather than X-rays to provide remarkably clear and detailed pictures of internal organs and tissues. It requires specialized equipment and expertise and allows evaluation of some body structures that may not be as visible with other imaging methods. The advantages of magnetic resonance imaging are the following:

(i) MRI has been shown to be sensitive for invasive breast cancer.

(ii) Its sensitivity is not impaired by dense parenchyma.

(iii) MRI is not associated with ionizing radiation.

For now, mostly because of the cost, breast MRI is not routinely used in a screening setting.

8.3 SOME STATISTICAL ISSUES

This section only covers some basic issues; more advanced topics are covered in Section 8.7.

8.3.1 The Response Rate

The response rate or test rate is the probability that the test result from a subject is positive, $T = +$, or $\pi t = \Pr(T = +)$. It can be seen that

$$\pi_t = \Pr(T = +) = \Pr(T = +, D = +) + \Pr(T = +, D = -)$$

$$\pi_t = \Pr(T = + | D = +)\Pr(D = +) + \Pr(T = + | D = -)\Pr(D = -)$$

$$\pi_t = S^{+}\pi + (1 - S^{-})(1 - \pi)$$

This simple result has some interesting implications:

(i) Since most diseases—such as all cancers, for example—are "rare" (low prevalence and incidence), specificity—not sensitivity—has a dominating effect on the response rate because of its coefficient, $(1 - \pi)$, as compared to the coefficient of sensitivity, π.

(ii) The response rate is a function of disease prevalence, π, making it problematic if we want to use it as an estimate of disease prevalence. More details on this topic are given in Section 8.4.

Example 8.6

Scenario	Sensitivity	Specificity	Prevalence	Response Rate
1A	0.9	0.9	0.002	0.1016
1B	0.9	0.9	0.005	
1C	0.9	0.9	0.200	
2A	0.9	0.8	0.002	0.2014
2B	0.9	0.8	0.005	
3A	0.8	0.9	0.002	0.1014

8.3.2 The Issue of Population Random Testing

One of the hot issues of biomedical ethics is if we should conduct random testing for diseases such as AIDS. Those against the practice often cite concerns about errors, privacy and confidentiality, and unwanted consequences (such as job loss). Those promoting the practice (e.g., policy makers) often want to know the magnitude of the problem in order to justify spending on research as well as interventions. The real emphasis should be about predictivity.

Example 8.7 The current estimate for AIDS for the United States is 0.3%. Let us use the sensitivity and specificity of the ELISA test (Weiss et al., 1985), $S^+ = 0.977$, $S^- = 0.926$.

(i) For a population such as the U.S. population with $\pi = 0.003$, we have

$$P^+ = \frac{(0.977)(0.003)}{(0.977)(0.003) + (0.074)(0.997)} = 0.038 \text{ or } 3.8\%$$

(ii) For a higher-risk subpopulation such as intravenous drug abusers or prisoners, say, $\pi = 0.20$, we have

$$P^+ = \frac{(0.977)(0.20)}{(0.977)(0.20) + (0.074)(0.80)} = 0.767 \text{ or } 76.7\%$$

That means the same truth holds: the higher the prevalence, the higher the positive predictive value. Some investigators imply that a good test must yield $P^+ \geq 50\%$, by either improving its characteristics (S^+ and S^-) or by selecting the population in which the test is used so that the background prevalence is higher. The question is: When does it make sense to screen diseases?

8.3.3 Screenable Disease Prevalence

From

$$P^+ = \frac{S^+\pi}{S^+\pi + (1-S^-)(1-\pi)}$$

we can solve for π:

$$\pi = \frac{(1-S^-)P^+}{S^+ + (1-S^-)P^+ - S^+P^+}$$

Then we could set a lower limit for P^+ to obtain our screenable disease prevalence.

■ **Example 8.8** The following table shows some typical results when we set, for example, $P^+ = 0.80$ or a positive predictive value at 80%.

		Sensitivity, S^+				
		0.5	0.9	0.95	0.98	0.99
	0.5					
	0.9		0.308		0.29	
Specificity, S^-	0.95			0.174		
	0.98		0.082		0.075	
	0.99					

That is, if $S^+ = S^- = 0.98$, we attain a positive predictive value of 80% if prevalence $\pi \geq 0.075$—much higher if the test is not that good. It is also noted that specificity has more influence on positive predictive value; therefore it has more influence on the determination of the screenable disease prevalence: $\pi \geq 0.082$ when $(s^- = 0.98,\ s^+ = 0.90)$ but $\pi \geq 0.29$ when $(s^- = 0.90,\ s^+ = 0.98)$.

■ **Example 8.9** The Pap test or Pap smear test (or cytology test) is an important part of women's health care. The smeared cells or cell suspension is placed on a glass slide, stained with a special dye (Pap stain), and viewed under a microscope. It is used to detect cervical cancer as well as some vaginal or uterine infections. As for cervical cancer, it is still not very sensitive, especially cases in the early stage. However, because it is highly specific (about 99%), its positive predictive value is high, making it suitable for case identification.

■ **Example 8.10** At the present time, the specificity and sensitivity for mammography are $S=0.966$ and $S^{+}=0.647$. Using

$$\pi = \frac{(1-S^{-})P^{+}}{(1-S^{-})P^{+} + S^{+}(1-P^{+})}$$

we can obtain the following screenable disease prevalence:

Predictive Value, $P+$	Screenable Prevalence
1%	53 per 100,000
2%	107 per 100,000
5%	276 per 100,000
10%	581 per 100,000

When comparing the screenable disease prevalence to breast cancer incidences from the SSER database, we find the following:

Age Group	Rate
35–39	59 per 100,000
40–44	119 per 100,000
45–49	194 per 100,000
50–54	254 per 100,000
55–59	313 per 100,000

We could draw the following conclusions:

(i) If screening starts at age 40:

Incidence rate is about 119 per 100,000.

Positive predictive value is 2%.

Negative predictive value is 99.96%.

(ii) If screening starts at age 50:

Incidence rate is about 254 per 100,000.

Positive predictive value is 5%.

Negative predictive value is 99.91%.

The question is: Can we justify starting the screening process at age 40 instead of 50? It is not an easy question to answer.

(i) Unfortunately, very often, neither maneuver—either improving its characteristics or selecting the population with higher prevalence—may be able to yield $P^+ \geq 50\%$. That may be reasonable but it is too much to ask: even tests that are useful clinically may not pass!
(ii) For HIV infection, perhaps one should only screen high-risk subpopulations, like intravenous drug abusers or prisoners. But what do we do for breast cancer? We know that early detection of breast cancer saves lives.

What about a retest?

(i) If screening starts age 40 and a woman is recalled, the chance of her having cancer would be about 2%. Another recall for biopsy would raise the predictive value to 28% (which is similar to ACS' data of about 20%—with perhaps lower specificity).
(ii) If screening starts at age 50 and a woman is recalled, the chance of her having cancer would be about 5%. Another recall for biopsy would raise the predictive value to 50%; that qualifies it as a "good" procedure as stipulated by some investigators.

8.3.4 An Index for Diagnostic Competence

Clearly, it is desirable that a diagnostic test or screening procedure should be highly sensitive and highly specific; both error rates should be small. Other things (cost, ease of application, etc.) being equal, a test with larger values of both sensitivity and specificity is obviously better. If not that clear-cut, one has to consider the relative costs associated with two forms of error; and if the two types of error are equally important, it may be desirable to have a single index to measure the diagnostic competence of the test. Candidates might include *overall agreement* and the *kappa statistic*, among others.

Overall Agreement

The simplest measure would be the overall agreement, Pr $(T = D)$. However, unlike sensitivity and specificity, the overall agreement is influenced by the disease prevalence.

$$\begin{aligned}
\Pr(T = D) &= \Pr(T = D = +) + \Pr(T = D = -) \\
\Pr(T = D) &= \Pr(T = + | D = +)\Pr(D = +) + \Pr(T = - | D = -)\Pr(D = -) \\
\Pr(T = D) &= \pi S^+ + (1-\pi)S^-
\end{aligned}$$

The probability of misclassification, $\Pr(T \neq D)$, depends on the disease prevalence too.

Kappa Statistic

Another alternative is the Kappa statistic. Kappa is a popular statistic often used to measure agreement between observers. It adjusts overall agreement for *chance agreement*.

$$\kappa = \frac{\{\text{Overall agreement}\} - \{\text{Chance agreement}\}}{1 - \{\text{Chance agreement}\}}$$

$$\kappa = \frac{[\Pr(T = D = +) + \Pr(T = D = -)] - [\Pr(T = +)\Pr(D = +) + \Pr(T = -)\Pr(D = -)]}{1 - [\Pr(T = +)\Pr(D = +) + \Pr(T = -)\Pr(D = -)]}$$

We can express kappa as a function of sensitivity, specificity, and prevalence; then fix sensitivity and specificity and prove that, similar to the case of overall agreement, kappa is a monotonic function of disease prevalence.

Youden Index

Both overall agreement and kappa statistics are functions of disease prevalence; that leaves one measure, *Youden's index* (Youden, 1950). If the two types of error are equally important, Youden's index J is defined as

$$J = 1 - (\alpha + \beta) = S^{+} + S^{-} - 1$$

Youden's index J has interesting characteristics: (i) it is based on a simple principle—small sum of errors (when neither one has priority), (ii) its value is larger when both sensitivity and specificity are high, and (iii) it does not depend on the disease prevalence. And there are other reasons for using it as well! For example, we might ask: "When does a process qualify as a test?"

To decide if a process is a test, the minimum criterion it must pass is that it detects disease better than by chance alone. A process can only qualify as a test if it selects diseased persons with higher probability than pure guessing: $P^{+} > \pi$. The decision therefore depends on J, Youden's index:

$$P^{+} = \frac{S^{+}\pi}{S^{+}\pi + (1 - S^{-})(1 - \pi)}$$

$$P^{+} > \pi \Longleftrightarrow S^{+} > 1 - S^{-} \Longleftrightarrow J > 0$$

More characteristics of Youden's index J are presented in Section 8.4.

■ **Example 8.11** Thermography is a way of measuring and mapping the heat from the breast with the use of a special camera; the idea is that changes in temperature may indicate disease. It is approved by the FDA as safe but not as a standalone screening test for breast cancer. According to the American Cancer Society (ACS), "studies have not proven this to be an effective screening tool for early detection of breast cancer and it is not a replacement for mammograms. It is not considered a reliable detection test since it can miss some cancers and can give a high false positive rate." The following are some historical data.

In a study of thermography, used to detect breast cancer in the early stages, Moskowitz et al. (1976) had experienced thermographers read films from people with and without breast cancer.

	Thermography		
Cancer	Positive	Negative	Unreadable
Yes	181	239	42
No	233	302	70

Discarding unreadable films, we have

$$s^+ = \frac{181}{181 + 239} = 0.43$$

$$s^- = \frac{302}{302 + 233} = 0.56$$

In other words, thermography does not qualify as a test because the condition $J > 0$ or $S^+ > 1 - S^-$ is violated; that agrees with the ACS position.

8.4 PREVALENCE SURVEYS

Decisions about choosing and using screening programs depend on two aspects:

1. Who benefits from the screening, the screenees or society? For example, a medical test is usually ordered for the benefit of the patient but HIV testing, for example, is usually performed not because the individual requested to be tested, but rather because blood banks routinely test all blood donations—for the benefit of others (i.e., society).
2. The cost and ease of application. Diagnostic procedures with high accuracy are obviously desirable but are frequently too expensive and/or too complicated to be used on a large scale; therefore a less sophisticated screening test—even with inferior accuracy—may be preferred.

There are two opposite strategies:

1. In many screening programs, when a positive test carries a high cost, emphasis may be put on high positive predictive value and specificity.

2. In the mass screening of blood donations for blood banks, however, society could be ill-served by emphasis on specificity. The cutoff for ELISA is usually set to have high sensitivity (positives by ELISA could be followed with a confirmatory Western blot). Emphasis is on negative predictive value and sensitivity; here a retest is needed.

We'll take the first approach. When details are needed, we focus on positive predictive value. If readers prefer the other approach to focus on negative predictivity, it could be taken and users can proceed similarly; the difference is only in resource allocation (sample sizes).

With our choice to focus on positive predictive value, estimation of disease prevalence often follows one of two different designs; we'll just call them Design 1 (and Design 1 Plus) and Design 2. In Design 1, we assume that sensitivity and specificity are known; in its more common version, Design 1 plus, sensitivity and specificity are unknown. In Design 2, a new screening test is used and calibrated against a known and more precise diagnostic test.

8.4.1 Known Sensitivity and Specificity

Design 1 is the most simple, most ideal, but less practical design. We have a screening test T; its sensitivity S^+ and specificity S^- have been independently established. It is less practical because we assume that sensitivity and specificity are known without errors. If such a test is available, then it does not make any difference if we focus on positive predictive value or negative predictive value. This design will be expanded into Design 1 Plus, where the assumed knowledge of sensitivity and specificity are relaxed. A *prevalence survey* is conducted in one target population in order to estimate the disease prevalence, $\pi = \Pr(D = +)$. Data are very simple: x of n subjects are found "positive" by the test.

A Preliminary Solution

A commonly used estimate of the disease prevalence is the frequency of positive tests,

$$p_t = \frac{x}{n}$$

$$\text{Var}(p_t) = \frac{\pi_t(1-\pi_t)}{n}$$

But this is an unbiased estimate of $\pi_t = \Pr(T = +)$, the response rate, whereas we want to estimate the disease prevalence, $\Pr(D = +)$. How good is it as an estimate of prevalence?

The estimate p_t depends not only on the disease prevalence (which it is supposed to estimate) but also on the characteristics of the test—the sensitivity and the specificity, S^+ and S^-. It can be shown that it is badly biased as an estimator of p. It is biased upward, usually overestimating the disease prevalence:

$$\begin{aligned}
\pi_t &= \Pr(T=+) = \Pr(T=+, D=+) + \Pr(T=+, D=-) \\
\pi_t &= \Pr(T=+|D=+)\Pr(D=+) + \Pr(T=+|D=-)\Pr(D=-) \\
\pi_t &= S^+\pi + (1-S^-)(1-\pi) \\
&= \pi + (1-S^-)(1-\pi) - (1-S^+)\pi \\
&= E(p_t) \\
\text{Bias} &= E(p_t) - \pi \\
&= (1-S^-)(1-p) - (1-S^+)p
\end{aligned}$$

For rare diseases, such as cancers, $(1-\pi)$ is much larger than π and $(1-S^-)$ and $(1-S^+)$ are often roughly equal; the bias is more likely positive—therefore we likely over estimate the disease prevalence using p_t.

■ **Example 8.12** Let us assume for $S^+ = S^- = 0.9$ and $\pi = 0.1$, that the bias $= (0.1)(0.9) - (0.1)(0.1) = 0.08$, which is more affected by specificity. In this example, the point estimate could be twice the true value; its expected value is 0.18 versus $\pi = 0.1$.

A Necessary Correction

A correction is needed because we want to estimate the disease prevalence, $\pi = \Pr(D=+)$, not the response or test rate, $\pi_t = \Pr(T=+)$. We should use

$$p = x^*/n$$

where x^* is the number of cases (true positives), instead of p_t. However, x^* is not available/observable. We can, however, proceed indirectly as follows—where J is Youden's index as introduced in Section 8.3.3:

$$\begin{aligned}
\pi_t &= \Pr(T=+) = \Pr(T=+, D=+) + \Pr(T=+, D=-) \\
\pi_t &= \Pr(T=+|D=+)\Pr(D=+) + \Pr(T=+|D=-)\Pr(D=-) \\
\pi_t &= S^+\pi + (1-S^-)(1-\pi) \\
\pi &= \frac{\pi_t + S^- - 1}{J}; \; J = S^+ + S^- - 1, \text{ leading to} \\
p &= \frac{p_t + S^- - 1}{J}
\end{aligned}$$

■ **Example 8.13** Let us assume for $S^+ = S^- = 0.9$ and $\pi = 0.1$, that we have

$$\begin{aligned}
\pi_t &= S^+\pi + (1-S^-)(1-\pi) \\
&= 0.18
\end{aligned}$$

Let us suppose that $x = 20$ out of $n = 100$ subjects are positive:

$$p_t = 0.20$$
$$p = \frac{p_t + S^- - 1}{J} = 0.125$$

■ **Example 8.14** Stamler et al. (1976) surveyed one million people and found 24.7% had diastolic blood pressure (DBP) > 90 mmHg and 11.6% had DBP 95 mmHg—using p_t, of course. Carey et al. (1976), using elevation of BP in three separate readings as the criterion for having hypertension (the "truth"), found $S^+ = 0.930$ and $S^- = 0.911$, good characteristics. Yet, correcting p_t to get p shows dramatic results: Stamler is 24.7% becomes 18.8% and 11.6% becomes 3.2%: estimates of p_t and p can differ by a factor of 4.

$$p = \frac{p_t + s^- - 1}{j}$$
$$S^+ = 0.930;\ S^- = 0.911$$
$$J = S^+ + S^- - 1 = 0.841$$
$$p_t = 0.116$$
$$p = \frac{0.116 + 0.911 - 1}{0.841} = 0.032$$

A correction, using p instead of p_t, is a substantial improvement; if S^+ and S^- are known a priori, then p is unbiased for π.

$$\pi = \frac{\pi_t + S^{-1} - 1}{J};\ J = S^+ + S^{-1} - 1$$
$$p = \frac{\pi_t + S^{-1} - 1}{J}$$
$$E(p) = \frac{\pi_t + S^{-1} - 1}{J}$$
$$= \pi$$

The standard error of the new and improved estimator p is

$$p = \frac{p_t + S^- - 1}{J}$$
$$\mathrm{Var}(p) = \frac{\mathrm{Var}(p_t)}{J^2}$$
$$SE(p) = \frac{1}{J}\sqrt{\frac{p_t(1 - p_t)}{n}}$$

This result indicates that the *precision* of estimation of the prevalence depends only on the size of Youden's index rather than on any function of sensitivity and specificity. This is a very important result, which justifies the value and the potential of Youden's index J, that the better test is the one with larger value of Youden's index.

8.4.2 Unknown Sensitivity and Specificity

This less ideal but more practical version of Design 1, called Design 1 Plus, is still a simple design; we have a screening test T but its sensitivity S^+ and specificity S^- are *not* known. A *prevalence survey* is conducted in one target population in order to estimate the disease prevalence, $\pi = \Pr(D = +)$. Data are still in the very simple form: x of n subjects found positive by the test. Sensitivity and specificity need to be estimated but there are not enough data from this prevalence survey to do so. Therefore sensitivity and specificity are estimated using two other independent samples; S^+ is estimated by the proportion s^+ from a sample of size n_1, and S^- is estimated by the proportion s^- from a sample of size n_0.

The solution is simple: we use the same estimator p of Design 1, where sensitivity and specificity S^+ and S^- are replaced by their estimates from the two independent samples, the proportions s^+ and s^-:

$$p = \frac{p_t + s^- - 1}{s^+ + s^- - 1}$$

Using a Taylor series expansion, we can approximate its expected value and its variance.

Sources of Bias

When sensitivity and specificity are unknown and are estimated using two other independent samples, p is no longer unbiased. As seen from the last two terms of the following formula, the bias comes from the estimation of sensitivity and specificity. However, the bias is much smaller as compared to what we have in Design 1; it is negligible if the other two samples n_1 and n_0 are both large.

$$E(p) = \pi + \frac{\pi}{J^2}\frac{S^+(1-S^+)}{n_1} - \frac{(1-\pi)}{J^2}\frac{S^-(1-S^-)}{n_0}$$

Sources of Variability

The first term of the variance is due to the prevalence survey itself; the last two terms are due to our need to estimate sensitivity and specificity.

$$\mathrm{Var}(p) = \frac{1}{J^2}\frac{p_t(1-p_t)}{n} + \frac{\pi^2}{J^2}\frac{S^+(1-S^+)}{n_1} + \frac{(1-\pi)^2}{J^2}\frac{S^-(1-S^-)}{n_0}$$

Priority

We already knew that positive predictive value, our chosen high priority, is much more affected by the value of specificity. Now compare the last two terms in the variance of p, Var(p): in common cases where both S^+ and S^- are high but π is low, the last term dominates. That means the contribution of variability in the estimate of the specificity is usually the dominant term in calculating the precision of the estimated disease prevalence; we thus need a larger n_0.

Optimal Allocation of Resources

The sensitivity and specificity have different effects on the estimated disease prevalence and its precision. If we regard the sum $m = n_1 + n_0$ as fixed and find the choices n_1 and n_0 that minimize the sum of the last two terms in Var(p), the result is as follows:

$$\frac{n_1}{n_0} \cong \frac{\pi}{1-\pi}\sqrt{\frac{S^+(1-S^+)}{S^-(1-S^-)}}$$

■ **Example 8.15** In setting up Design 1 Plus, suppose we have $S^+ = S^- = 0.93$ approximately, and $\pi = 0.05$. Then n_0 should be 19 times as large as n_1 in order to minimize the variance of p:

$$\frac{n_1}{n_0} \cong \frac{\pi}{1-\pi}\sqrt{\frac{S^+(1-S^+)}{S^-(1-S^-)}} = \frac{(0.05)}{(0.95)}\sqrt{\frac{(0.93)(0.07)}{(0.93)(0.07)}} = \frac{1}{19}$$

Estimation of Positive Predictivity

When disease prevalence π has been estimated, say, using Design 1 Plus, let us turn our attention to the next targets—predictive values P^+ and P^-. With our chosen priority, we focus on P^+ estimated by $p+$. Since

$$P^+ = \frac{S^+\pi}{S^+\pi + (1-S^-)(1-\pi)} = \frac{S^+\pi}{\pi_t}$$

we have, as its estimator,

$$p^+ = \frac{s^+}{j}\left[1 - \frac{(1-s^-)}{p_t}\right]$$

with S^+ and S^- being estimated from two independent samples by proportions s^+ and s^-.

Theorem 8.1 (Gastwirth, 1987): *When n, n_1, and n_0 are all large, the sampling distribution of p^+ is approximately normal with mean P^+ and variance*

$$\mathrm{Var}(p^+) = \left\{\frac{S^+(1-S^-)}{J\pi_t}\right\}^2 \frac{\pi_t(1-\pi_t)}{n} + \left\{\frac{\pi(1-S^-)}{J\pi_t}\right\}^2 \frac{S^+(1-S^+)}{n_1} + \left\{\frac{(1-\pi)S^+}{J\pi_t}\right\}^2 \frac{S^-(1-S^-)}{n_0}$$

The following are some implications, similar to those we have in the estimation of the disease prevalence.

1. *Priority.* Compare the last two terms in Var(p^+). In common cases where both S^+ and S^- are high but π is low, the last term dominates. That means the contribution of variability in the estimate of the specificity is usually the dominant term in calculating the precision of the estimated positive predictive value; we thus need a larger n_0. This is very similar to the result concerning precision of the estimated disease prevalence.

2. *Optimal Resource Allocation.* The sensitivity and specificity have different effects on positive predictive value and the precision of its estimate. If we regard the sum $m = n_1 + n_0$ as fixed and find the choices n_1 and n_0 that minimize the sum of the last two terms in Var(p^+), result is as follows:

$$\frac{n_1}{n_0} \cong \frac{\pi}{1-\pi}\sqrt{\frac{(1-S^+)(1-S^+)}{S^+S^-}}$$

■ **Example 8.16** In setting up Design 1 Plus, suppose we have $S^+ = S^- = 0.93$ approximately, and $\pi = 0.05$. Then n_0 should be 253 times as large as n_1 in order to minimize the variance of p^+:

$$\frac{n_1}{n_0} \cong \frac{\pi}{1-\pi}\sqrt{\frac{(1-S^+)(1-S^-)}{S^+S^-}} = \frac{(0.05)}{(0.95)}\sqrt{\frac{(0.07)(0.07)}{(0.93)(0.93)}} \cong \frac{1}{253}$$

■ **Example 8.17** The ELISA test for HIV is used to screen donated blood for blood banks. An evaluation of ELISA yielded the following estimates (Weiss et al., 1985): $s^+ = 0.977$ ($n_1 = 88$) and $s^- = 0.926$ ($n_0 = 297$). It is obvious that it was not an optimal sample size.

Tables 8.1 and 8.2 present estimated positive predictive value and its standard error for prevalence surveys with $n = 500$ and $n = 10{,}000$.

In the two tables we treated the sensitivity and specificity as fixed (column 3) and as estimated (column 4). The results show that when the prevalence is large, say, 40% or above, positive predictive value is high and its standard error is small—even for $n = 500$. On the other hand, when prevalence is low, 5% or below, positive predictive value falls below 50% and its standard error gets larger—regardless of the size of the

TABLE 8.1. Sample Size $n = 500$

Prevalence	p^+	S Fixed SE(p^+)	S Estimated SE(p^+)	Percentage of Var(p^+) Due to Estimation of S^-
0.50	0.930	0.007	0.017	84.9
0.40	0.898	0.009	0.025	85.2
0.20	0.768	0.024	0.057	82.1
0.10	0.595	0.049	0.103	77.0
0.05	0.410	0.082	0.154	72.0
0.03	0.290	0.106	0.190	69.0
0.01	0.118	0.143	0.243	65.2

TABLE 8.2. Sample Size $n = 10{,}000$

Prevalence	p^+	S Fixed SE(p^+)	S Estimated SE(p^+)	Percentage of Var(p^+) Due to Estimation of S^-
0.50	0.930	0.001	0.016	98.5
0.40	0.898	0.002	0.023	98.9
0.20	0.768	0.005	0.052	99.0
0.10	0.595	0.041	0.091	98.5
0.05	0.410	0.018	0.132	98.1
0.03	0.290	0.024	0.160	97.8
0.01	0.118	0.032	0.199	97.4

prevalence survey; most of its sampling variability is due to the estimation of the specificity.

The overall lesson is that if sensitivity and specificity are unknown and if resources are limited, resources are best spent getting a good estimate of specificity; the sample size n_0 is even more important than the size of the main survey.

8.4.3 Prevalence Survey with a New Test

Design 2, involving a new screening test, is a little complicated but more practical and more often used in real prevalence surveys. We have a *reference* screening test R; its sensitivity and specificity have been independently established. The test itself is too expensive or too complicated for use on a large scale. We also have a new screening test T, which is easy to use. The test's characteristics S^+ and S^- are unknown but this test T has been calibrated against R for accuracy. A prevalence survey is conducted in one target population in order to estimate the disease prevalence, π. Data are still in the very simple form: x of n subjects found positive by the new test.

Calibration of the New Test

Diagnostic procedures with high accuracy are obviously desirable but are frequently too expensive and/or too complicated to be used on a large scale; therefore a less sophisticated screening test—even with inferior accuracy—may be preferred. For a new test, its accuracy measures S^+ and S^- must be determined (and weighed against its costs). Those can be estimated if the test is applied to individuals whose true disease states are known; but this may be difficult. The new test is customarily evaluated against a standard or reference test (with its own errors).

In this calibration, the basic question is: Do the tests select the diseased and the healthy equally well? The approach is to apply both tests, the reference R and the new T, simultaneously to the same group of individuals—independent of the main intended prevalence survey.

Calibration Key Parameters

Let R and T denote the test results by the reference test and the new test, respectively. The key parameters are two conditional probabilities:

$$C^+ = \Pr(T = +|R = +) \quad \text{and} \quad C^- = \Pr(T = -|R = -)$$

These conditional probabilities are defined similar to the sensitivity and specificity but are called *copositivity* and *conegativity*. In Design 2, the copositivity and conegativity, C^+ and C^-, are assumed known (i.e., determined before and independent of the prevalence survey).

Estimation of Disease Prevalence

An estimator of the disease prevalence can be obtained in three steps.

Step 1: Applying the solution of Design 1, if there is an estimate p_r of $\pi_r = \Pr(R = +)$, then an estimate of $\pi = \Pr(D = +)$ is

$$p = \frac{p_r + S_r^- - 1}{J_r}; \text{ where } J_r = S_r^+ + S_r^- - 1$$

Step 2: Applying the solution of Design 1 again but, this time, to obtain an estimate of $\pi_r = \Pr(R = +)$, we then have

$$p_r = \frac{p_t + C^- - 1}{J_t}; \text{ where } J_t = C^+ + C^- - 1$$

Step 3: Substitute p_r from Step 2 into the right-hand side of the equation in Step 1. The result is

$$\text{Point estimate: } p = \frac{p_t + C^- - 1 + J_t(S^- - 1)}{J_r J_t} \quad \text{where } p_t = \frac{x}{n}$$

$$\text{Standard error: } SE(p) = \frac{1}{J_r J_t}\sqrt{\frac{p_t(1 - p_t)}{n}}$$

It is interesting to note that the standard error depends on the product of the two Youden indices.

8.5 THE RECEIVER OPERATING CHARACTERISTIC CURVE

Diagnostic tests have been presented as always having dichotomous outcomes. In some cases, the result of the test may be binary, but in many cases it is based on the dichotomization of a continuous separator. To deal with a continuous separator, we need a well-known graph called the *receiver operating characteristic curve* or *ROC curve*. If the idea, in the developmental stage, was to classify people as "diseased" (condition present) or "healthy" (condition absent) based on a certain continuous measurement (from blood or urinary components), then we need to *dichotomize* the measurement: if the measurement is "high," the subject is classified as "diseased"—if it's "low," the subject is "healthy." But the basic question is: "How high is high?" or "How low is low?"

8.5.1 The ROC Function and ROC Curve

Let us first consider the following simple but plausible model where the separator Y is normally distributed with the same variance, but different means; it can be seen from Figure 8.7 that no matter where you "cut," both errors result! More important, specificity and sensitivity are functions of the cutpoint y.

Assumption

In the case of many diseases, the larger values of the separator Y are associated with the diseased population (also called *population of the cases*) and smaller values are associated with the control or nondiseased (or healthy) population (e.g., blood glucose for diabetes, PSA for prostate cancer, antibodies for infections). For many others, the smaller values of the separator Y are associated with the diseased population and larger values are associated with the nondiseased population (e.g., static admittance for otitis media, TSH for hyperthyroidism).

We will assume, without loss of generality, that larger values of Y are associated with the diseased population.

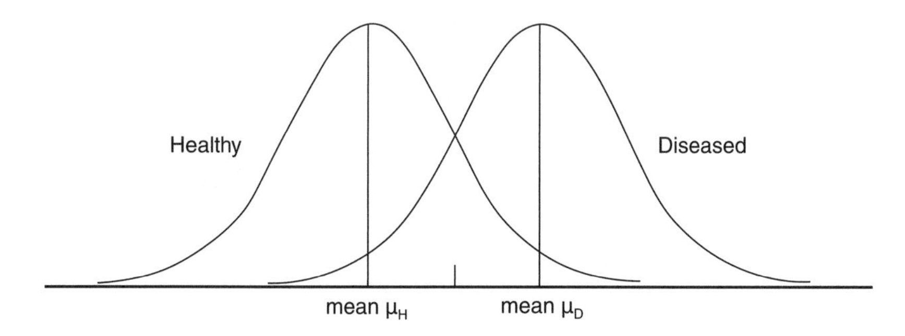

FIGURE 8.7. Normal distributions with the same variance.

Sensitivity and Specificity

With our assumption that larger values of Y are associated with the diseased population, the sensitivity, $\Pr(T=+|D=+)$, associated with a cutpoint $Y=y$ is

$$\begin{aligned} S^+(y) &= \Pr(Y > y|D = +) = \text{True-positive rate} \\ &= 1-\Pr(Y \le y|D = +) = 1-F^+(y) \end{aligned}$$

where $F^+(y) = \Pr(Y \le y|D = +)$ is the cumulative distribution function (cdf) of Y for the diseased population (or population of cases).

With the same assumption that larger values of Y are associated with the diseased population, the specificity, $\Pr(T=-|D=-)$, associated with a cutpoint $Y=y$ is

$$\begin{aligned} S^-(y) &= \Pr(Y \le y|D = -) = F^-(y) \\ 1-S^-(y) &= 1-F^-(y) = \text{False-positive rate} \end{aligned}$$

where $F^-(x)$ is the cumulative distribution function (cdf) of Y for the nondiseased or healthy population.

ROC Function and ROC Curve

A function R from [0, 1] to [0, 1] that maps false-positive rate to true-positive rate, $(1 - F^-(y))$ to $(1 - F^+(y))$, is called the *ROC function*:

$$R[1-F^-(y)] = 1-F^+(y) \quad \text{or} \quad R[1-S^-(y)] = S^+(y)$$

The graph of $R(.)$ is called the *ROC curve* (Figure 8.8). The ROC curve, the graph of sensitivity, $S^+(y)$, versus (1 – specificity), $(1 - S^-(y))$, is generated as the cutpoint y moves through its range of possible values. More information can be found in Lusted (1971a,b) and Swets (1979).

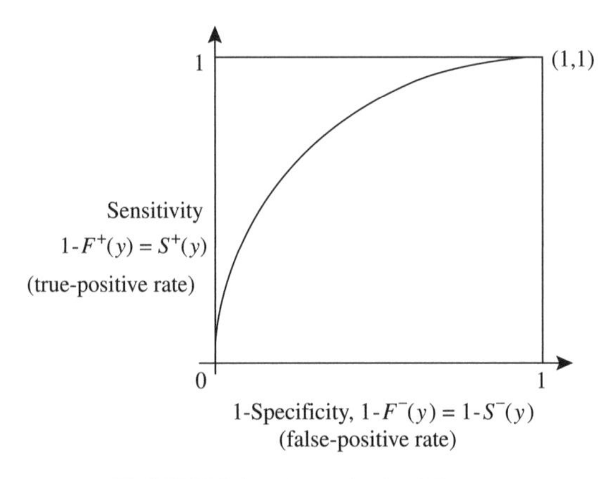

FIGURE 8.8. The typical ROC cuve.

8.5.2 Some Parametric ROC Models

Let $S(t)$ be the *survival function*, $S(t) = (1 - \text{cdf})$. We have

$$\begin{aligned} S_D(t) &= 1 - F^+(t) \\ S_H(t) &= 1 - F^-(t) \\ R[S_H(t)] &= S_D(t) \\ R(u) &= S_D[S_H^{-1}(u)];\ u \in [0, 1] \end{aligned}$$

The following are two possibilities: the bi-normal and bi-log-logistic ROC curves.

The Bi-normal ROC Curve

When test results, or separator Y values, are assumed to be normally distributed in both diseased and nondiseased populations, we have the *bi-normal ROC curve*:

$$\begin{aligned} &Y_1 \in N(\mu_1, \sigma_1^2),\ Y_0 \in N(\mu_0, \sigma_0^2) \\ &ROC(t) = \Phi\left[\frac{\mu_1 - \mu_0}{\sigma_1} + \frac{\sigma_0}{\sigma_1}\Phi^{-1}(t)\right] \end{aligned}$$

where slope is equal to 1 if the variances are equal.

The Bi-log-logistic ROC Curve

An alternative to the popular normal distribution is the *logistic distribution*; the standard logistic curve looks very much like the standard normal curve but with thicker tails:

$$\text{Standard logistic density:}\quad f(x) = \frac{e^x}{(1 + e^x)^2}$$

$$\text{General logistic density:}\quad f(x) = \frac{\dfrac{1}{\sigma}\exp\left(\dfrac{x - \mu}{\sigma}\right)}{\left[1 + \exp\left(\dfrac{x - \mu}{\sigma}\right)\right]^2}$$

where μ is the mean, and σ is standard deviation.

If $\ln(X)$ is distributed as logistic, X is distributed as log-logistic; the log-logistic distribution is similar to the log-normal distribution but with thicker tails, so it fits better for real, nonnegative measurements.

$$\begin{aligned} S(t) &= \frac{1}{1 + (\rho t)^\nu} \\ \rho &= e^{-\mu},\ \text{where } \mu \text{ is the mean} \\ \nu &= \frac{1}{\sigma},\ \text{where } \sigma \text{ is standard deviation} \end{aligned}$$

When test results, or separator Y values, are assumed to be distributed as log-logistic in both diseased and nondiseased populations, we have the *bi-log-logistic ROC curve*:

$$S(t) = \frac{1}{1+(\rho t)^{\nu}}$$

$$\sigma_{\mathrm{D}} = \sigma_{\mathrm{H}} = \sigma$$

$$R(u) = \frac{u}{u+(1-u)\exp\left(-\frac{\mu_{\mathrm{D}}-\mu_{\mathrm{H}}}{\sigma}\right)}$$

8.5.3 Estimation of the ROC Curve

The next question is how to estimate the ROC curve given two independent samples: $\{y_{0i};\ i=1,\ldots, n_0\}$ and $\{y_{1j};\ j=1,\ldots, n_1\}$ from n_0 controls and n_1 cases.

The Empirical Estimator

The simplest way to estimate $R(.)$ is to replace cumulative distribution functions (cdfs) $F^{+}(y)$ and $F^{-}(y)$ by their empirical estimates $p^{+}(y)$ and $p^{-}(y)$; $p^{+}(y)$ is the portion of the n_1 observations y_{1j} of the cases which are less than or equal to y, and $p^{-}(y)$ is defined similarly. This is a nonparametric estimate and $\{1-p^{-}(y), 1-p^{+}(y)\}$ is an unbiased estimator of $\{1-F^{-}(y), 1-F^{+}(y)\}$ but, as Bamber (1975) put it, "the sample ROC (for continuous Y) can never be anything but a finite set of points." If there are no ties in the combined sample of y_{0i} and y_{1j} there are $(n_0 n_1)$ points.

Steck (1971) actually made an attempt to connect the dots, turning them into a step function. He combined two samples and arranged them in the usual increasing order. He described the empirical estimator as "a random walk from the bottom-left corner (0, 0) to the top-right corner (1,1)—and read the combined order sample from largest to smallest—whose next step is $1/n_1$ up or $1/n_0$ to the right according to whether the next observation in the ordered combined sample is a case's measurement (y_1) or a control's measurement (y_0)."

■ **Example 8.18** Let us consider, for simplicity, a small data set with two cases and three controls. Assume the numerical values of these five measurements are ranked as follows:

$$y_{01} < y_{02} < y_{11} < y_{03} < y_{12}$$

The empirical estimate of the ROC curve is presented in the following figure; it is similar to an empirical cumulative distribution function (cdf) of size 2 with weight $\frac{1}{2}$ at point 0 and point $\frac{1}{3}$.

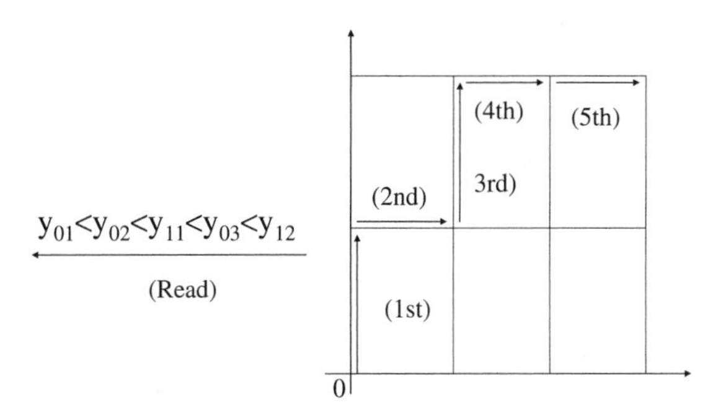

A Rank-Based Estimator

Le (1997) reformulated the empirical estimator by defining a *score* u_j for each case's observation y_{1j}:

$$u_j = \frac{n_0 - (S_j - R_j)}{n_0}$$

where R_j is the rank of y_{1j} among the cases and S_j is the rank of y_{1j} among observations in the pooled sample. Each u represents the percentage of control observations greater than the case observation. By treating $\{u_j\}$ as a *pseudo-sample*, $R(.)$ is estimated by the empirical step function:

$$R_{n_0}(u) = \frac{\text{Number of } u_j \leq u}{n_0}; \text{ for } 0 \leq u \leq 1$$

$R_{n_0}(.)$ converges in probability to $R(.)$. When there are no ties in the combined sample, it is identical to the empirical estimator. However, the "whole" curve is defined and it has two advantages:

1. There is an easy option to handle ties: use of *midranks* can handle graded or ordinal separators.
2. It provides a pseudo-sample, the $\{u_j\}$, which can be used for other purposes; for example, comparison of ROC curves

■ **Example 8.19** Consider the same information as in Example 8.18: $\{y_{01} < y_{02} < y_{11} < y_{03} < y_{12}\}$ with $n_1 = 2$ and $n_0 = 3$. We have $u_1 = 0$, and $u_2 = \frac{1}{3}$, leading to

$$R_2(u) = \begin{cases} \frac{1}{2} \text{ for } 0 \leq u < \frac{1}{3} \\ 1 \text{ for } \frac{1}{3} \leq u \leq 1 \end{cases}$$

The result is identical to that in Example 8.18, an empirical cumulative distribution function (cdf) of size 2 with weight $\frac{1}{2}$ at points 0 and $\frac{1}{3}$.

8.5.4 Index for Diagnostic Accuracy

The ROC curve is a graphical device to show all possible combinations of sensitivity and specificity but, for simplicity, it is desirable to reduce an entire curve to a single quantitative index of diagnostic accuracy. Possibilities include the difference between means of Y for the two populations, those with disease and those without; and the ratio of variances. However, the most popular one has been the area under the ROC curve.

Area Under the ROC Curve

The area under the curve has a powerful interpretation and it is related to other well-known statistics, making it easier to learn its statistical properties. Suppose that an observation Y_1 is randomly sampled from the diseased population and another random observation Y_0 is independently sampled from the nondiseased population. Let $\Pr(Y_1 > Y_0)$ denote the probability of the event that the Y_1 observation is larger than the Y_0 observation. We have

$$A = \Pr(Y_1 > Y_0)$$

$$A = \int (1 - F^+) d(1 - F^-)$$

$$A = \int_0^1 R(u) du$$

$$A = \text{Area under ROC curve}$$

The area under the ROC curve measures the size and the importance of the difference between two populations, those with the disease and those without it. It tells how accurately a given diagnostic test differentiates two populations by giving the probability of correct ranking: the probability of separating a case from a control—a measure of *separation power*. If Y_0 and Y_1 are identically distributed, $A = \frac{1}{2}$; the maximum value of 1.0 is attained if and only if the cases' distribution lies entirely above that of controls. We can keep the practice of graphing sensitivity versus $(1 - \text{specificity})$ in forming the ROC curve. If the larger values of the separator Y are associated with the diseased population and smaller values are associated with the nondiseased population, as we are assuming, the curve is above the main diagonal joining (0, 0) and (1, 1) and $\frac{1}{2} \leq A \leq 1$. If the smaller values of the separator Y are associated with the diseased population and larger values are associated with the nondiseased population, the curve is below the main diagonal and $0 \leq A \leq \frac{1}{2}$.

Area Under ROC Curve and Mann–Whitney Statistic

Given two independent samples, $\{y_{0i};\ i = 1, \ldots, n_0\}$ and $\{y_{1j}, j = 1, \ldots, n_1\}$ from n_0 controls and n_1 cases, there are $n_0 n_1$ ways of pairing a Y_0 observation with a Y_1 observation. The Mann–Whitney statistic U counts the number of pairs (y_{0i}, y_{1j}), out of $n_0 n_1$ possible pairs, where $y_{0i} < y_{1j}$; and statistic U is related to Wilcoxon's rank sum W. From the interpretation of the area A under the ROC curve, we have (U is the Mann–Whitney statistics, which is related to the Wilcoxon's rank sum W)

$$A = \Pr(Y_1 > Y_0) = U / n_0 n_1$$

Area Under ROC Curve and Wilcoxon Statistic

Since the Mann–Whitney statistic U is related to the Wilcoxon's rank sum statistic W, the area A under the ROC curve is related to the statistic W. We can also derive such a relationship directly using the area under the $R_{n0}(.)$ curve, which is formed from the scores, u_j's, of Section 8.5.2, where W_1 is the sum of the ranks for the cases.

$$\text{Area} = (u_2-u_1)\frac{1}{n_1} + (u_3-u_2)\frac{2}{n_1} + \cdots + (u_{n_1}-u_{n_1-1})\frac{n_1-1}{n_1} + (1-u_{n_1})$$

$$A = 1-\frac{u_1+u_2+\cdots+u_{n_1}}{n_1} = 1-u$$

$$u_j = \frac{n_0-(S_j-R_j)}{n_0}$$

leading to

$$A = \frac{W_1-\frac{1}{2}n_1(n_1+1)}{n_0 n_1}$$

8.5.5 Estimation of Area Under ROC Curve

A point estimate of the area under the ROC curve is obvious from its relationship to the Mann–Whitney and Wilcoxon statistics.

$$A = U/n_0 n_1$$

$$A = \frac{1}{n_0 n_1}\left\{W_1-\tfrac{1}{2}n_1(n_1+1)\right\}$$

The variance of this estimator also seems straightforward:

$$\text{Var}(A) = \text{Var}(W_1)/(n_0 n_1)^2$$

However, few people are familiar with the variance of Wilcoxon's statistic W, because we usually use it only under the null hypothesis. Exact variance and standard error are in fact complicated, but there are good approximations.

The exact standard error of the area A under the ROC curve is given as follows.

(i) Q_2 is the probability that y-values for two randomly selected cases will both be smaller than the y-value for a randomly selected control.

(ii) Q_1 is the probability that the y-value for one randomly selected case will be smaller than both y-values for two randomly selected controls.

(iii) Q_1 and Q_2 are estimated by the proportions of triplets satisfying the required properties.

Then

$$SE(A) = \sqrt{\frac{A(1-A) + (n_0-1)(Q_1-A^2) + (n_1-1)(Q_2-A^2)}{n_0 n_1}}$$

Hanley and McNeil (1982), using simulation, showed that standard error of A is only very slightly influenced by the distribution of the separator Y. Then they provided, by assuming that X is distributed as a negative exponential, very simple approximations for Q_1 and Q_2:

$$Q_1 \cong \frac{2A^2}{1+A}$$

$$Q_2 \cong \frac{A}{2-A}$$

■ **Example 8.20** In medical imaging studies, the following common rating method is generally used. Images from diseased and nondiseased subjects are thoroughly mixed, then presented in this random order to a reader who is asked to rate each on a discrete ordinal 5-point scale ranging from definitely normal to definitely abnormal. The following table gives the rating of 109 CT images (data from Hanley and McNeil, 1982).

	Rating by Reader					
Status	Definitely Normal (1)	Probably Normal (2)	Questionable (3)	Probably Abnormal (4)	Abnormal (5)	Total
Normal	33	6	6	11	2	58
Abnormal	3	2	2	11	33	51
Total	36	8	8	22	35	109

From these data, with a five-category ordinal outcome, we have

$$A = \frac{\{(33)(2+2+11+33)+(6)(2+11+33)+(6)(11+33)+(11)(33)\}+(1/2)\{(33)(3)+(6)(2)+(11)(11)+(2)(33)\}}{(51)(58)}$$

$$A = 0.893$$

The probability is 0.893 that an abnormal subject has a higher rating; or the probability is 0.107 that an abnormal subject has a smaller rating. In this example, a higher rating is associated with abnormality.

Using the Hanley–McNeil approximation, we can obtain

$$Q_1 \cong \frac{2A^2}{1+A} = \frac{(2)(0.893)^2}{1+0.893} = 0.843$$

$$Q_2 \cong \frac{A}{2-A} = \frac{0.893}{2-0.893} = 0.807$$

leading to

$$SE(A) = \sqrt{\frac{(0.893)(1-0.893)+(58-1)(0.843-0.893^2)+(51-1)(0.807-0.893^2)}{(58)(51)}}$$
$$= 0.031$$

A 95% confidence interval for A is $0.893 \pm (1.96)(0.031) = (0.832, 0.954)$.

Using the Hanley–McNeil approximation for the variance, confidence intervals can be formed as seen in Example 8.20. In fact, there are several different ways to obtain standard error and/or confidence intervals but numerical results are very similar. For example, Ury (1972) suggested an approximate $(1-\alpha)$ 100% confidence interval, obtained by applying the *Chebyshev inequality*, which is as follow:

$$A \pm \frac{1}{\sqrt{4\min(n_0, n_1)(1-\alpha)}}$$

■ **Example 8.21** For the same data of Example 8.20, the approximate 95% confidence interval for A by Ury (1972) is very similar to the result in Example 8.20:

$$0.893 \pm \frac{1}{\sqrt{(4)(51)(0.95)}} = 0.893 \pm 0.072 = (0.821, 0.965)$$

8.6 THE OPTIMIZATION PROBLEM

Consider this trivial example. A man has a physical examination; he's concerned that his prostate-specific antigen (PSA) level is "kind of high." Does he have prostate cancer? The truth would be confirmed by a biopsy. But it's a very painful process. At the very least, could we say if he needs a biopsy? Then, suppose prostate cancer has been confirmed; the next concern is whether the cancer has spread to neighboring lymph nodes. Knowledge would dictate an appropriate treatment strategy. Again, the truth would be confirmed by performing a laparotomy, but any surgery involves risks. The question is whether we can accurately predict nodal involvement without surgery.

Possible predictors include acid phosphatase level in blood serum, among others. Is the level "high enough"? In both stages of this example, a decision is needed. Unless we can dichotomize the measurement (PSA or acid phosphatase), discoveries about the role of prostate-specific antigen (PSA) and the role of acid phosphatase level in blood serum will remain laboratory findings—without real applications. Therefore finding an optimal cutpoint is an important step and is part of translational research.

For a continuous marker/predictor such as PSA or acid phosphatase level in blood serum, the basic question is: "How high is high?" or "How low is low?" Keep in mind that, for any continuous separator Y, no matter where you cut, both errors result! More importantly, the sizes of these errors depend on the cutpoint $Y = y$. In practice, cutpoints are formed arbitrarily because we fail to form and justify a criterion or criteria. We need an *optimal cutpoint*; but what do we mean by optimal? There may be more than one solution.

8.6.1 Basic Criterion: Youden's Index

Youden's index (Youden, 1950) is defined simply as

$$J = 1-(\alpha+\beta) = S^{+} + S^{-} - 1$$

where α is the size of Type I error (false-positive) and β is the size of Type π error (false- negative). It is not influenced by the disease prevalence and its value is large when both sensitivity and specificity are high. There are a number of other important reasons that make Youden's index J even more special.

Process Qualification

Suppose we want to decide if a "process" is qualified as a "test" (used for diagnosis).

The minimum criterion a process must pass to qualify as a test is that it detects disease better than by chance alone. A process can only qualify as a test if it selects diseased persons with higher probability, that is, $P^{+} = \Pr(D = + | T = +) > (= \Pr(D = +)$; thus knowledge $(T = +)$ helps.

$$P^{+} > \pi \Longleftrightarrow J > 0$$

In other words, Youden's index J is a measure of basic association: the disease D and the test T are independent if and only if $J = 0$.

Diagnostic Power

In using a diagnostic marker or test, the gains are

$$[P^{+} - \pi] \quad \text{and} \quad [P^{-} - (1-\pi)]$$

Youden's index J is some kind of weighted average of those gains:

$$\frac{1}{\pi(1-\pi)J} = \frac{1}{P^{+}-\pi} + \frac{1}{P^{-}-(1-\pi)}$$

Measure of Precision

As seen in Section 8.4, suppose we have a screening test T; its sensitivity S^+ and specificity S^- have been independently established. A prevalence survey is conducted in order to estimate the disease prevalence. The data show x of n subjects found positive by the test; the disease prevalence $\pi = \Pr\ (D = +)$ is estimated by

$$p = \frac{p_t + S^- - 1}{J}$$

The variance and standard error of this estimate are

$$\mathrm{Var}(p) = \frac{\mathrm{Var}(p_t)}{J^2}$$

$$SE(p) = \frac{1}{J}\sqrt{\frac{p_t(1-p_t)}{n}}$$

The *precision* of estimation of the prevalence depends only on Youden's index; the larger the index, the smaller the standard error.

Because of the above reasons, one basic criterion would be to maximize youden's index of the dichotomized test in the determination of an optimal cutpoint for a continuous marker. With this strategy, when using the resulting dichotomized test in a prevalence survey, we would obtain an estimate with minimal error. Of course, there are other gains too.

8.6.2 Possible Solutions

In the determination of an optimal cutpoint for a continuous marker by maximizing Youden's index of the dichotomized test, there are several ways to get there.

Empirical Approach

We could pool the two samples, the cases and the controls, and arrange the data in the combined sample in increasing order. At each midway between two data points, calculate the sensitivity s^+ and specificity s^-; then Youden's index $J = s^+ + s^- - 1$. We then locate the cutpoint corresponding to the maximum value of J. Of course, the solution is simple but it's hard to determine its standard error.

Nonparametric Approach

The ROC function $R(.)$ maps $U = 1 - S^-$ on the horizontal axis to $V = S^+$ on the vertical axis, $V = R(U)$. Youden's index $(J = S^+ + S^- - 1 = R(U) - U)$ is maximized when $0 = R'(U) - 1$, or $R'(U) = 1$. The process could be as follows:

Step 1: Smooth the empirical estimate by any smoothing technique (e.g., lowess).
Step 2: Locate the point with slope = 1 to obtain specificity.
Step 3: Go to the control sample to get the cutpoint.

It may require lots of data and it's still very hard to determine standard error.

Semi-nonparametric Approach

We are still looking for the point on the curve with slope = 1, but we first fitt the empirical data with a smooth curve $Y = R(U|\theta)$ because it takes less data to do a better job than nonparametric smoothing (we need a model but can check for goodness-of-fit). The two components to be chosen are (i) a meaningful parameter θ and (ii) a functional form for $R(.)$.

Since what we have on the axes of the ROC curve are two survival functions, one possibility is the *proportional hazards model* (PHM), also called *Lehmann's alternatives*. The results are

$$1-F^{+}(y) = (1-F^{-}(y))^{\theta};\ \text{or}\ v = R(u) = 1-(1-u)^{\theta},\ 0 \leq u \leq 1$$

$$\text{Area under the ROC curve:} \quad A = \frac{\theta}{\theta+1}$$

$$\text{Youden's index:} \quad J(u) = v-u = 1-u-(1-u)^{\theta}$$

The semi-nonparametric approach can be summarized in four steps;

Step 1: Model the ROC function by the PHM:

$$R(u) = 1-(1-u)\theta,\ 0 \leq u \leq 1$$

Step 2: Maximize Youden's index,

$$J = 1-u-(1-u)\theta$$

to obtain $u = F^{-}(y)$.

Step 3: Solve for the optimal cutpoint: $y = (F^{-})^{-1}(u)$.

Step 4: In the result, θ is estimated from $A = \theta/(\theta + 1)$; and the area A is obtained from the Wilcoxon rank sum:

$$\frac{dJ(u)}{du} = -1 + \theta(1-u)^{\theta-1} = 0$$

$$u = 1 - \frac{1}{\theta^{1/(\theta+1)}}$$

$$\theta = \frac{A}{1-A}$$

$$A = \frac{1}{n_0 n_1}\{W_1 - \tfrac{1}{2}n_1(n_1+1)\}$$

The value u obtained is the optimal value for the cumulative distribution function (cdf) of the control group; knowing the value of u, and having the sample of controls, leads to the optimal cutpoint for the marker Y. In implementation, we could first get the rank sum W_0 by using a packaged computer program such as SAS. The next step is calculating the area A under the ROC curve, then θ, the parameter of the PHM. We then calculate u, which is (1- cdf) of the controls. Finally, we calculate the optimal cutpoint. Of course, it is still not easy to obtain the standard error, but it is possible by the delta method.

■ **Example 8.22** For the same data as in Example 8.20, we have

$$A = \text{Area} = 0.014$$

$$\theta = \frac{A}{1-A} = 0.014 \text{ approximately}$$

$$u = 1 - \frac{1}{\theta^{1/(\theta-1)}} = F^{-}(x) = 0.986$$

$F^{-}(3)$ is 0.986; the optimal cutpoint is between 4 and 5 (56/58 = 0.966). Classify as "abnormal" only those with a rating of 5 (resulting test is 99% specific but only 63% sensitive). The result is similar to the ACS recommendation to classify as "abnormal/tumor" only those mammograms with a rating of 5 = definitely abnormal.

■ **Example 8.23** There were 53 patients with prostate cancer: 20 of them with nodal involvement and 33 without. We examined the level of acid phosphatase in blood serum ($\times 100$). Data are reproduced from Miller et al. (1980) and are as follows:

Patients Without Nodal Involvement: 40, 40, 46, 47, 48, 48, 49, 49, 50, 50, 50, 50, 50, 52, 52, 55, 55, 56, 59, 62, 62, 63, 65, 66, 71, 75, 76, 78, 83, 95, 98, 102, 187.

Patients with Nodal Involvement: 48(6), 49(9), 51(16), 56(21.5), 67(30), 67 (30), 67(30), 70(32,5), 70(32.5), 72(35), 76(37.5), 78(40), 81(41), 82(42.5), 82(42.5), 84(45), 89(46), 99(49), 126(51), 136(52). Numbers in parentheses are the ranks in the combined sample; midranks are used for tied observations.

We found the following results:

$$\text{Rank sum for cases: } W_1 = 689$$

$$\text{Area under the ROC curve: } A = 0.726$$

$$\theta = 0.378$$

$$u_{\text{op}} = 0.209$$

Leading to an optimal cutpoint of 0.75, subjects with a level of acid phosphatase in blood serum greater than 0.75 are classified as "involved"; this corresponds to a specificity of 0.791 (which is $1 - u_{\text{op}}$) and a sensitivity of 0.554.

Parametric Approach

In general, a parametric approach can be summarized as follows:

$$\begin{aligned} S_{\text{D}}(t) &= 1 - F^{+}(t) = S^{+} \\ S_{\text{H}}(t) &= 1 - F^{-}(t) = 1 - S^{-} \\ R[S_{\text{H}}(t)] &= S_{\text{D}}(t) \\ R(u) &= S_{\text{D}}[S_{\text{H}}^{-1}(u)];\ u \in [0,1] \\ J &= R(u) - u;\ u = 1 - S^{-} \end{aligned}$$

Then solving to get the optimal location,

$$\begin{aligned} J' &= R'(u) - 1 \\ &= 0 \Longleftrightarrow 1 = R'(u) \end{aligned}$$

For example, when the separator X is distributed as log-logistic in both populations—the population of cases and the population of controls—the survival function of the log-logistic distribution given by

$$\begin{aligned} S(t) &= \frac{1}{1 + (\rho t)^{\nu}} \\ \rho &= e^{-\mu}, \text{ where } \mu \text{ is mean} \\ \nu &= \frac{1}{\sigma}, \text{ where } \sigma \text{ is standard deviation} \end{aligned}$$

When we further assume that the two distributions have equal variances, the ROC function can easily be expressed as

$$\sigma_{\text{D}} = \sigma_{\text{H}} = \sigma$$

Then

$$\begin{aligned} R(u) &= \frac{u}{u + (1-u)\exp\left(-\frac{\mu_{\text{D}} - \mu_{\text{H}}}{\sigma}\right)} \\ &= \frac{u}{u + (1-u)\beta} \end{aligned}$$

The term involved in the last equation is

$$\beta = \exp\left(-\frac{\mu_D - \mu_H}{\sigma}\right)$$

$$0 < \beta < 1 \text{ for } \mu_D > \mu_H$$

Applying the above method of optimization, we obtain

$$R(u) = \frac{u}{u + (1-u)\beta}$$

$$R'(u) = \frac{\beta}{[u + (1-u)\beta]^2}$$

$$R'(u) = 1 \Longleftrightarrow u = \frac{-\beta + \sqrt{\beta}}{1-\beta}$$

$$\text{Optimal:} \quad S^- = 1-u = S^+ = \frac{1-\sqrt{\beta}}{1-\beta}$$

where

$$\beta = \exp\left(-\frac{\mu_D - \mu_H}{\sigma}\right) = \exp(-d)$$

where d is the *effect size*, the difference of population means divided by the common standard deviation. In other words, the effect size determines the screening value of a biomarker.

■ **Example 8.24** The following table shows the screening values and values of sensitivity and specificity at the optimal cutpoint of a biomarker for various effect sizes:

d	$S^- = S^+$
1	62%
2	73%
3	82%
4	88%

8.7 STATISTICAL CONSIDERATIONS

In this section, we cover three basic issues: evaluation of screening tests, comparison of screening tests, and consideration of subjects' characteristics.

8.7.1 Evaluation of Screening Tests

In an evaluation, we want to see if certain test is "acceptable," that is, meeting certain prespecified criteria. For example, we may want to accept and use only tests with specificity of at least 90%. A test's evaluation is simpler when the endpoint is binary, while evaluation is more complicated for continuous endpoints.

Evaluation of Tests on Binary Scale

For diagnostic or screening tests with binary endpoint, for example, the presence or absence of a bacterium (infections) or some specific DNA sequence (genetic tests), their evaluations are simple. The evaluations are simple because all parameters, such as sensitivity and specificity, are proportions or functions of proportions, making the evaluation process very simple—or at least straightforward.

For example, when we have n_0 as the number of controls and n_{00} as the number of controls tested negative,

$$\begin{aligned} S^- &= \Pr(T = - | D = -) \\ s^- &= \frac{n_{00}}{n_0} \\ \text{Var}(s^-) &= S^-(1 - S^-)/n_0 \\ SE(s^-) &= \sqrt{(s^-)(1 - s^-)/n_0} \end{aligned}$$

Therefore if we want to accept and use only tests with specificity of at least 90%, we try one-sample one-sided test using

$$z = \frac{s^- - 0.9}{SE(s^-)}$$

Evaluation of Tests on Continuous Scale

For diagnostic tests with continuous endpoint, for example, blood glucose level or antibody level measured by ELISA, diagnostic power is represented by the area under the ROC curve. Estimation of this parameter has been covered previously; however, results are more often used for comparison—not for evaluation.

For the case of a continuous endpoint, we can also impose a condition; say, to define a "good test" for a particular disease as one that detects 90% or more of the diseased population while misclassifying no more than 5% of the well. Here an optimal dichotomization is not needed; the question is if we could find a cutpoint, any cutpoint, so that the resulting dichotomized test meets the condition. Given data from the developmental stage, we can construct a z-test as follows. Let W_{95} be the 95th percentile of the distribution for controls (or Well) and D_{10} be the 10th percentile of the distribution for the cases (or Diseased). Assume that large values of the test are associated with the disease, as in the case of blood glucose. Then, the (alternative) hypothesis we wish to test is

$$H_0: \ \theta = D_{10} - W_{95} > 0$$

The rationale is as follows. Let C be the cutpoint; in order to misclassify no more than 5% of the well, we must have

$$\text{Condition 1:} \ C \geq W_{95}$$

Similarly, in order to detects 90% or more of the cases, we must have

$$\text{Condition 2:} \ C \leq D_{10}$$

In order to satisfy both Conditions 1 and 2, we must have $D_{10} \geq W_{95}$ or as seen in H_0.

Let (μ_0, σ_0) and (μ_1, σ_1) be the mean and standard deviation of the populations of controls and cases, respectively. Then we have

$$W_{95} = \mu_0 + 1.645\sigma_0$$

$$D_{10} = \mu_1 - 1.282\sigma_1$$

$$\theta = \mu_1 - 1.282\sigma_1 - \mu_0 - 1.645\sigma_0$$

Given two independent samples, one from the cases and one from the controls, we have

$$\hat{\theta} = \bar{y}_1 - 1.282 s_1 - \bar{y}_0 - 1.645 s_0$$

$$\text{Var}(\hat{\theta}) = \frac{\sigma_1^2}{n_1}\left\{1 + \frac{(1.282)^2}{2}\right\} + \frac{\sigma_0^2}{n_0}\left\{1 + \frac{(1.645)^2}{2}\right\}$$

$$SE(\hat{\theta}) = \sqrt{\frac{s_1^2}{n_1}\left\{1 + \frac{(1.282)^2}{2}\right\} + \frac{s_0^2}{n_0}\left\{1 + \frac{(1.645)^2}{2}\right\}}$$

$$z = \frac{\hat{\theta}}{SE(\hat{\theta})} \quad \text{versus} \ N(0, 1) \ \text{under} \ H_0$$

The z-score is referred to the percentiles of the standard normal distribution for a decision to reject or not to reject the null hypothesis.

8.7.2 Comparison of Screening Tests

In a comparison, we want to see if two tests, usually a new versus a more established one, have the same performance using a statistical test or tests of significance. The comparison is easy; the more difficult problem is how to express the "level of difference" if the two screening tests do not have the same performance (i.e., statistical test is significant).

Designs for Comparison of Tests

Decisions about which test or tests to recommend for widespread use and which to abandon, assuming that more than one are acceptable, are made on the basis of research studies that compare the accuracies of the tests.

(i) If each study subject is tested by all tests, referred to as *paired design*, even more than two tests are under consideration.

(ii) If each study subject is tested by one test, we refer to the design as *unpaired*.

Use of Unpaired Design

In the design stage of unpaired design, we follow the same design principles of multiarm randomized clinical trials. Those include well-defined inclusion–exclusion criteria and clear a priori definition of disease and test result—including measurement scale, preparation of study protocol, and randomization to ensure that study arms are balanced with regard to factors affecting test performance and/or result. Blinding—if feasible—may be needed to ensure integrity of disease and test assessments. An analysis plan must also be in place.

The data layout is as follows, with four groups of test subjects:

Cases	Test Result		
Test	Negative ($T = -$)	Positive ($T = +$)	Total
A	$n_{10}(A)$	$n_{11}(A)$	$n_1(A)$
B	$n_{10}(B)$	$n_{11}(B)$	$n_1(B)$

Controls	Test Result		
Test	Negative ($T = -$)	Positive ($T = +$)	Total
A	$n_{00}(A)$	$n_{01}(A)$	$n_0(A)$
B	$n_{00}(B)$	$n_{01}(B)$	$n_0(B)$

Data analysis to compare the two screening tests is straightforward. We can perform two separate chi-squared tests, one for cases and one for controls; for an overall level of α, each test is performed at $\alpha/2$. If the difference between two diagnostic tests is found to be statistically significant, the level of difference should be summarized and presented; this part is more complicated than the statistical tests themselves. The two commonly used parameters are the ratio of two sensitivities (RS^+) and the ratio of two specificities (RS^-); these are ratios of independent proportions. Variances are calculated as follows:

$$RS^{+}(A,B) = \frac{S_A^+}{S_B^+}$$

$$\text{Var}\{\ln RS^{+}(A,B)\} = \frac{1-S_A^+}{n_1(A)} + \frac{1-S_B^+}{n_1(B)}$$

$$RS^{-}(A,B) = \frac{S_A^-}{S_B^-}$$

$$\text{Var}\{\ln RS^{-}(A,B)\} = \frac{1-S_A^-}{n_0(A)} + \frac{1-S_B^-}{n_0(B)}$$

■ **Example 8.25** The following is a simple data set from the book by Pepe (2004). This is a randomized study of chronic villus sampling (CVS: Test *B*) versus early amniocentesis (EA: Test *A*) for fetus abnormality (Disease *D*).

Cases	Test Result		
Test	Negative ($T = -$)	Positive ($T = +$)	Total
EA	6	116	122
CVS	13	11	124

Controls	Test Result		
Test	Negative ($T = -$)	Positive ($T = +$)	Total
EA	4844	34	4878
CV	4765	111	4876

Results of the two chi-squared tests are

$$\text{Cases}: \chi^2 = \frac{246\{(6)(111)-(13)(116)\}^2}{(19)(127)(122)(124)} = 4.78$$

$$\text{Controls}: \chi^2 = \frac{9754\{(4844)(11)-(4765)(34)\}^2}{(9609)(145)(4878)(4876)} = 41.54$$

Both tests are statistically significant, especially the comparison of specificities from the two control groups. The differences are summarized into the ratio of sensitivities and the ratio of specificities:

$$RS^{+} = \exp\left\{\ln\frac{0.951}{0.895} \pm (2.28)\sqrt{\frac{1-0.951}{122} + \frac{1-0.895}{124}}\right\} = (0.981,\ 1.151)$$

$$RS^{-} = \exp\left\{\ln\frac{0.993}{0.977} \pm (2.28)\sqrt{\frac{1-0.993}{4878} + \frac{1-0.977}{4876}}\right\} = (1.007,\ 1.024)$$

Use of Paired Design

The design and analysis of unpaired designs are simple but, if feasible, paired designs are more desirable. However, paired designs are only valid if tests do not interfere with each other; be cautious because interference can be subtle. Also pay attention to cooperation of the subjects; *order* should be randomized if feasible. The following are advantages of paired designs:

(i) They are more efficient because the impact of between-subject variability is minimized.
(ii) Possibilities of confounding are eliminated.
(iii) One can examine characteristics of subjects where tests yield different results; this can lead to insight about test performance and sometimes strategies for improvement.
(iv) One can assess the value of applying combinations of tests compared to single tests.

The data layout with two groups of test subjects—one group of cases and one group of controls—is as follows:

	Test A Result		
Test B Result	Negative	Positive	Total
Negative	a_1	b_1	$n_{10}(B)$
Positive	c_1	d_1	$n_{11}(B)$
Total	$n_{10}(A)$	$n_{11}(A)$	n_1

	Test A Result		
Test B Result	Negative	Positive	Total
Negative	a_0	b_0	$n_{00}(B)$
Positive	c_0	d_0	$n_{01}(B)$
Total	$n_{00}(A)$	$n_{01}(A)$	n_0

Data analysis to compare the two screening tests is also straightforward, as in the case of unpaired designs. We can perform two separate McNemar's chi-squared tests, one for the set of cases and one for the set of controls; for an overall level of α, each test is performed at $\alpha/2$. If the difference between two diagnostic tests is found to be statistically significant, the level of difference should be summarized and presented; this part is more complicated than the statistical tests themselves. The two commonly used parameters are the ratio of two sensitivities (RS^+) and the ratio of two specificities (RS^-). However, these are no longer ratios of independent proportions; the method becomes even a little more complicated.

$$S_A^+ = n_{11}(A)/n_1$$

$$S_B^+ = n_{11}(B)/n_1$$

$$RS^+(A,B) = \frac{b_1 + d_1}{c_1 + d_1}$$

$$\text{Var}\left\{\ln RS^+(A,B)\right\} = \frac{b_1 + c_1}{(b_1 + d_1)(c_1 + d_1)}$$

$$RS^-(A,B) = \frac{b_0 + a_0}{c_0 + a_0}$$

$$\text{Var}\left\{\ln RS^-(A,B)\right\} = \frac{b_0 + c_0}{(b_0 + a_0)(c_0 + a_0)}$$

(For more details, see Cheng and Macaluso (1997, pp. 104–106).)

■ **Example 8.26** The following is a simple data set from the book by Pepe (2004). A paired study of exercise stress test (EST: Test *B*) versus chest pain history (CPH: Test *A*) for diagnosing coronary artery disease (Disease *D*):

Cases	Test A Result		
Test B Result	Negative	Positive	Total
Negative	25	183	208
Positive	29	786	815
Total	54	969	1023

Controls	Test A Result		
Test B Result	Negative	Positive	Total
Negative	151	176	327
Positive	46	69	115
Total	197	245	442

Results of the two chi-squared tests are

$$\text{Cases}: \chi^2 = \frac{(183-29)^2}{183+29} = 111.87$$

$$\text{Controls}: \chi^2 = \frac{(176-46)^2}{176+46} = 76.13$$

Both tests are statistically significant, especially more so with the comparison of sensitivities from the two groups of cases. The differences are summarized into the ratio of sensitivities and the ratio of specificities:

$$RS^{+}(A,B) = \frac{183+786}{29+783} = 1.189$$

$$\text{Var}\{\ln RS^{+}(A,B)\} = \frac{183+29}{(183+786)(29+786)} = (0.016)^2$$

$$RS^{-}(A,B) = \frac{176+69}{46+69} = 2.130$$

$$\text{Var}\{\ln RS^{-}(A,B)\} = \frac{176+46}{(176+69)(46+69)} = (0.089)^2$$

leading to confidence intervals for the two ratios:

$$RS^{+} = \exp\{\ln(1.189) \pm (2.28)(0.016)\} = (1.146, 1.232)$$

$$RS^{-} = \exp\{\ln(2.130) \pm (2.28)(0.089)\} = (1.738, 2.609)$$

8.7.3 Consideration of Subjects' Characteristics

Various factors can influence the performance of a diagnostic test: the environment in which it is performed, the characteristics of a technician, and especially the characteristics of the test subject. It may be important to identify and understand the influence of such factors in order to optimize conditions for using a test. In general, regression analysis can be used to make inferences.

The following are only a few examples:

(i) The ability of mammography to detect breast cancer depends on the woman's age: younger women have denser breasts, which renders the mammogram more difficult to read and to interpret.

(ii) Men and women differ in their abilities to perform physical exercise; gender should be considered in evaluating an exercise stress test.

(iii) The experience of the pathologist makes a difference in reading histologic slides.
(iv) Serum levels of PSA tend to be higher in older men than in younger men; it has been proposed that criteria for a positive PSA test should be more stringent in older men (i.e., higher cutpoint).
(v) Levels of PSA also tend to be high in men with enlarged (but not cancerous) prostate glands.

In general, it is important to identify both optimal and suboptimal conditions. Insights can lead to modifications of a test to improve its performance. Even if we find, in research, that a factor does not influence test performance, we may be able to relax the conditions under which the test is performed. (So "negative findings" are beneficial as well.)

For diagnostic tests with binary endpoint, for example, the presence or absence of a bacterium (infections) or some specific DNA sequence (genetic tests), we can model the *true-positive probability* ($\Pr(T=+|\mathrm{D}=+)$, or sensitivity) or the false-positive probability ($\Pr(T=+|\mathrm{D}=-)$, or $(1-\text{specificity})$). Note that, in both, the event $(T=+)$ is random but the effects of a covariate may be different; for example, a comorbidity may only affect $\Pr(T=+|D=+)$ but not $\Pr(T=+|D=-)$.

We need to model a probability, $\Pr(T=+|\mathrm{D}=+)$ or $\Pr(\mathrm{T}=+|\mathrm{D}=-)$, as a function of one or a set of several covariates—each is binary or continuous. Some prefer to model the log of the probability as a linear function of covariates; some prefer the conventional use of logistic regression (see more details in Section 4.1). In the first choice, to model the log of the probability as a linear function of covariates,

$$\ln P = \beta_0 + \beta_1 x$$

The advantage of the first approach is easy interpretation of model parameters—the probability is changed by a multiple constant. The disadvantages are that fitted probabilities may exceed 1.0, and the unit exponential distribution is defined only on the positive range.

When we have a common set of covariates, we can consider fitting a composite model to the combined data instead of fitting separate models to $\Pr(T=+|D=+)$ and $\Pr(T=+|D=-)$. The advantage is that one can test if a covariate is common to both models by including an *interaction term* (product of covariate and D). If so, a combined model requires estimation of fewer parameters, leading to greater precision. Suppose $X=0/1$ is a binary covariate and D is coded as $(+=1, -=0)$. Then in the combined model,

$$\ln\{\Pr(T=+)\} = \beta_0 + \beta_1 X + \beta_2 D + \beta_3 XD$$

β_1 represents the effect of X on false-positive probability, whereas $(\beta_1 + \beta_3)$ represents the effect of X on true-positive probability (sensitivity); β_2 only tells the difference of test responses from the diseased and the nondiseased populations.

In the second choice for diagnostic tests with binary endpoint, we apply the logistics regression model, but issues are similar.

Of course, if the endpoint of a test is on the continuous scale, we can simply use regular regression analysis with the usual normal error regression model.

EXERCISES

8.1 Reexamine the case of Example 8.4, where we assume that there is a good screening procedure (98% sensitive and 97% specific) and let us consider two other examples—a low-risk subpopulation (prevalence is 0.5%) and a higher-risk subpopulation (prevalence is 10%). In each case, calculate the positive predictive value and the negative predictive value.

8.2 Prove that, in general:

(**a**) The higher the prevalence, the higher the positive predictive value.

(**b**) The higher the prevalence, the lower the negative predictive value.

8.3 Complete the table in Example 8.3. Using the results of the last row, draw your conclusion concerning the effect of S^+ on π when S^- is high.

8.4 Express the kappa statistic (κ) as a function of S^+, S^-, and π. Is it an increasing or decreasing function of π?

8.5 Prove that $P^- > (1 - \pi)$ if and only if $J > 0$.

8.6 Prove that, in the 2-by-2 cross-classification of $D(+, -)$ versus $T(+, -)$, the odds ratio is equal to 1 if and only if $J = 0$.

8.7 Let us assume that we know $\pi = 0.1$, and we have estimates $s^+ = s^- = 0.9$. Using sample sizes in two independent investigations with $n_1 = 100$ and $n_0 = 300$, find the bias for p in Design 1 Plus and compare the result to that of p_t for Design 1 in Example 8.12.

8.8 Assume the same numbers as in Exercise 8.7, and that a prevalence survey is conducted with a sample size $n = 1000$. Find the percentage of $\text{Var}(p^+)$ due to the prevalence survey, and the estimations of s^+ and s^-.

8.9 Refer to the data set on prostate cancer in Example 8.23. Form the ROC curve and estimate the area under it, including 95% confidence intervals.

8.10 Model the ROC curve according to the proportional hazards model (also called Lehmann's alternatives):

$$1 - F^+(y) = \{1 - F^-(y)\}^\theta; \text{ or } R(u) = 1 - \{1 - u\}^\theta, 0 \le u \le 1$$

Show how θ is related to the area under the curve. Then show how to determine specificity, sensitivity, and cutpoint $Y = y$ so as to maximize Youden's index.

8.11 In a previous study (Anderson et al., 2001) of environmental tobacco smoke, we compared two groups of nonsmoking women: $n_1 = 23$ women who had male partners who smoked in the home and $n_0 = 22$ women who had male partners who did not smoke. Urine samples were obtained and analyzed and the comparison was based on a number of chemicals, among then cotinine (a metabolite of nicotine, in nmol/mL) and NNAL and its glucuronide, NNAL-Gluc (NNAL and NNAL-Gluc are metabolites of the tobacco-specific lung carcinogen called NNK, in pmol/mL). Data (cotinine, NNAL + NNAL-Gluc) are given below (ND is for "not detectable," the limit of detection for cotinine is 0.003 nmol/mL and for NNAL and NNAL-Gluc is 0.005 pmol/mL; one case has a missing value for NNAL + NNAL-Gluc):

Nonexposed Women. (ND, ND), (ND, ND), (ND, ND), (ND, ND), (ND, ND), (ND, 0.008), (0.003, ND), (0.003, 0.015), (0.006, ND), (0.007, ND), (0.007, ND), (0.007, 0.018), (0.008, ND), (0.008, ND), (0.009, ND), (0.01, ND), (0.012, ND), (0.016, ND), (0.017, ND), (0.019, 0.047), (0.025, ND), and (0.03, ND).

Exposed Women. (ND, 0.067), (0.003, 0.009), (0.003, 0.012), (0.007, 0.039), (0.008, ND), (0.008, 0.010), (0.008, 0.011), (0.009, ND), (0.011, 0.037), (0.017, 0.072), (0.018,—), (0.021, 0.083), (0.036, 0.022), (0.037, 0.032), (0.042, 0.063), (0.046, ND), (0.053, 0.210), (0.076, 0.041), (0.099, 0.018), (0.101, 0.031), (0.111, 0.018), (0.122, 0.282), and (0.200, 0.027).

Determine the optimal cutpoint for cotinine and the corresponding sensitivity and specificity.

8.12 Using the same data set as in Exercise 8.11, determine the optimal cutpoint for NNAL + NNAL-Gluc and the corresponding sensitivity and specificity.

8.13 Compare the test Obtained in Exercise 8.11 and the test obtained in Exercise 8.12. Calculate the 95% confidence intervals for RS^+ and RS^-.

8.14 Use the data (Example 8.25) for chronic villus sampling (CVS: Test B) versus early amniocentesis (EA: Test A) for fetus abnormality. Calculate the joint 80% confidence intervals for RS^+ and RS^-.

8.15 Use the data (Example 8.26) for exercise stress test (EST: Test B) versus chest pain history (CPH: Test A) for diagnosing coronary artery disease. Calculate the joint 80% confidence intervals for RS^+ and RS^-.

8.16 Use the data (Example 8.26) for exercise stress test (EST: Test B) versus chest pain history (CPH: Test A) for diagnosing coronary artery but treat the design as unpaired. Compare the two diagnostic tests.

9

TRANSITION FROM CATEGORICAL TO SURVIVAL DATA

Methodology discussed in this text has been directed toward the analyses of categorical data: this chapter is an exception. The topics covered here—basic survival analysis and Cox's proportional hazards regression—were developed to deal with survival data; they are included here for a number of reasons. First, the borderline between categorical and survival data is rather vague, especially for beginning students. Survival analysis is focused on the occurrence of an event, such as death—a binary outcome. Therefore, for beginners, it may be confused with the type of data that require the logistic regression analysis of Chapter 4.

The basic difference is that, first, for survival data, studies have staggered entry and subjects are followed for varying lengths of time; they do not have the same probability for the event to occur even if they have identical characteristics, a basic assumption of the logistic regression model. Second, statistical tests for the comparison of survival distributions are special forms of the Mantel–Haenszel method of Section 2.4. In addition, the coverage of Cox's proportional hazards regression would enrich or supplement Chapter 4 because, for certain special cases, this model and the conditional logistic regression model correspond to the same likelihood function and are analyzed using the same computer program (Le and Lindgren, 1988). Finally, the inclusion of some introductory survival analysis at the end of this book is a natural extension; most methods used in survival analysis are generalizations of those for categorical data. For those students in applied fields, such as epidemiology, access to this methodology would be beneficial because most may not be adequately prepared for the level of sophistication of a full course in survival analysis (Le, 1997a). Sections 9.1 and 9.2 introduce some basic concepts and techniques of survival analysis; Cox's regression models are covered in Sections 9.3 and 9.4. The new Section 9.5 is focused on competing risks.

9.1 SURVIVAL DATA

In prospective studies, the important feature is not only the outcome event, such as death, but the time to that event, the survival time. In order to determine the survival time T, three basic elements are needed: (i) a time origin or starting point, (ii) an ending event of interest, and (iii) a measurement scale for the passage of time.

For example, Figure 9.1 shows the lifespan T from birth (starting point) to death (ending event) in years (measurement scale).

The time origin or starting point should be defined precisely but it need not be birth; it could be the start of a new treatment (randomization date in a clinical trial) or the admission to a hospital or a nursing home. The ending event should also be defined precisely but it need not be death; a nonfatal event such as the relapse of a disease (e.g., leukemia) or the relapse from a smoking cessation program or discharge to the community from a hospital or a nursing home can satisfy the definition and is an acceptable choice. The use of calendar time in health studies is common and meaningful; however, other choices for a time scale are justified—for example, hospital cost (in dollars) from admission (starting point) to discharge (ending event).

The distribution of the survival time T from enrollment or starting point to the event of interest, considered as a random variable, is characterized by either of two equivalent functions: the survival function and the hazard function.

The survival function, denoted $S(t)$, is defined as the probability that an individual survives longer than t units of time:

$$S(t) = \Pr(T > t)$$

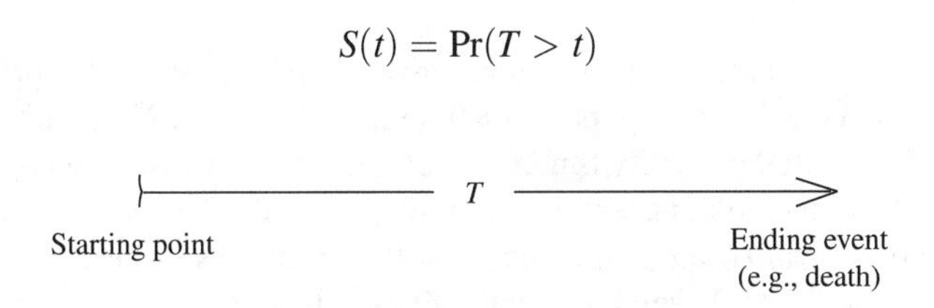

FIGURE 9.1. The survival time.

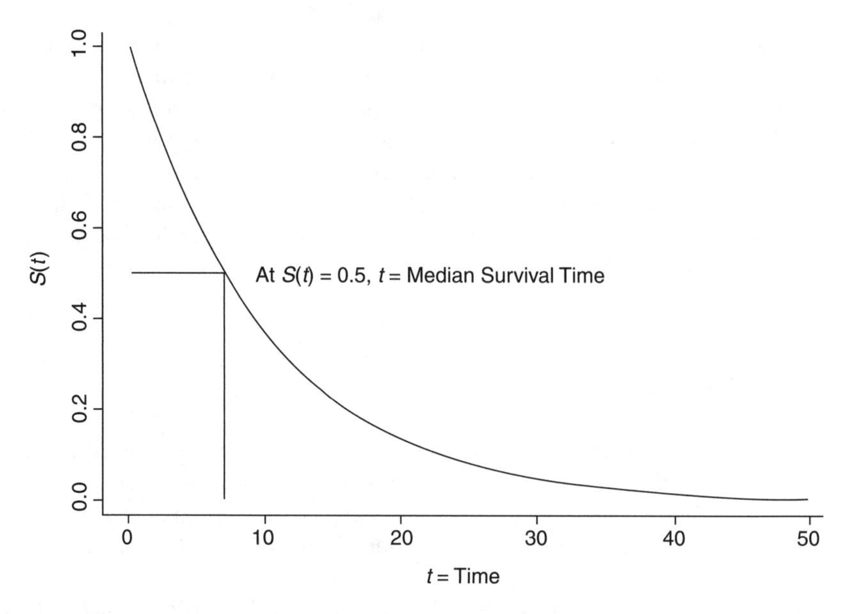

FIGURE 9.2. A Typical survival curve.

$S(t)$ is also known as the survival rate; for example, if times are in years, then $S(2)$ is the 2-year survival rate, $S(5)$ is the 5-year survival rate, and so on. The graph of $S(t)$ versus t is called the *survival curve*, as shown in Figure 9.2.

The hazard or risk function $\lambda(t)$ gives the instantaneous failure rate assuming that the individual has survived to time t:

$$\lambda(t) = \lim_{\delta \to 0} \frac{\Pr(t \le T \le t + \delta)}{\delta}$$

That means, for a small time increment δ, the probability of an event occurring during time interval $(t, t + \delta)$ is given approximately by

$$\lambda(t) \cong \frac{\Pr(t \le T \le t + \delta)}{\delta}$$

In other words, the hazard or risk function $\lambda(t)$ approximates the proportion of subjects dying or having events per unit time around time t. Note that this hazard function differs from the density function represented by the usual histogram; in the case of $\lambda(t)$, the numerator is a conditional probability. The hazard $\lambda(t)$ is also known as the *force of mortality* and is a measure of the proneness to failure as a function of age of the individual. When a population is subdivided into two subpopulations, say, E (for "exposed") and E' (for "nonexposed"), by the presence or absence of a certain characteristic (an exposure such as smoking), each subpopulation corresponds to a hazard or risk function and the ratio of two hazard functions,

$$RR(t) = \frac{\lambda(t;E)}{\lambda(t;E')}$$

is called the *relative risk* of exposure to factor E. In general, the relative risk $RR(t)$ is a function of time and measures the magnitude of an effect; when it remains constant, $RR(t) = \rho$, we have a *proportional hazards model* (PHM):

$$\lambda(t;E) = \rho\lambda(t;E')$$

with the risk of the nonexposed subpopulation serving as the baseline.

This is a multiplicative model; another way to express this model is

$$\lambda(t) = \lambda_0(t)e\beta x$$

where $\lambda_0(t)$ is $\lambda(t,E')$—the hazard function of the unexposed subpopulation—and the indicator (or covariate) x is defined as

$$x = \begin{cases} 0 & \text{if unexposed} \\ 1 & \text{if exposed} \end{cases}$$

The *regression coefficient* β represents the relative risk on the log scale. This model works with any covariate X—continuous or categorical; the above binary covariate is a very special case. Of course, the model can be extended to include several covariates; it is usually referred to as *Cox's regression model* or *proportional hazard regression model.*

A special source of difficulty in the analysis of survival data is the possibility that some individuals may not be observed for the full time to failure or event. The so-called random censoring arises in medical applications with animal studies, epidemiological applications with human studies, or clinical trials. In these cases, observation is terminated before the occurrence of the event. In a clinical trial, for example, patients may enter the study at different times; then each is treated with one of several possible therapies after a randomization. We want to observe their lifetimes from enrollment, but censoring may occur in one of the following forms:

- Loss to follow-up—the patient may decide to move elsewhere.
- Dropout—therapy may have such bad effects and it is necessary to discontinue.
- Termination of the study (for data analysis at a predetermined time.
- Death due to a cause not under investigation (e.g., suicide).

Figure 9.3 shows a description of a typical clinical trial.

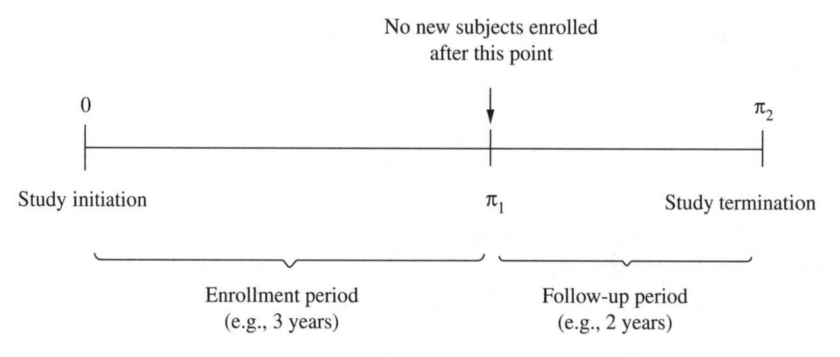

FIGURE 9.3. A clinical trial.

To make it amenable to statistical analysis, we make the crucial assumption that, conditionally on the values of any explanatory variables (or covariates), the prognosis for any individual who has survived to a certain time t should not be affected if the individual is censored at t. That is, an individual who is censored at t should be representative of all those subjects with the same values of the explanatory variables who survive to t. In other words, survival condition and reason of loss are independent; under this assumption, there is no need to distinguish the above four forms of censoring.

We assume that observations available on the failure time of n individuals are usually taken to be independent. At the end of the study, our sample consists of n pairs of numbers (t_i, d_i). Here d_i is an indicator variable for survival status ($d_i = 0$ if individual i is censored; $d_i = 1$ if individual i failed) and t_i is the time to failure (or event, if $d_i = 1$) or the censoring time (if $d_i = 0$); t_i is also called the *duration time*. We may also consider, in addition to t_i and d_i, values on a set of k covariates (or exploratory variables) associated with individual i representing cofactors such as age, sex, and treatment.

9.2 INTRODUCTORY SURVIVAL ANALYSIS

This section briefly introduces a popular method for the estimation of the survival function and a family of statistical tests for the comparison of survival distributions.

9.2.1 Kaplan–Meier Curve

Here we introduce the *product-limit* (PL) method of estimating the survival rates; this is also called the Kaplan–Meier method (Kaplan and Meier, 1958).

Let

$$t_1 < t_2 < \cdots < t_k$$

be the distinct observed death times in a sample of size n from a homogeneous population with survival function $S(t)$ to be estimated ($1 = k = n$). Let n_i be the number of subjects at risk at a time just prior to t_i $(1 = i = k)$; these are cases or subjects in the sample whose duration time is at least t_i, and d_i is the number of deaths at t_i. The survival function $S(t)$ is estimated by

$$\hat{S}(t) = \prod_{t_i \leq t} \left(1 - \frac{d_i}{n_i}\right)$$

which is called the *product-limit estimator* or Kaplan–Meier estimator with a 95% confidence interval given by

$$\hat{S}(t)\exp[\pm 1.96 s(t)]$$

where the standard error $s(t)$ is the square root of the variance (Greenwood, 1926):

$$s^2(t) = \sum_{t_i \leq t} \frac{d_i}{d_i(n_i - d_i)}$$

■ **Example 9.1** The remission times of 42 patients with acute leukemia were reported from a clinical trial undertaken to assess the ability of a drug called 6-mercaptopurine (6-MP) to maintain remission (Freireich et al., 1963; data were taken from Cox and Oakes, 1984). Each patient was randomized to receive either 6-MP or placebo. The study was terminated after 1 year; patients have different follow-up times because they were enrolled sequentially at different times. Times in weeks were as follows:

6-MP Group. 6, 6, 6, 7, 10, 13, 16, 22, 23, 6+, 9+, 10+, 11+, 17+, 19+, 20+, 25+, 32+, 32+, 34+, 35+.

Placebo Group. 1, 1, 2, 2, 3, 4, 4, 5, 5, 8, 8, 8, 8, 11, 11, 12, 12, 15, 17, 22, 23.

A $t+$ denotes a censored observation, that is, the case was censored after t weeks without a relapse. For example, 10+ is a case enrolled 10 weeks before study termination and still remission-free at termination.

According to the product-limit method, survival rates for the 6-MP group are calculated by constructing a table such as Table 9.1 with five columns. To obtain the Kaplan–Meier estimate of $S(t)$, multiply all values in column 4 up to and including t. From Table 9.1, we have, for example (see Figure 9.4):

Seven-week survival rate is 80.67%.

Twenty two-week survival rate is 53.78%.

We have a 95% confidence interval for $S(7)$, the 7-week survival rate: (0.6804, 0.9565).

TABLE 9.1. Survival Rates for 6-MP Group of Example 9.1

(1)	(2)	(3)	(4)	(5)
t	*n*	*d*	1-*d*/*n*	*S*(*t*)
6	21	3	0.8571	0.8571
7	17	1	0.9412	0.8067
10	15	1	0.9333	0.7529
13	12	1	0.9167	0.6902
16	11	1	0.9091	0.6275
22	7	1	0.8571	0.5378
23	6	1	0.8333	0.4482

Note: A SAS program would include these instructions:

```
PROC LIFETEST METHOD=KM;
TIME WEEKS*RELAPSE(0);
```

where `WEEKS` is the variable name for duration time, `RELAPSE` the variable name for survival status, `0` in `RELAPSE(0)` is the coding for censoring used in the data description, and `KM` stands for Kaplan–Meier method.

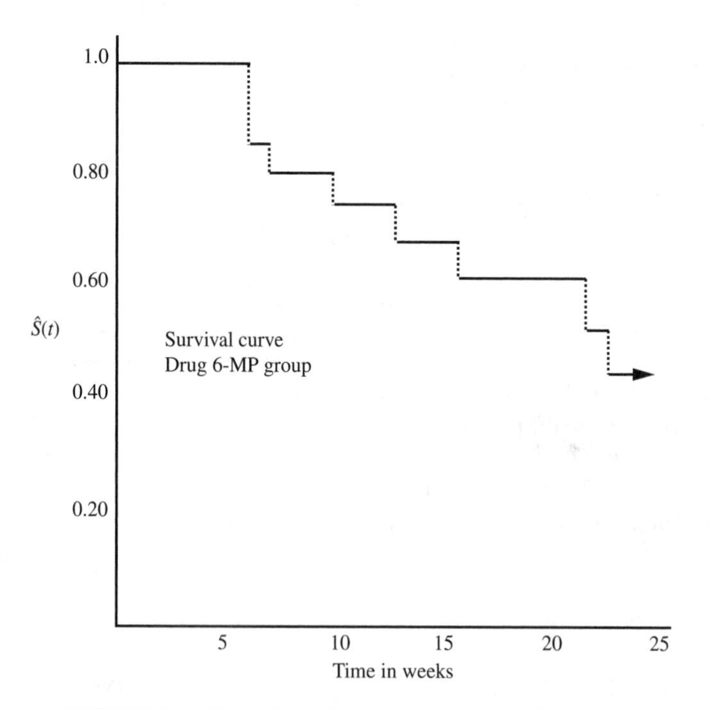

FIGURE 9.4. The estimated survival curve for 6-MP group.

9.2.2 Comparison of Survival Distributions

Suppose that there are n_1 and n_2 individuals corresponding to two treatment groups 1 and 2, respectively. The study provides two samples of survival data:

$$\{(t_{1i}, \delta_{1i});\ i = 1, 2, \ldots, n_1\} \quad \text{and} \quad \{(t_{2j}, \delta_{2j}); j = 1, 2, \ldots, n_2\}$$

In the presence of censored observations, tests of significance can be constructed as follows:

Step 1: Pool data from two samples together and let

$$t_1 < t_2 < \cdots < t_m;\ m \leq d \leq n_1 + n_2$$

be the distinct times with at least one event at each (d is the total number of deaths; $m \leq d$ because we may have more than one death at the same time point or time points).

Step 2: At ordered time t_i, $1 = i = m$, the data may be summarized into a 2×2 table:

Sample	Dead	Alive	Total
1	d_{1i}	a_{1i}	n_{1i}
2	d_{2i}	a_{2i}	n_{2i}
Total	d_i	a_i	n_i

where

n_{1i} = number of subjects from sample 1 who were at risk just before t_i;

n_{2i} = number of subjects from sample 2 who were at risk just before t_i;

$n_i = n_{1i} + n_{2i}$;

d_{1i} = number of deaths at time t_i from sample 1;

d_{2i} = number of deaths at time t_i from sample 2;

$d_i = d_{1i} + d_{2i}$;

$a_{1i} = n_{1i} - d_{1i}$;

$a_{2i} = n_{2i} - d_{2i}$;

$a_i = n_i - d_i$ = number of subjects from both samples who were at risk just before t_i but survived past time t_i.

In this form, the null hypothesis of equal survival functions implies the independence of "sample" and "status" in the above cross-classification 2×2 table. Therefore under the null hypothesis, the *expected value* of d_{1i} (the observed value at the top left corner) is

$$E_0(d_{1i}) = \frac{d_i n_{1i}}{n_i}$$

The variance is estimated using the hypergeometric model:

$$\text{Var}_0(d_{1i}) = \frac{n_{1i} n_{2i} a_i d_i}{n_i^2 (n_i - 1)}$$

Step 3: After constructing one 2×2 table for each uncensored observation, the evidence against the null hypothesis can be summarized in the following statistic:

$$\theta = \sum_{i=1}^{m} w_i [d_{1i} - E_0(d_{1i})]$$

where w_i is some weight associated with the 2×2 table at t_i; some tables may provide more information than others. We have, under the null hypothesis,

$$E_0(\theta) = 0$$

$$\text{Var}_0(\theta) = \sum_{i=1}^{m} w_i^2 \, \text{Var}_0(d_{1i})$$

$$= \sum_{i=1}^{m} w_i^2 \frac{n_{1i} n_{2i} a_i d_i}{n_i^2 (n_i - 1)}$$

The evidence against the null hypothesis is summarized in the standardized statistic:

$$z = \frac{\theta}{\sqrt{\text{Var}_0(\theta)}}$$

which is referred to the standard normal percentiles for a statistical decision. We may also refer the squared value of z to percentiles of the chi-squared distribution with 1 degree of freedom.

There are two important special cases:

(i) We may choose the size of the table as its weight:

$$w_i = n_i$$

This choice gives the generalized Wilcoxon test (also called the Gehan–Breslow test) (Gehan,1965a,b; Breslow, 1970); it is reduced to the Wilcoxon test in the absence of censoring.

(ii) We may choose to assign the same weights to all tables:

$$w_i = 1$$

This choice gives the log-rank test (also called the Cox–Mantel test; it is similar to the Mantel–Haenszel procedure for the combination of several 2×2 tables in the analysis of stratified categorical data) (Mantel and Haenszel, 1959; Cox, 1972; Tarone and Ware, 1977).

There are a few other interesting issues:

1. Which test should we use? The generalized Wilcoxon statistic puts more weight on the beginning observations, and because of that its use is more powerful in detecting the effects of short-term risks. On the other hand, the log-rank statistic puts equal weight on each observation and therefore, by default, is more sensitive to exposures with a constant relative risk (proportional hazards effect; in fact, we have derived the log-rank test as a score test using the proportional hazards model). Because of these characteristics, applications of both tests may reveal not only whether or not an exposure has any effect, but also the nature of the effect, short-term or long-term.

2. Because of the way the tests are formulated (terms in the summation are not squared), they are only powerful when one risk is greater than the other at all times. Otherwise, some terms in this sum are positive, while some other terms are negative and they cancel each other out. For example, the tests are virtually powerless for the case of crossing survival curves; in this case the assumption of proportional hazards is severely violated.

3. Some cancer treatments (e.g., bone marrow transplantation) are thought to have cured patients within a short time of initiation. Then, instead of all patients having the same hazard, a biologically more appropriate model, the cure model, assumes that an unknown proportion $(1 - p)$ are still at risk whereas the remaining proportion (p) have essentially no risk. If the aim of the study is to compare the cure proportions p, then neither the generalized Wilcoxon nor log-rank tests are appropriate (low power). One may simply choose a time point t far enough for the curves to level off, then compare the estimated survival rates by referring to percentiles of the standard normal distribution:

$$z = \frac{\hat{S}_2(t) - \hat{S}_1(t)}{\sqrt{\text{Var}[\hat{S}_2(t)] + \text{Var}[\hat{S}_1(t)]}}$$

Estimated survival rates and their variances are obtained as Section 9.2.1 (Kaplan–Meier procedure).

■ **Example 9.2** Refer back to the clinical trial to evaluate the effect of 6-mercaptopurine (6-MP) to maintain remission from acute leukemia (Example 9.1). The results of the tests indicate a highly significant difference between survival patterns of the two groups. The generalized Wilcoxon test shows a slightly larger statistic, indicating that the difference is slightly larger at earlier times; however, the log-rank test is almost equally significant, indicating that the use of 6-MP has a long-term effect (the effect does not wear off). (See Figure 9.5).

$$\text{Generalized Wilcoxon:} \quad X^2 = 13.46 \text{ with 1 degree of freedom, } p < 0.0001$$
$$\text{Log-rank:} \quad X^2 = 16.79 \text{ with 1 degree of freedom, } p = 0.0002$$

Note: A SAS program would include these instructions:

```
PROC LIFETEST METHOD=KM;
TIME WEEKS*RELAPSE(0);
STRATA DRUG;
```

where `KM` stands for Kaplan–Meier method, `WEEKS` is the variable name for duration time, `RELAPSE` is the variable name for survival status, `0` (in `RELAPSE(0)`) is the coding for censoring, and `DRUG` is the variable name specifying groups to be compared.

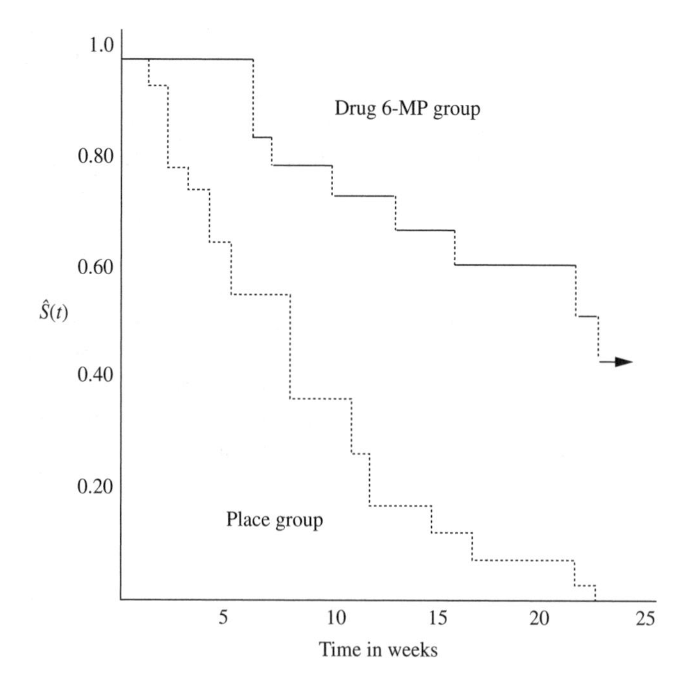

FIGURE 9.5. The relative positions of the two survival curves.

The above tests are applicable to the simultaneous comparison of several samples; when k groups are to be compared, the chi-squared tests, both the log-rank and generalized Wilcoxon, have $(k-1)$ degrees of freedom.

9.3 SIMPLE REGRESSION AND CORRELATION

In this section we discuss the basic ideas of simple regression analysis when only one predictor or independent variable is available for predicting the survival of interest; parts of this have been briefly introduced near the end of Section 9.1. The following example is used for illustration in this and the next section as well.

■ **Example 9.3** For a group of patients who died of acute myelogenous leukemia, they were classified into the two subgroups according to the presence or absence of a morphologic characteristic of white cells. Patients termed "AG positive" were identified by the presence of Auer rods and/or significant granulature of the leukemic cells in the bone marrow at diagnosis. For the "AG negative" patients these factors were absent.

AG-Positive Group, $n = 17$		AG-Negative Group, $n = 16$	
WBC	Time (weeks)	WBC	Time (weeks)
2,300	65	4,400	56
750	156	3,000	65
4,300	100	4,000	17
2,600	134	1,500	7
6,000	16	9,000	16
10,500	108	5,300	22
10,000	121	10,000	3
17,000	4	19,000	4
5,400	39	27,000	2
7,000	143	28,000	3
9,400	56	31,000	8
32,000	26	26,000	4
35,000	22	21,000	3
100,000	1	79,000	30
100,000	1	100,000	4
52,000	5	100,000	43
100,000	65		

Leukemia is a cancer characterized by an overproliferation of white blood cells; the higher the white blood count (WBC), the more severe the disease. Data in the table clearly suggest that there is such a relationship and thus, when predicting a leukemia patient's survival time, it is realistic to make a prediction dependent on WBC count (and any other covariates that are indicators of the progression of the disease).

9.3.1 Model and Approach

The association between two random variables X and T, the second of which—the suvival time T—may be only partially observable due to right censoring, has been the focus of many investigations starting with the historical breakthrough by Cox (1972). Cox's regression model or the proportional hazards model (PHM) expresses a loglinear relationship between X and the hazard function of T, as briefly introduced near the end of Section 9.1.

$$\begin{aligned}\lambda(t|X=x) &= \lim_{\delta\to 0}\frac{\Pr[t \le T < t+\delta | t \le T, X = x]}{\delta}\\ &= \lambda_0(t)\exp(\beta x)\end{aligned}$$

In this model, $\lambda_0(t)$ is an unspecified baseline hazard, that is, hazard at $X=0$, and β is an unknown regression coefficient. The estimation of β and subsequent analyses are performed as follows.

Denote the ordered distinct death times by

$$t_1 < t_2 < \cdots < t_m$$

and let R_i be the risk set just before time t_i, n_i the number of subjects in R_i, D_i the death set at time t_i, d_i the number of subjects (i.e., deaths) in D_i, and C_i the collection of all possible combinations of subjects from R_i; each combination—or subset of R_i—has d_i members, and D_i is itself one of these combinations. For example, if three subjects (A, B, and C) are at risk just before time t_i and two of them (A and B) die at t_i, then

$$\begin{aligned}R_i &= \{A, B, C\},\ n_i = 3\\ D_i &= \{A, B\},\ d_i = 2\\ C_i &= \{\{A, B\} = D_i, \{A, C\}, \{B, C\}\}\end{aligned}$$

The number of elements in C_i is the number of ways to choose d_i subjects out of n_i subjects,

$$\binom{n_i}{d_i}$$

Cox (1972) suggests using the following as a likelihood function, called the *marginal* or *partial likelihood function*; basically, it is a *conditional approach*—conditioned on the events observed in the sample:

$$\begin{aligned} L &= \prod_{i=1}^{m} \Pr(D_i | R_i, d_i) \\ &= \prod_{i=1}^{m} \frac{\exp(\beta s_i)}{\sum_{C_i} \exp(\beta s_u)} \\ s_i &= \sum_{D_i} x_j \\ s_u &= \sum_{D_u} x_j \\ D_u &\in C_i \end{aligned}$$

An alternative likelihood, proposed by Peto (1972), is

$$L = \prod_{i=1}^{m} \frac{\exp(\beta s_i)}{\left[\sum_{R_i} \exp(\beta x_u)\right]^{d_i}}$$

which seems to work reasonably well when the number of ties is not excessive and, therefore, has become rather popular.

9.3.2 Measures of Association

We first consider the case of a binary covariate with the conventional coding

$$x_i = \begin{cases} 0 & \text{if subject } i \text{ is not exposed} \\ 1 & \text{if subject } i \text{ is exposed} \end{cases}$$

Here, the term "exposed" may refer to a risk factor such as smoking, or a patient's characteristic such as race (white/nonwhite) or sex (male/female).

It can be seen that, from the proportional hazards model,

$$\lambda(t;\ \text{nonexposed}) = \lambda_0(t)$$

$$\lambda(t; \text{exposed}, x = 1) = \lambda_0(t)e^{\beta}$$

$$\frac{\lambda(t; \text{nonexposed})}{\lambda(t; \text{exposed}, x = 1)} = e^{\beta}$$

The ratio of the two risk functions represents the relative risk (RR) of the exposure, exposed versus nonexposed. In other words, the regression coefficient β is the value of the relative risk on the log scale.

Similarly, we have for a continuous covariate X and any value x of X,

$$\lambda(t; X=x) = \lambda_0(t)e^{\beta x}$$
$$\lambda(t; X=x+1) = \lambda_0(t)e^{\beta(x+1)}$$
$$\frac{\lambda(t; X=x+1)}{\lambda(t; X=x)} = e^{\beta}$$

That is, the regression coefficient β represents the relative risk (RR) due to one unit increase in the value of X, $(X=x+1)$ versus $(X=x)$: for example, a systolic blood pressure of 114 mmHg versus 113 mmHg. For m units increase in the value of X, say, $(X=x+m)$ versus $(X=x)$, the corresponding relative risk is $\exp(m\beta)$.

The regression coefficient β can be estimated iteratively using the first and second derivatives of the partial likelihood function. From the results, we can obtain a point estimate of the relative risk and its 95% confidence interval—first on the log scale and then exponentiating the endpoints of the interval.

It should be noted that the relative risk, used as a measure of association between survival time and a covariate, depends on the coding scheme for a binary factor and, for a continuous covariate X, the scale with which to measure X. For example, if we use the following coding for a factor,

$$x_i = \begin{cases} -1 & \text{if not exposed} \\ 1 & \text{if exposed} \end{cases}$$

then

$$\lambda(t) = \begin{cases} \lambda_0(t)e^{-\beta} & \text{if not exposed} \\ \lambda_0(t)e^{\beta} & \text{if exposed} \end{cases}$$

leading to

$$\begin{aligned} RR &= \frac{\lambda(t|\text{exposed})}{\lambda(t|\text{nonexposed})} \\ &= \exp(-2\beta) \end{aligned}$$

and its 95% confidence interval: $\exp[2(\hat{\beta} \pm 1.96 SE(\hat{\beta}))]$. Of course, the estimate of β and its standard error, under the new coding scheme, is only half of that under the former scheme; therefore the estimate of the relative risk remains unchanged.

The following example, however, shows the obvious effect of measurement scale in the case of a continuous measurement.

■ **Example 9.4** Refer to the data for patients with acute myelogenous leukemia in Example 9.3 and suppose we want to investigate the relationship between survival time of AG-positive patients and white blood cell (WBC) count in two different ways, using either (i) $X = \text{WBC}$ or (ii) $X = \log(\text{WBC})$.

(i) For $X = \text{WBC}$, we find $\beta = 0.0000167$, from which, the relative risk for a patient with WBC = 100,000 versus another patient with WBC = 50,000 would be

$$\begin{aligned} RR &= \exp[(100,000 - 50,000)(0.0000167)] \\ &= 2.31 \end{aligned}$$

(ii) For $X = \log(\text{WBC})$ – logarithm of base 10, we find $\beta = 0.612331$, from which, the relative risk for a patient with WBC = 100,000 versus another patient with WBC = 50,000 would be

$$\begin{aligned} RR &= \exp[\log(100,000) - \log(50,000)](0.612331) \\ &= 1.53 \end{aligned}$$

The above results show that the relative risks are different for two different choices of the scale of measurement for X and this causes an obvious problem of choosing an appropriate measurement scale. Of course, we assume a linear model and one choice of X would fit better than the other (there are methods for checking this assumption).

Note: A SAS program would include these instructions:

```
PROC PHREG DATA = CANCER;
MODEL WEEKS*DEATH(0) = WBC;
```

where `CANCER` is the name assigned to the data set, `WEEKS` the variable name for duration time, `DEATH` the variable name for survival status, and `0` in `DEATH(0)` is the coding for censoring.

9.3.3 Tests of Association

The null hypothesis to be considered is

$$H_0\colon \beta = 0$$

The reason for interest in testing whether or not $\beta = 0$ is that $\beta = 0$ implies there is no relation between survival time T and the covariate X under investigation. The decision is made using

$$z = \frac{\hat{\beta}_i}{SE(\hat{\beta}_i)}$$

In performing this test, we refer the value of the z-statistic to percentiles of the standard normal distribution. For the case of a categorical covariate, the test based on the score statistic of Cox's regression model is identical to the log-rank test of Section 9.2.2.

9.4 MULTIPLE REGRESSION AND CORRELATION

The effect of some factor on survival time may be influenced by the presence of other factors through effect modifications (i.e., interactions). Therefore in order to provide a more comprehensive prediction of the future of patients with respect to duration, course, and outcome of a disease, it is very desirable to consider a large number of factors and sort out which ones are most closely related to diagnosis. In this section, we discuss a multivariate method for risk determination. This method, which is multiple regression analysis, involves a linear combination of the explanatory or independent variables; the variables must be quantitative with particular numerical values for each patient. Information concerning possible factors is usually obtained as a subsidiary aspect from clinical trials that were designed to compare treatments. A covariate or prognostic patient characteristic may be binary or dichotomous, categorical, or continuous (categorical factors are represented by dummy variables). Examples of dichotomous covariates are sex and presence or absence of a certain comorbidity. Categorical covariates include race and different grades of symptoms; these can be covered by the use of dummy or indicator variables. Continuous covariates include patient age and blood pressure. In many cases, data transformations (e.g., taking the logarithm) may be desirable.

9.4.1 Proportional Hazards Models with Several Covariates

Suppose we want to consider k covariates simultaneously. The proportional hazards model of the previous section can easily be generalized and expressed as

$$\begin{aligned}
\lambda(t|X) &= \lim_{\delta \to 0} \frac{\Pr[t \leq T < t+\delta | t \leq T, X = (x_1, \ldots, x_k)]}{\delta} \\
&= \lambda_0(t) \exp\left(\sum_{i=1}^{k} \beta_i x_i\right)
\end{aligned}$$

$\lambda_0(t)$ is an unspecified baseline hazard, that is, hazard at $X = 0$ for all covariates, and the β's are unknown regression coefficients. In order to have a meaningful baseline hazard, it may be necessary to "center" continuous covariates about their means:

$$x^* = x - \bar{x}$$

so that $\lambda_0(t)$ is the hazard function associated with a "typical patient" (i.e., a hypothetical one having all covariates at their average values in the sample).

The estimation of β and subsequent analyses are performed similar to the univariate case using Cox's partial likelihood function of Section 9.3.1 with obvious extension from one to several covariates:

$$L = \prod_{i=1}^{m} \Pr(D_i|R_i, d_i)$$

$$= \prod_{i=1}^{m} \frac{\exp\left(\sum_j \beta_j s_{ji}\right)}{\sum_{C_i} \exp\left(\sum_j \beta_j s_{ju}\right)}$$

$$s_{ji} = \sum_{D_i} x_j$$

$$s_{ju} = \sum_{Du} x_j$$

$$D_u \in C_i$$

Also similar to the univariate case, $\exp(\beta_i)$ represents (i) the relative risk associated with an exposure if X_i is binary, or (ii) the relative risk due to one unit increase in X_i if X_i is continuous. After an estimate of β_i and its standard error have been obtained, a 95% confidence interval for the above relative risk is given by

$$\exp[\hat{\beta}_i \pm 1.96SE(\hat{\beta}_i)]$$

These results are necessary in the effort to identify important prognostic or risk factors. Of course, before such analyses are done, the problem and the data have to be examined carefully. If some of the variables are highly correlated, then one or fewer of the correlated factors are as likely to be good predictors as all of them; information from other similar studies also has to be incorporated so as to drop some of these correlated explanatory variables. The use of products, such as x_1x_2, and higher power terms, such as $x_1{}^2$, may be necessary and can improve the goodness-of-fit. It is important to note that we are assuming a *linear* regression model in which, for example, the relative risk due to one unit increase in the value of a continuous X ($X = x + 1$ versus $X = x$) is independent of x. Therefore if this linearity seems to be violated, the incorporation of powers of X should be seriously considered. The use of products will help in the investigation of possible effect modifications. Finally, we confront the messy problem of missing data; most packaged computer programs would delete the patient if one or more covariate values are missing.

9.4.2 Testing Hypotheses in Multiple Regression

Once we have fitted a multiple proportional hazards regression model and obtained estimates for the various parameters of interest (regression coefficients), we want to answer questions about the contributions of various factors to the prediction of the future of the patient. There are three types of such questions:

(i) An overall test: Taken collectively, does the entire set of explanatory or independent variables contribute significantly to the prediction of survivorship?

(ii) Test for the value of a single factor: Does the addition of one particular variable of interest add significantly to the prediction of survivorship over and above that achieved by other independent variables?

(iii) Test for contribution of a group of variables: Does the addition of a group of variables add significantly to the prediction of survivorship over and above that achieved by other independent variables?

Overall Regression Tests We now consider the first question stated above concerning an overall test for a model containing k factors, say,

$$\begin{aligned}\lambda(t|X) &= \lim_{\delta\to 0}\frac{\Pr[t \le T < t+\delta | t \le T, X = (x_1,\ldots,x_k)]}{\delta}\\ &= \lambda_0(t)\exp\left(\sum_{i=1}^{k}\beta_i x_i\right)\end{aligned}$$

The null hypothesis for this test may be stated as: "all k independent variables considered together do not explain the variation in survival times." In other words,

$$H_0\colon\ \beta_1 = \beta_2 = \cdots = \beta_k = 0$$

Two likelihood-based statistics can be used to test this "global" null hypothesis; each has an asymptotic chi-squared distribution with k degrees of freedom under H_0.

(i) *Likelihood Ratio Test:*

$$X^2_{\text{LR}} = 2[\ln L(\hat{\beta}) - \ln L(\mathbf{0})]$$

(ii) *Score Test:*

$$X^2_{\text{S}} = \left[\frac{\delta \ln L(\mathbf{0})}{\delta\boldsymbol{\beta}}\right]\left[-\frac{\delta^2 \ln L(\mathbf{0})}{\delta\boldsymbol{\beta}^2}\right]^{-1}\left[\frac{\delta \ln L(\mathbf{0})}{\delta\boldsymbol{\beta}}\right]$$

Both statistics are provided by most standard computer programs such as SAS and they are asymptotically equivalent, yielding identical statistical decisions most of the time.

Tests for a Single Variable Let us assume that we now wish to test whether the addition of one particular independent variable of interest adds significantly to the prediction of survivorship over and above that achieved by other factors already present in the model. The null hypothesis for this test may be stated as: "factor X_i does not have any value added to the prediction of survivorship given that other factors are already included in the model." In other words,

$$H_0\colon\ \beta_i = 0$$

To test such a null hypothesis, one can perform a likelihood ratio chi-squared test, with 1 degree of freedom, similar to that for the above global hypothesis:

$$X^2_{\text{LR}} = 2[\ln L(\hat{\boldsymbol{\beta}}; \text{all } X\text{'s}) - \ln L(\hat{\boldsymbol{\beta}};\ \text{all other } X\text{'s with } X_i \text{ deleted})]$$

A much easier alternative method uses

$$z = \frac{\hat{\beta}_i}{SE(\hat{\beta}_i)}$$

In performing this test, we refer the value of the z-statistic to percentiles of the standard normal distribution. This is equivalent to Wald's chi-squared test as applied to one parameter.

Contribution of a Group of Variables This testing procedure addresses the more general problem of assessing the additional contribution of two or more factors to the prediction of survivorship over and above that made by other variables already in the regression model. In other words, the null hypothesis is of the form where $1 < m < k$.

$$H_0\colon\ \beta_1 = \beta_2 = \cdots = \beta_m = 0$$

To test such a null hypothesis, one can perform a likelihood ratio chi-squared test, with m degrees of freedom:

$$X^2_{\text{LR}} = 2[\ln L(\hat{\boldsymbol{\beta}}; \text{all } X\text{'s}) - \ln L(\hat{\boldsymbol{\beta}}; \text{all other } X\text{'s with } m\, X\text{'s under investigation deleted})]$$

As with the above z-test for a single variable, this procedure is very useful for assessing the importance of potential explanatory variables. In particular, it is often used to test whether a similar group of variables, such as demographic characteristics, is important for the prediction of survivorship; these variables have some trait in common. Another application would be a collection of powers and/or product terms (referred to as interaction variables). It is often of interest to assess the interaction effects collectively before trying to consider individual interaction terms in a model as previously suggested. In fact, such use reduces the total number of tests to be performed and this, in turn, helps to provide better control of overall Type I error rates, which may be inflated due to multiple testing.

Stepwise Procedure

In applications, our major interest is to identify important prognostic factors. In other words, we wish to identify from many available factors a small subset of factors that relate significantly to the length of survival time of patients. In that identification process, of course, we wish to avoid a Type I (false-positive) error. In a regression analysis, a Type I error corresponds to including a predictor that has no real relationship to survivorship; such an inclusion can greatly confuse the interpretation of the regression results. In a standard multiple regression analysis, this goal can be achieved by using a strategy that adds into or removes from a regression model one factor at a time according to a certain order of relative importance. The details of this stepwise process for survival data are similar to those for logistic regression in Chapter 4.

Stratification

The proportional hazards model requires that, for a covariate X—say, an exposure—the hazards functions at different levels, (t; exposed) and $\lambda(t$; nonexposed), are proportional. Of course, sometimes there are factors, for which the different levels produce hazard functions that deviate markedly from proportionality. These factors may not be under investigation themselves, especially those of no intrinsic interest, those with a large number of levels and/or those where interventions are not possible. But these factors may act as important confounders, which must be included in any meaningful analysis so as to improve predictions concerning other covariates. Common examples include sex, age, and neighborhood. To accommodate such confounders, an extension of the proportional hazards model is desirable. Suppose there is a factor that occurs on q levels and for which the proportional hazards model may be violated. If this factor is under investigation as a covariate, then the model and subsequent analyses are not applicable. However, if this factor is not under investigation and is considered only as a confounder so as to improve analyses and/or predictions concerning other covariates, then we can treat it as a stratification factor. By doing that, we will get no results concerning this factor (which are not wanted), but in return we do not have to assume that the hazards functions corresponding to different levels are proportional (which may be severely violated). Suppose the stratification factor Z has q levels; this factor is not clinically important itself but adjustments are still needed in efforts to investigate other covariates. We define the hazards function for an individual in the jth stratum (or level) of this factor as

$$\lambda_j(t|X) = \lambda_{0j}(t)\exp\left(\sum_{i=1}^{k}\beta_i x_i\right)$$

for $j = 1, 2, \ldots, q$, where $\lambda_{0j}(t)$ is an unspecified baseline hazard for the jth stratum and X represents other k covariates under investigation (excluding the stratification itself). The baseline hazards functions are allowed to be arbitrary and are completely unrelated (and, of course, not proportional to each other). The basic additional assumption here, which is the same as that in the *analysis of covariance* (ANCOVA), requires that the β's are the same across strata (i.e., the so-called parallel lines assumption, which is testable).

In the analysis, we identify distinct times of events for the jth stratum and form the partial likelihood $L_j(\beta\text{'s})$ as in the previous sections. The overall partial likelihood of β's is then the product of those q stratum-specific likelihoods:

$$L(\beta\text{'s}) = \prod_{j=1}^{q} L_j(\beta\text{'s})$$

Subsequent analyses, finding maximum likelihood estimates as well as using score statistics, are straightforward. For example, if the null hypothesis $\beta = 0$ for a given covariate is of interest, the score approach would produce a stratified log-rank test. An important application of stratification, the analysis of epidemiologic matched studies resulting in the conditional logistic regression model, was presented in Chapter 5.

■ **Example 9.5** Refer to the myelogenous leukemia data of Example 9.3. Patients were classified into the two groups according to the presence or absence of a morphologic characteristic of white blood cells and the primary covariate is white blood cell (WBC) count. Using

$$\begin{aligned} X_1 &= \ln(WBC) \\ X_2 &= AG\text{-}group \\ &= \begin{cases} 0 & \text{if negative} \\ 1 & \text{if positive} \end{cases} \end{aligned}$$

we fit the following model with one interaction term:

$$\lambda(t|X) = \lambda_0(t)\exp(\beta_1 x_1 + \beta_2 x_2 + \beta_3 x_1 x_2)$$

From the results, it can be seen that the interaction effect is almost significant at the 5% level ($p = 0.0732$); that is, the presence of the morphologic characteristic modifies substantially the effect of WBC count. (See Table 9.2)

TABLE 9.2. Investigation of Effect Modification

Factor	Coefficient	Standard Error	z-Statistic	p-Value
WBC Count	0.14654	0.17869	0.821	0.4122
AG-group	−5.85637	2.75029	−2.129	0.0332
Product	0.49527	0.27648	1.791	0.0732

9.4.3 Time-Dependent Covariates and Applications

In prospective studies, since subjects are followed over time, values of many independent variables or covariates may be changing; covariates such as patient age, blood pressure, even treatment. In general, covariates are divided into two categories: fixed and time dependent. A covariate is time dependent if the difference between covariate values from two different subjects may be changing with time. For example, sex and age are fixed covariates; a patient's age is increasing by one a year but the difference in age between two patients remains unchanged. On the other hand, blood pressure is an obvious time-dependent covariate. The following are three important groups of time-dependent covariates.

Examples

(i) Personal characteristics whose measurements are periodically made during the course of a study. Blood pressure fluctuates; so do cholesterol level and weight. Smoking and alcohol consumption habits may change.

(ii) Cumulative exposure. In many studies, exposures such as smoking are often dichotomized; subjects are classified as exposed or unexposed. But this may be oversimplified, leading to loss of information; the length of exposure may be important. As time goes on, a nonsmoker remains a nonsmoker but "years of smoking" for a smoker increases.

(iii) Another important group consists of "switching treatments." In a clinical trial, a patient may be transferred from one treatment to another due to side effects or even by the patient's request. Organ transplants form another category with switching treatments; when a suitable donor is found, a subject is switched from nontransplanted group to transplanted group. The case of intensive care units is even more complicated, where a patient may be moving in and out more than once.

Implementation

Recall that in the analysis using the proportional hazards model, we order the death times and form the partial likelihood function:

$$
\begin{aligned}
L &= \prod_{i=1}^{m} \Pr(D_i | R_i, d_i) \\
&= \prod_{i=1}^{m} \frac{\exp\left(\sum_j \beta_j s_{ji}\right)}{\sum_{C_i} \exp\left(\sum_j \beta_j s_{ju}\right)} \\
s_{ji} &= \sum_{D_i} x_j \\
s_{ju} &= \sum_{D_u} x_j \\
D_u &\in C_i
\end{aligned}
$$

where R_i is the risk set just before time t_i, n_i the number of subjects in R_i, D_i the death set at time t_i, d_i the number of subjects (i.e., deaths) in D_i, and C_i the collection of all possible combinations of subjects from R_i; each combination — or subset of R_i — has d_i members, and D_i is itself one of these combinations. For example, if three subjects (A, B, and C) are at risk just before time t_i and two of them (A and B) die at t_i, then

$$R_i = \{A, B, C\}, \quad n_i = 3$$
$$D_i = \{A, B\}, \quad d_i = 2$$
$$C_i = \{\{A, B\} = D_i, \{A, C\}, \{B, C\}\}$$

In this approach, we try to explain why an event or events occurred to a subject or subjects in D_i while all subjects in R_i are equally at risk, and why an event or events did not occur to a subject or subjects in another subset $D_u \in C_i$ of the same size. This explanation, through the use of s_{ji} and s_{ju}, is based on the covariate values measured at time t_i. Therefore this needs some modification in the presence of time-dependent covariates because events at time t_i; should be explained by values of covariates measured at that particular moment. Blood pressure, for example, measured years before may become irrelevant.

First, notations are expanded to handle time-dependent covariates. Let x_{jil} be the value of factor x_j measured from individual l at time t_i; then the above likelihood function becomes

$$L = \prod_{i=1}^{m} \Pr(D_i | R_i, d_i)$$
$$= \prod_{i=1}^{m} \frac{\exp\left(\sum_j \beta_j s_{jii}\right)}{\sum_{C_i} \exp\left(\sum_j \beta_j s_{jiu}\right)}$$
$$s_{jii} = \sum_{l \in D_i} x_{jil}$$
$$s_{ju} = \sum_{v \in D_u} x_{jiv}$$
$$D_u \in C_i$$

From this new likelihood function, applications of subsequent steps (estimation of β's, formation of test statistics, and the estimation of the baseline survival function) are straightforward. In practical implementation, most standard computer programs have somewhat different procedures for two categories of time-dependent covariates: those that can be defined by a mathematical equation (external) and those measured directly from patients (internal). The former categories are much easier implemented.

A Simple Test of Goodness-of-Fit

Treatment of time-dependent covariates leads to a simple test of goodness-of-fit. Consider the case of a fixed covariate, denoted by X_1. Instead of the basic proportional hazards model,

$$\begin{aligned}\lambda(t|X_1 = x_1) &= \lim_{\delta \to 0} \frac{\Pr[t \le T < t+\delta | t \le T, X_1 = x_1]}{\delta} \\ &= \lambda_0(t)\exp(\beta_1 x_1)\end{aligned}$$

we can define an additional time-dependent covariate X_2, $X_2 = X_1 t$, the product of x and time t, and consider the expanded model:

$$\begin{aligned}\lambda(t|X_1 = x_1) &= \lim_{\delta \to 0} \frac{\Pr[t \le T < t+\delta | t \le T, X_1 = x_1]}{\delta} \\ &= \lambda_0(t)\exp(\beta_1 x_1 + \beta_2 x_2)\end{aligned}$$

and examine the significance of

$$H_0\colon \beta_2 = 0$$

The reason for interest in testing whether or not $\beta_2 = 0$ is that $\beta_2 = 0$ implies a goodness-of-fit of the proportional hazards model for the factor under investigation, $X = X_1$. Of course, in defining the new covariate X_2, t could be replaced by any function of t; a commonly used one is $\log(t)$.

This simple approach results in a test of a specific alternative to the proportionality. The computational implementation here is very much similar to the case of cumulative exposures; however, X_1 may be binary or continuous. We may even investigate the goodness-of-fit for several variables simultaneously.

■ **Example 9.6** Refer to the data set in Example 7.1, where the remission times of 42 patients with acute leukemia were reported from a clinical trial undertaken to assess the ability of a drug called 6-mercaptopurine (6-MP) to maintain remission. Each patient was randomized to receive either 6-MP or placebo. The study was terminated after 1 year; patients have different follow-up times because they were enrolled sequentially at different times. Times in weeks were as follows:

6-MP Group. 6, 6, 6, 7, 10, 13, 16, 22, 23, 6+, 9+, 10+, 11+, 17+, 19+, 20+, 25+, 32+, 32+, 34+, 35+.

Placebo Group. 1, 1, 2, 2, 3, 4, 4, 5, 5, 8, 8, 8, 8, 11, 11, 12, 12, 15, 17, 22, 23.

A $t+$ denotes a censored observation, that is, the case was censored after t weeks without a relapse. For example, 10+ is a case enrolled 10 weeks before study termination and still remission-free at termination.

Since the proportional hazards model is often assumed in the comparison of two survival distributions such as in this example (also see Example 9.2), it is

TABLE 9.3. Investigation of Goodness-of-Fit

Factor	Coefficient	Standard Error	z-Statistic	p-Value
X_1	−1.5539	0.8108	−1.917	0.0553
$X_2 = X_1 t$	−0.0075	0.0693	−0.108	0.9142

desirable to check it for validity (if the proportionality is rejected, it would lend support to the conclusion that this drug does have some cumulative effects).

Let X_1 be the group indicator variable (=0 if Placebo, and 1 if treated by 6-MP drug) and $X_2 = X_1 t$, the product of x and time t, which also represents treatment weeks (time t is recorded in weeks). In order to judge the validity of the proportional hazards model with respect to X_1, it is the effect of this newly defined covariate X_2 that we want to investigate.

We fit the following model:

$$\begin{aligned}\lambda(t|X_1 = x_1) &= \lim_{\delta \to 0} \frac{\Pr[t \leq T < t+\delta | t \leq T, X_1 = x_1]}{\delta} \\ &= \lambda_0(t)\exp(\beta_1 x_1 + \beta_2 x_2)\end{aligned}$$

and from the results, it can be seen that the *accumulation effect* or *lack of fit*, represented by X_2 is insignificant; in other words, there is not enough evidence to be concerned about the validity of the proportional hazards model. (See Table 9.3)

Note: A SAS program would include these instructions:

```
PROC PHREG DATA = CANCER;
MODEL WEEKS*RELAPSE(0) = DRUG TESTEE;
TESTEE = DRUG*WEEKS;
```

where `WEEKS` is the variable name for duration time, `RELAPSE` is the variable name for survival status, `0` in `RELAPSE(0)` is the coding for censoring, `DRUG` is the 0/1 group indicator (i.e., X_1) and `TESTEE` is the newly created variable (i.e., X_2). In SAS, `TESTEE`—as a time-dependent covariate—is defined after the `MODEL` statement.

9.5 COMPETING RISKS

Competing risk is a rather advanced topic. To do a thorough job in this area, one needs a stronger background in survival analysis—a lot more than just reading or learning from this introductory chapter. However, we decided to include an introductory section because the topic is very important and very popular in the conduct and analysis of translational research.

The primary objective of a Phase II clinical trial is to screen for antitumor activity; agents that are found to have substantial antitumor activity and an appropriate spectrum of toxicity are likely incorporated into combinations to be evaluated for patient benefit in controlled, randomized Phase III clinical trials. The secondary aim of Phase II clinical trials is often concerned with patient survivorship; other secondary aims include pharmacokinetics and duration of response. For Phase II trials, which are often small (20–30 patients) and short term, lasting just a year or two, 1-year survival rate is often used a benchmark to measure treatment success (e.g., versus historical data).

If survival is of high priority, disease-free survival is a popular endpoint; overall survival is often recommended as another endpoint. In some problems, cause-specific deaths may be identified; if so, we may be interested in studying cause-specific mortality. In doing so, we have to deal with competing risks. A competing risk is defined as risk for an event whose occurrence precludes the occurrence of another event, which may be the primary event under investigation. Even in early-phase clinical trials—for example, in a nonrandomized Phase II clinical trial, we often have to face the task of estimating a failure probability in the presence of a competing risk. For example, in the presence of all other causes: "What is the probability that a patient would die from cancer within 5 years? The followings are a few other examples.

(i) Loss to follow-up is not a competing risk (we can't observe death during a clinical trial because the patient moved; but the patient would die later).
(ii) In a cancer study, death due to a noncancer cause (e.g., suicide) is a competing risk.
(iii) In a study where the primary event is "Relapse," death without sign of relapse is a competing risk.
(iv) In a study of graft versus host disease, after a bone marrow transplant, death is a competing risk.

Why do competing risks affect the failure probability of the primary risk under investigation? Failures or events from the competing risks reduce the number of patients at risk of failure from the cause of interest; if not handled properly, we would overestimate the failure rate due to the primary risk under investigation.

9.5.1 Redistribution to the Right Method

In Section 9.2.1, we introduced the product-limit (PL) method of estimating the survival rates; this is also called the Kaplan–Meier method.

Let

$$t_1 < t_2 < \cdots < t_k$$

be the distinct observed death times in a sample of size n from a homogeneous population with survival function $S(t)$ to be estimated $(1 = k = n)$. Let n_i be the number of subjects at risk at a time just prior to $t_i (1 = i = k)$; these are cases or subjects in the sample whose duration time is at least t_i; and d_i is the number of deaths at t_i. The survival function $S(t)$ is estimated by

$$\hat{S}(t) = \prod_{t_i \leq t} \left(1 - \frac{d_i}{n_i}\right)$$

which is called the product-limit estimator or Kaplan–Meier curve. Basically, the approach in the Kaplan–Meier method was estimating the *survival probability* at each time with events; that is, the "height" of the survival curve at that time point. Actually, there are two options in forming such a step-function survival curve; the alternative to the Kaplan–Meier method is to estimate the "jump" at each time with events. If we could do so, the Kaplan–Meier method follows.

■ **Example 9.7** Consider a simple, hypothetical data set $\{3, 6+, 8, 11\ +, 17\}$, and define the "jump" $J(t_i)$ at time t_i as

$$J(t_i) = S(t_i^-) - S(t_i)$$

the drop from one step to the next step on the step-function survival curve. We have the following for the above small hypothetical data set:

t	n	d	$1 - d/n$	$S(t)$	$J(t)$
3	5	1	0.8	0.8	$1.0 - 0.80 = 0.20$
8	3	1	0.667	0.533	$0.8 - 0.533 = 0.267$
17	1	1	0	0	0.533

Note that we can form the curve if we know the "height" $S(t_i)$ at all time points with events. Equivalently, we can also form the curve if we know the "jump" $J(t_i)$ at all time points with events. How do we find these jumps in the presence of censoring and without calculating all the heights first? Before we can answer this question, let us revisit the estimation of the survival curve.

Let first consider data without censoring. Given an uncensored sample of survival times $\{t_i\}$, $i = 1, 2, \ldots, n$, we can estimate $S(t)$ by simply counting the proportion of survival times greater than t:

$$S_n(t) = \frac{\text{Number of } t_i > t}{n}$$

In other words, the empirical estimate of $S(t)$, applied to data without censoring, puts a mass (or jump) of $1/n$ at each of the observed times; say, $1/n = 1/5$ at each time point of $\{3, 6, 8, 11, 17\}$. Now, consider a data set censoring $\{3, 6+, 8, 11+, 17\}$. What do we do to afford the two censored subjects at $t = 6$ and $t = 11$?

Efron (1967) introduced another method for computing the Kaplan–Meier or product-limit estimate, called the *redistribution to the right algorithm*; it can be briefly explained as follows:

Step 1: For the data set $\{3, 6+, 8, 11+, 17\}$, a death did not occur at $t = 6$ but some time later; the most likely places are the later observed times (to the right of $t = 6$).

It is reasonable to redistribute its mass or jump (=1/5) equally to $t=8$, 11, and 17. Their jumps (at $t=6$, $t=11$, and $t=17$) each become

$$1/5+(1/3)(1/5)=4/15 \text{ or } 0.267$$

Step 2: The new mass 4/15 at 11, calculated as above, is redistributed further to the right because there is no death at $t=11$; the jump at $t=17$ becomes 4/15 + 4/15 = 0.534.

In summary, for the sample {3, 6+, 8, 11+, 17}:

(i) the jump at $t=3$ is 1/5 = 0.200;
(ii) the jump at $t=8$ is 1/5 + (1/3)(1/5) = 0.267; and
(iii) the jump at $t=17$ is 0.267 + 0.267 = 0.534.

With these jumps, we can construct the survival curve. The heights are:

(i) 1–0.200 = 0.800 at $t=3$;
(ii) 0.800–0.267 = 0.533 at $t=8$; and
(iii) 0.533–0.533 = 0 at $t=17$.

This agrees with the results in the above table. Efron's redistribution to the right algorithm provides all the jumps; so it is equivalent to the Kaplan–Meier method. The basic principle is that "the weight at each censored time is redistributed equally to future time points—including points without events which will be redistributed again themselves later. The method can be summarized as follows:

(i) $J(t_1)=1/n$ (if the smallest time is an event).
(ii) $J(t_i)=J(t_{i-1})[1+c_{i-1}/n_i]$, where c_{i-1} is the number of subjects censored in interval $(t_{i-1}, t_i]$.

9.5.2 Estimation of the Cumulative Incidence

We can now return to our task of studying competing risks, first as a special, interesting application of redistribution to the right method. As can be seen from Section 9.5.1, there are two different approaches to the estimation of the survival function. One is a likelihood-based method—the Kaplan–Meier method—and the other is based on the redistribution to the right principle. The latter one can easily be modified to handle data with competing risk. That simple modification is as follows.

Consider the same example—{3, 6+, 8, 11+, 17}—but assume that subject $t=11+$ was really censored (i.e., no event at the study termination after following for a period of $t=11$) but subject 6+ died by a different cause—say, heart disease—while we studied cancer mortality. The question is: When calculating the

Kaplan–Meier estimates—either by the Kaplan–Meier method or Efron's redistribution to the right method—should we consider patients who failed from a competing risk as being censored at the time of failure? The answer is "no": they already died (say, from heart diseases), and are not there to die later from cancer; the weights should *not* be redistributed. We would overestimate probabilities of failure at future times if some censored observation was not *really* censored. And that's why the identification of events by competing risks is very important if we want to study cause-specific mortality. In the rule for calculating the jumps—(i) $J(t_1) = 1/n$, (if smallest time is an event) and (ii) $J(t_i) = J(t_{i-1})[1 + c_{i-1}/n_i]$, where c_{i-1} is the number of subjects censored in interval $(t_{i-1}, t_i]$—in the presence of competing risks, events by other causes are not included in c_{i-1}.

The general settings in translational or clinical research can be formulated as follows. A patient may die (i.e., having an event) by one of k causes. Let T be the patient's survival time from a time origin to an event (by any cause). Let $S(t) = \Pr(T > t)$, for $t > 0$, be the *overall survival function* and $\lambda_i(t)$ be the ith cause-specific hazard:

$$\lambda_i(t) = \lim_{\delta \to 0} \frac{\Pr[t \leq T < t + \delta, \text{cause} = i | t \leq T]}{\delta}$$

Since the risk to the patient is the sum of the risks by all causes,

$$S(t) = \exp\left[-\int_0^t \left\{ \sum_{i=1}^{k} \lambda_i(x) \right\} dx \right]$$

The quantity $\{1 - S(t)\}$ is like the overall cumulative distribution function (cdf); however, instead of estimating the overall survival function $S(t)$, we are often interested in cause-specific (cumulative) mortality. Usually, only one of the causes is of primary interest, say, cause i (e.g., death by cancer). Let $F_i(t)$ be the cause-specific failure probability up to time t by cause i; that is, the probability of failure (or event) of type i will occur before time t in the presence of competing risks (other causes):

$$F_i(t) = \int_0^t \lambda_i(x) S(x) dx$$

This parameter of interest is called the *cumulative incidence* (up to time t). The term "incidence" can be seen or explained from its components:

(i) $S(x)$ represents the probability of surviving (having no event) at time x.

(ii) $\lambda_i(x)dx$ represents the probability of having an event at time i in interval $(x, x + dx)$.

It is similar to the estimation of the survival function. This new and important parameter—the cumulative incidence—can be estimated in two different ways: one is a likelihood-based method, similar to the Kaplan–Meier method, and the other is based on the redistribution to the right principle. A brief discussion follows.

Redistribution to the Right Method

Efron's redistribution to the right algorithm would provide a very simple tool. Without competing risks, the basic principle is the weight at each censored time is redistributed equally to future time points. In this basic scenario, there *is* redistribution *to* censored times (and, by definition) there *is* redistribution *from* censored times. When calculating the jumps, in the presence of competing risk, patients who failed from a competing risk are NOT treated as being censored at the time of failure; jumps associated with these time points are NOT redistributed. That is, we still use the same formula for calculating the jumps; however, there *is* redistribution *to* all censored times—including times having events of a different cause—but *no* redistribution *from* times having events of a different and competing cause or source.

Finally, we obtain the cause-specific failure probability up to time t, or cumulative incidence to time t by the cause I under investigation (say, cancer), by adding up all the jumps up to that time point.

■ **Example 9.8** Consider the data set $\{4, 5, 6+, 7, 15+\}$, where $t+$ denotes a censored observation (i.e., terminated but no event). In addition, suppose we know that $t = 5$ is the time to death of a subject by cause #2—not under investigation (i.e., a competing cause). Suppose time is in months and we want to estimate the (cause-specific) cumulative incidence by 1 year (or $t = 12$). A simple solution is obtained as follows:

(i) A weight of 1/5 is assigned to each of the five subjects; for example, $t = 4$.
(ii) The assigned weight 1/5 assigned to $t = 5$ was *not* redistributed because that subject died from a competing cause.
(iii) The assigned weight 1/5 assigned to $t = 6+$ *was* redistributed, 1/10 each, to $t = 7$ and $t = 15+$; the weight at each becomes 1/5 + 1/10.

The cumulative incidence by 1 year (or 12 months) is obtained by adding the weights assigned to $t = 4$ and $t = 7$:

$$1/5 + (1/5 + 1/10) = 0.50 \text{ or } 50\%$$

Efron's approach allows us to estimate (cause-specific) cumulative incidence; it is easy to do and easy to understand but it does not provide us a needed standard error.

A Likelihood-Based Solution

A formal approach, a likelihood-based solution, to the estimation of the cumulative incidence could be formed as follows. Consider those time points "with event(s)" (by any of the k causes): $t_1 < t_2 < \cdots < t_m$. At t_j, let n_j be the number of subjects "at risk" (from any cause of failure) and $d_j = (\Sigma d_{ij}$ be the total number of events; d_{ij} is the number of event(s) by cause i at time t_j. In other words, we decompose the number of events at time t_j into events by different causes at that time. In order to simplify, we consider having two types of risks: type 1 is under investigation and all competing risks are grouped into type 2. The function or parameter to be estimated is

$$F_1(t) = \int_0^t \lambda_1(x)S(x)dx \hat{=} \sum_{j:t_j \le t} \lambda_{1j}S(t_{j-1})$$

We use all data, regardless of causes of events, to estimate the survival function by either method, the Kaplan–Meier method or redistribution to the right; however, only cause-specific deaths d_{ij} are involved in the estimation of cause-specific hazard λ_{ij}. For estimation of cause-specific hazards, the likelihood function is

$$L = \prod_{j=1}^{m} \left[\prod_{i=1}^{k} \lambda_{ij}^{d_{ij}}\right]\left[1 - \sum_{i=1}^{k} \lambda_{ij}\right]^{n_j - d_j}$$

If there is only one cause ($k = 1$), the above marginal likelihood function becomes

$$L = \prod_{j=1}^{m} [\lambda_j^{d_j}][1 - \lambda_j]^{n_j - d_j}$$

leading to the usual Kaplan–Meier estimates of hazard points of Section 9.2.1. In the general case ($k > 1$), the MLE result is

$$\hat{\lambda}_{ij} = \frac{d_{ij}}{n_j}$$

For example, with $k = 2$,

$$F_1(t) = \int_0^t \lambda_1(x)S(x)dx \hat{=} \sum_{j:t_j \le t} \lambda_{1j}S(t_{j-1})$$

The resulting estimate for this cumulative incidence is

$$\hat{F}_1(t) = \sum_{j:t_j \le t} \frac{d_{1j}}{n_j}\hat{S}(t_{j-1})$$

Variance and standard error can be obtained by the delta method (error propagation). There is a formula for a good approximation (Korn and Dorey, 1991); one can and should write a SAS "macro" to implement this procedure.

■ **Example 9.9** Consider data set {4, 5, 6+, 7, 15+}, where $t+$ denotes a censored observation (i.e., terminated but no event); in addition, suppose we know that $t = 5$ is time to death of a subject by cause #2—not under investigation (i.e., a competing cause). Suppose time is in months and we want to estimate the (cause-specific) cumulative incidence by 1 year (or $t = 12$). A solution (50%) was found in Example 9.8 using the redistribution to the right method. Let us confirm that using the likelihood-based method.

t	n	d	$1 - d/n$	$S(t)$
4	5	1	0.80	0.8
5	4	1	0.25	0.6
7	2	1	0.50	0.3

A point estimate of the cumulative incidence by 1 year is the same 50%:

$$\left[\frac{1}{5}\right](1.0) + \left[\frac{0}{4}\right](0.8) + \left[\frac{1}{2}\right](0.6) = 0.2 + 0.3$$

It should be noted that, in this example, if $t = 5$—which is an event by a competing risk—is treated as being censored, we would have the following:

t	n	d	$1 - d/n$	$S(t)$
4	5	1	0.80	0.8
7	2	1	0.50	0.4

The estimated survival probability at 1 year is 40% and the cumulative incidence at 1 year is 60%—instead of 50%. In other words, if we treat events by competing risks as censored observations, we overestimate the cumulative incidence, as previously noted.

9.5.3 Brief Discussion of Proportional Hazards Regression

Competing risks data are inherent to clinical medical research in which response to treatment can be classified in terms of failure or death from the disease process under investigation, say, cancer, from other disease processes, or from non-disease-related competing causes.

With data from explanatory covariates available, standard survival analysis involves modeling the hazards function (proportional hazards model):

$$\begin{aligned}\lambda(t|X) &= \lim_{\delta \to 0} \frac{\Pr[t \leq T < t+\delta | t \leq T, X = (x_1, \ldots, x_k)]}{\delta} \\ &= \lambda_0(t) \exp\left(\sum_{i=1}^{k} \beta_i x_i\right)\end{aligned}$$

as seen in Sections 9.3.1 and 9.4.1. The results of Cox's proportional hazards regression lead to direct interpretation in terms of survival probabilities because the hazards function and the survival function are well connected:

$$S(t) = \int_0^t \lambda(x)dx$$

With data from explanatory covariates available, standard analysis for competing risks data may be proceeded similarly by modeling the cause-specific hazards function:

$$\lambda_i(t|X = x) = \lim_{\delta \to 0} \frac{\Pr[t \leq T < t+\delta, \text{cause} = i | t \leq T, X = x]}{\delta}$$

The extension of Cox's proportional hazards model, its partial likelihood function, and subsequent inferences are possible (Larson, 1984; Prentice et al., 1978). The problem is that the results of this analysis imply the effects of covariates on the cause-specific hazards function but no corresponding translation in terms of survival probabilities for the disease process under investigation. We have the overall survival function,

$$S(t) = \exp\left[-\int_0^t \left\{\sum_{i=1}^{k} \lambda_i(x)\right\} dx\right]$$

but the cause-specific survival function is not defined for the disease process under investigation. Conventionally, the cumulative incidence of Section 9.5.2 has become the marker of choice, which has provided information about survivorship in the presence of competing risk. Therefore we like to draw inferences concerning the effects of covariates on the cumulative incidence, which is intuitively appealing and more easily explained to nonstatisticians. However, many authors have noted that the effect of a covariate on the cause-specific hazards function of a particular risk may be very different from the effect of the same covariate on the corresponding cumulative incidence (Gray, 1988; Pepe, 1991). In the most extreme case, a covariate may have strong influence on the cause-specific hazards function but almost no effect on the cumulative incidence. This leads to the need for extending Cox's proportional hazards model for use in the presence of competing risks.

Gray (1988) constructed K-sample tests for differences in the cumulative incidence function and Fine and Gray (1999) extended it into a regression approach for data analysis with competing risks; however, these methods are beyond the scope of this book on categorical data. Basically, it is to assume, for some known increasing function $g(.)$, the model

$$g\{F_1(t)|X)\} = h_0(t)\exp\left(\sum_{i=1}^{k}\beta_i x_i\right)$$

where $h_0(t)$ is an unspecified increasing function. One of the possibilities is to consider

$$g(u) = \log\{-\log(1-u)\}$$

which is akin to the popular proportional hazards model. One important step in the analysis is the construction of "subsets at risk" because those individuals who have already failed from causes other than the cause under investigation prior to time t are not "at risk" at time t.

EXERCISE

9.1 *Pneumocystis carinii* pneumonia (PCP) is the most common opportunistic infection and a life-threatening disease in HIV-infected patients. Many North Americans with HIV have one or two episodes of PCP during the course of their HIV infection. PCP is a consideration factor in mortality, morbidity, and expense; and recurrences are common. We have the following for our data set in Table 9.4.

- Treatments, coded as A and B.
- Patient characteristics: Baseline CD4 count, Sex (1 = Male), Race (1 = White, 2 = Black, 3 = Others).
- Weight (lb).
- Homosexuality (1 = Yes, 0 = No).
- PCP recurrence indicator (1 = Yes).
- PDate or time to recurrence (months).
- Die or Survival indicator (1 = Yes).
- DDate or time to death (or to date last seen for survivors; in months).

TABLE 9.4. AIDS Data

Treatment	CD4	Sex	Race	Weight	Homosexual	PCP	Pdate	Die	DDate
B	2	1	1	142	1	1	11.9	0	14.6
B	139	1	2	117	0	0	11.6	1	11.6
A	68	1	2	149	0	0	12.8	0	12.8
A	12	1	1	160	1	0	7.3	1	7.7
B	77	1	1	120	1	0	18.1	1	18.1
A	56	1	1	158	0	0	14.7	1	14.7
B	208	1	2	157	1	0	24.0	1	24.0
A	40	1	1	122	1	0	16.2	0	16.2
A	53	1	2	125	1	0	26.6	1	26.6
A	28	1	2	130	0	1	14.5	1	19.3
A	162	1	1	124	0	0	25.8	1	25.8
B	163	1	2	130	0	1	14.5	1	19.3
A	65	1	1	120	0	0	19.4	0	19.4
A	247	1	1	167	0	0	23.4	0	23.4
B	131	1	1	160	0	0	2.7	0	2.7
A	25	1	1	130	1	0	20.1	1	20.2
A	118	1	1	155	0	0	17.3	0	17.3
B	21	1	1	126	0	0	6.0	1	6.0
B	81	1	2	168	1	0	1.6	0	1.6
A	89	1	1	169	1	0	29.5	0	29.5
B	172	1	1	163	1	0	24.2	1	16.5
B	21	1	1	164	1	1	4.9	0	22.9
A	7	1	1	139	1	0	14.8	1	14.8
B	94	1	1	165	1	0	29.8	0	29.0
B	14	1	2	170	0	0	21.9	1	21.9
A	38	1	1	170	1	0	18.8	1	18.8
A	73	1	1	140	1	0	20.5	1	20.5
B	25	1	2	190	0	0	13.1	1	13.1
A	13	1	3	121	1	0	21.4	0	21.4
B	30	1	1	145	1	0	21.0	0	21.0
A	152	1	3	124	1	0	19.4	0	19.4
B	68	1	3	150	0	0	17.5	1	17.4
A	27	1	1	128	1	0	18.5	0	18.5
A	38	1	1	159	1	1	18.3	1	18.3
B	265	1	3	242	0	0	11.1	1	11.1
B	29	0	3	130	0	0	14.0	0	14.3
A	73	1	3	130	1	0	11.1	0	11.1
B	103	1	1	164	1	0	11.1	0	11.1
B	98	1	1	193	1	0	10.2	0	10.2
A	120	1	1	170	1	0	5.2	0	5.2
B	131	1	2	184	0	0	5.5	0	5.5
A	48	0	1	160	0	0	13.7	1	13.7
B	80	1	1	115	1	0	12.0	0	12.0
A	132	1	3	130	1	0	27.0	0	27.0
A	54	1	1	148	1	0	11.7	0	11.7

(continued)

TABLE 9.4 (*Continued*)

Treatment	CD4	Sex	Race	Weight	Homosexual	PCP	Pdate	Die	DDate
B	189	1	1	198	1	0	24.5	0	24.5
B	14	1	2	160	0	0	1.3	0	1.3
A	321	1	1	130	1	1	18.5	0	18.6
B	148	1	1	126	1	0	22.8	0	22.8
B	54	1	1	181	1	1	14.5	0	15.7
A	17	1	1	152	1	0	19.8	0	19.8
A	37	1	3	120	1	0	16.6	0	16.8
B	71	0	1	136	0	0	16.5	0	16.5
A	9	1	3	130	1	0	8.9	1	8.9
A	22	1	2	190	1	0	8.5	0	8.5
A	43	1	1	134	1	0	17.9	1	17.9
B	103	1	1	110	1	0	20.3	0	20.3
A	146	1	1	213	1	0	20.5	0	20.5
A	92	1	1	128	1	0	14.2	0	14.2
B	218	1	1	163	1	0	1.9	0	1.9
A	100	1	1	170	1	0	14.0	0	14.0
B	148	1	1	158	1	1	15.4	0	16.1
A	76	1	1	149	0	1	15.9	1	23.4
B	44	1	2	124	1	0	7.3	0	7.3
B	30	1	1	181	1	0	6.6	1	6.6
B	260	1	1	165	1	1	7.5	1	18.0
B	40	1	1	204	0	0	21.0	1	21.0
A	90	1	1	149	1	1	17.0	0	21.8
A	120	1	1	152	0	0	21.8	0	21.8
B	80	1	1	199	1	1	20.6	0	20.6
A	170	1	1	141	1	0	18.6	1	18.6
A	54	1	1	148	1	0	18.6	0	18.6
A	151	1	1	140	1	0	21.2	0	21.2
B	107	1	1	158	1	0	22.5	1	22.5
A	9	1	1	116	1	0	18.0	1	18.0
B	79	1	3	132	0	0	22.6	1	22.6
A	72	1	1	131	1	0	19.9	0	19.9
A	100	1	1	182	1	0	21.2	0	21.2
B	16	1	2	106	1	0	18.3	0	18.3
B	10	1	1	168	1	0	24.7	0	24.7
A	135	1	1	149	1	0	23.8	0	23.8
B	235	1	1	137	0	0	22.7	0	22.7
B	20	1	1	104	0	0	14.0	0	14.0
A	67	1	2	150	0	0	19.4	1	19.4
B	7	1	1	182	1	0	17.0	1	17.0
B	139	1	1	143	1	0	21.4	0	21.4
B	13	1	3	132	0	0	23.5	0	23.5
A	117	1	1	144	1	0	19.5	1	19.5
A	11	1	2	111	1	0	19.3	1	19.3

TABLE 9.4 (*Continued*)

Treatment	CD4	Sex	Race	Weight	Homosexual	PCP	Pdate	Die	DDate
B	280	1	1	145	1	0	11.6	1	11.6
A	119	1	1	159	1	1	13.1	0	19.3
B	9	1	1	146	1	1	17.0	1	18.2
A	30	1	2	150	0	0	20.9	0	20.9
B	22	1	1	138	1	1	1.1	1	10.0
B	186	1	3	114	1	0	17.2	0	17.2
A	42	1	1	167	1	0	19.2	0	19.2
B	9	1	2	146	1	0	6.0	1	6.0
B	99	1	1	149	0	1	14.8	0	14.8
A	21	1	1	141	1	0	17.7	0	17.7
A	16	1	2	116	0	0	17.5	0	17.5
B	10	1	1	143	1	1	8.3	1	8.3
B	109	1	1	130	1	1	12.0	0	12.1
B	72	1	1	137	0	0	12.8	0	12.8
B	582	1	1	143	1	0	15.7	0	15.7
A	8	1	2	134	1	0	9.3	1	9.3
A	69	1	1	160	0	0	10.1	0	10.1
A	57	1	1	138	1	0	10.2	0	10.2
A	47	1	1	159	1	0	9.1	0	9.1
A	149	1	3	152	0	0	9.8	0	9.8
B	229	1	3	130	1	0	9.4	0	9.4
A	9	1	1	165	1	0	9.2	0	9.2
A	10	1	1	162	0	0	9.2	0	9.2
A	78	1	1	145	1	0	10.2	0	10.2
B	147	1	1	180	1	0	9.0	0	9.0
B	126	1	1	124	1	0	5.5	0	5.5
A	19	1	2	192	0	0	6.0	0	6.0
A	142	1	1	170	1	0	17.3	1	17.3
B	277	0	1	140	0	0	17.0	0	17.0
B	80	1	1	130	1	0	15.0	0	15.0
A	366	1	1	150	1	0	14.9	0	14.9
A	76	1	1	180	1	0	9.2	0	9.2
A	13	1	1	171	1	0	30.2	0	30.2
B	17	1	1	276	0	0	15.8	1	15.8
B	193	1	1	164	1	0	22.5	1	22.5
A	108	1	1	161	0	0	24.0	0	24.0
B	41	1	1	153	0	0	23.9	0	23.9
A	113	1	1	131	0	0	21.4	0	21.4
B	1	1	2	136	0	0	19.6	0	19.6
A	47	1	1	168	1	0	18.2	1	18.2
B	172	1	2	195	1	0	10.3	1	10.3
A	247	1	1	123	1	0	16.2	0	16.2
B	21	1	2	124	0	0	9.7	0	9.7
A	113	1	1	131	0	0	21.4	0	21.4
B	1	1	2	136	0	0	19.6	0	19.6

(*continued*)

TABLE 9.4 (*Continued*)

Treatment	CD4	Sex	Race	Weight	Homosexual	PCP	Pdate	Die	DDate
A	47	1	1	168	1	0	18.2	1	18.2
B	172	1	2	195	1	0	10.3	1	10.3
A	247	1	1	123	1	0	16.2	0	16.2
B	21	1	2	124	0	0	9.7	0	9.7
B	38	1	2	160	1	0	14.7	0	14.7
B	50	1	1	127	1	0	13.6	0	13.6
A	4	1	2	218	0	0	12.9	0	12.9
A	150	1	1	200	1	0	11.7	1	11.7
A	97	1	2	156	0	0	11.9	0	11.9
B	312	1	1	140	1	0	10.6	0	10.6
B	35	1	1	155	1	0	11.0	0	11.0
A	100	1	1	157	1	0	9.2	0	9.2
A	69	1	1	126	0	0	9.2	0	9.2
A	124	1	2	135	1	0	6.5	0	6.5
B	25	1	1	162	1	0	16.0	1	16.0
A	61	0	2	102	0	0	18.5	1	18.5
B	102	1	1	177	1	1	11.3	0	17.4
A	198	1	2	164	0	0	23.2	0	23.2
B	10	1	1	173	0	0	8.4	1	8.4
A	56	1	1	163	1	0	11.9	0	11.9
A	43	1	1	134	1	0	9.2	0	9.2
B	202	1	2	158	0	0	9.2	0	9.2
B	102	1	1	177	1	1	11.3	0	17.4
A	198	1	2	164	0	0	23.2	0	23.2
B	10	1	1	173	0	0	8.4	1	8.4
A	56	1	1	163	1	0	11.9	0	11.9
A	43	1	1	134	1	0	9.2	0	9.2
B	202	1	2	158	0	0	9.2	0	9.2
A	31	1	1	150	1	0	9.5	1	9.5
B	243	1	1	136	1	0	22.7	0	22.7
B	40	1	1	179	1	0	23.0	0	23.0
A	365	1	1	129	0	0	17.9	0	17.9
A	29	1	2	145	0	1	0.6	1	2.6
A	97	1	1	127	0	0	13.7	0	13.7
B	314	1	3	143	0	0	12.2	1	12.2
B	17	1	1	114	0	1	17.3	0	17.7
A	123	1	1	158	0	0	21.5	0	21.5
B	92	1	1	128	0	0	6.0	0	6.0
A	39	0	2	150	0	0	10.8	0	10.8
A	87	1	1	156	1	0	28.3	0	28.3
A	93	1	1	170	0	0	23.9	0	23.9
A	4	0	2	104	0	0	21.0	0	21.0
A	60	1	1	150	0	0	6.3	0	6.3
B	20	0	1	133	0	0	17.3	1	17.3

TABLE 9.4 (*Continued*)

Treatment	CD4	Sex	Race	Weight	Homosexual	PCP	Pdate	Die	DDate
A	52	1	3	125	0	0	12.0	0	12.0
B	78	0	1	99	0	0	16.7	0	16.7
B	262	1	2	192	0	0	12.7	0	12.7
A	19	1	2	143	1	0	6.0	1	6.0
A	85	1	1	152	0	0	10.8	0	10.8
B	6	1	1	151	1	0	13.0	0	13.0
B	53	1	2	115	0	0	8.9	0	8.9
A	386	1	1	220	1	0	27.6	0	27.6
A	12	1	1	130	1	0	26.4	1	26.4
B	356	0	1	110	0	0	27.8	0	2.7.8
A	19	1	1	187	0	0	28.0	0	28.0
A	39	1	2	135	0	0	2.9	0	2.9
B	9	1	1	139	0	1	6.9	0	6.9
B	44	1	2	112	0	0	23.1	0	23.1
B	7	1	1	141	1	0	15.9	1	15.9
A	34	1	1	110	1	1	0.4	1	6.1
B	126	1	1	155	1	1	0.2	0	6.9
B	4	1	1	142	1	0	20.3	0	20.3
A	16	1	1	154	0	0	14.7	1	14.7
A	22	1	1	121	1	0	21.4	0	21.4
B	35	1	1	165	1	0	21.2	0	21.2
A	98	1	1	167	1	0	17.5	0	17.5
A	357	1	1	133	0	0	16.6	0	16.6
B	209	1	1	146	1	0	15.6	0	15.6
A	138	1	1	134	1	0	8.8	0	8.8
B	36	1	1	169	1	0	3.4	0	3.4
A	90	1	1	166	0	0	30.0	0	30.0
B	51	1	1	120	0	1	26.0	0	26.0
B	25	1	2	161	0	0	29.0	0	29.0
A	17	1	1	130	0	0	7.3	1	7.3
A	73	1	1	140	0	0	20.9	0	20.9
A	123	1	1	134	1	0	17.4	0	17.4
B	161	1	1	177	1	1	19.3	1	19.3
B	105	1	1	128	1	1	3.7	0	23.5
A	74	1	2	134	1	0	24.8	0	24.8
A	7	1	1	130	0	1	10.3	1	10.3
B	29	0	1	97	0	0	13.1	0	23.9
A	84	1	1	217	1	0	24.8	0	24.8
B	9	1	1	158	1	0	23.5	0	23.5
A	29	1	1	160	1	0	23.5	0	23.5
B	24	1	1	136	1	0	19.4	0	19.4
B	715	1	2	126	1	0	15.7	0	15.7
A	147	1	1	170	1	1	9.8	0	16.8
A	162	1	1	137	0	0	11.0	0	11.0
B	35	1	1	150	1	0	11.9	0	11.9

(*continued*)

TABLE 9.4 (*Continued*)

Treatment	CD4	Sex	Race	Weight	Homosexual	PCP	Pdate	Die	DDate
B	14	1	1	153	1	0	8.2	1	8.2
B	227	1	1	150	1	0	9.5	0	9.5
B	137	1	1	145	1	0	9.0	0	9.0
A	48	1	1	143	0	0	8.3	0	8.3
A	62	1	1	175	1	0	6.7	0	6.7
A	47	1	1	164	1	0	5.5	0	5.5
B	7	1	1	205	0	0	6.9	0	6.9
B	9	1	1	121	0	1	19.4	0	23.9
B	243	1	2	152	0	1	12.0	1	12.0
A	133	1	1	136	0	0	23.1	0	23.1
A	56	1	1	159	1	0	23.2	0	23.2
A	11	1	1	157	0	0	8.7	1	8.7
A	94	1	2	116	0	0	15.1	1	15.1
A	68	1	1	185	1	0	21.3	0	21.3
A	139	1	1	145	1	0	19.1	0	19.1
B	15	0	1	114	0	0	17.4	0	17.4
B	22	1	2	125	1	0	4.4	1	4.4

Consider each of these endpoints: relapse (treating death as censoring), death (treating relapse as censoring) and death or relapse, whichever comes first. For each endpoint, answer these questions:

(a) Estimate the survival function for white, homosexual men.

(b) Estimate the survival functions, one for each treatment.

(c) Compare the two treatments. Are the treatments different in the short-term or long-term?

(d) Compare men versus women.

(e) Taken collectively, do the covariates contribute significantly to prediction of survival?

(f) Fit the multiple regression model to obtain estimates of individual regression coefficients and their standard errors. Draw your conclusion concerning the conditional contribution of each factor.

(g) Within the context of the multiple regression model in (b), does treatment alter the effect of CD4?

(h) Focus on treatment as the primary factor. Taken collectively, was this main effect altered by any other covariates?

(i) Within the context of the multiple regression model in (b), is the effect of CD4 linear?

(j) Do treatment and CD4, individually, fit the proportional hazards model?

BIBLIOGRAPHY

Agresti A. (1990). *Categorical Data Analysis*. Hoboken, NJ: Wiley.

Ahlquist DA, McGill DB, Schwartz S, and Taylor WF. (1985). Fecal blood levels in health and disease: a study using HemoQuant. *New England Journal of Medicine* 314:1422–1428.

Ahn C. (1998). An evaluation of Phase I cancer clinical designs. *Statistics in Medicine* 17:1537–1549.

Alonzo TA, Pepe MS, and Moskowitz CS. (2002). Sample size calculations for comparative studies of medical tests for detecting presence of disease. *Biostatistics* 3:421–432.

Anderson KE, Carmella SG, Ye M, Bliss RL, Le C, Murphy L, and Hecht SS. (2001). Metabolites of a tobacco-specific lung carcinogen in the urine of nonsmoking women exposed to environmental tobacco smoke in their homes. *Journal of the National Cancer Institute* 93:378–381.

Arbuck SG. (1996). Workshop on phase I study design. *Annals of Oncology* 7:567–573.

Armitage P. (1975). *Sequential Medical Trials*. Oxford, UK: Blackwell.

Armitage P. (1977). *Statistical Methods in Medical Research*, pp. 433–438. Hoboken, NJ: Wiley.

Armitage P, McPherson K, and Rowe BC. (1969). Repeated significance tests on accumulating data. *Journal of the Royal Statistical Society, Series A* 132:235–244.

Bamber D. (1975). The area above the ordinal dominance graph and the area below the receiver operating characteristic graph. *Journal of Mathematical Psychology* 12:387–415.

Begg CB. (1991). Advances in statistical methodology for diagnostic medicine in the 1980s. *Statistics in Medicine* 10:1887–1895.

Begg CB and McNeil B. (1988). Assessment of radiologic tests:control of bias and other design considerations. *Radiology* 167:565–569.

Berkowitz GS. (1981). An epidemiologic study of pre-term delivery. *American Journal of Epidemiology* 113:81–92.

Berry G. (1983). The analysis of mortality by the subject-years method. *Biometrics* 39:173–184.

Bhattacharyya GK, Karandinos MG, and DeFoliart GR. (1979). Point estimates and confidence intervals for infection rates using pooled organisms in epidemiologic studies. *American Journal of Epidemiology* 109:124–131.

Bishop YMM, Fienberg SE, and Holland PW. (1975). *Discrete Multivariate Analyses: Theory and Practice*. Cambridge, MA: MIT Press.

Bishop YMM, Fienberg SE, and Holland PW. (1988). *Discrete Multivariate Analysis*, pp. 393–396. Cambridge, MA: MIT Press.

Blot WJ, Harrington M, Toledo A, et al. (1978). Lung cancer after employment in shipyards during World War II. *New England Journal of Medicine* 299:620–624.

Breslow N. (1970). A generalized Kruskal–Wallis test for comparing *K* samples subject to unequal patterns of censorship. *Biometrika* 57:579–594.

Breslow N. (1982). Covariance adjustment of relative-risk estimates in matched studies. *Biometrics* 38:661–672.

Breslow NE and Day NE. (1980). *Statistical Methods in Cancer Research, Volume I: The Analysis of Case–Control Studies*. Lyons, France: International Agency for Research on Cancer.

Brown BW. (1980). Prediction analyses for binary data. In: *Biostatistics Casebook*, edited by RG Miller, B Efron, BW Brown, and LE Moses, pp. 3–8. Hoboken, NJ: Wiley.

Buck AA and Gart JJ. (1966). Comparison of a screening test and a reference test in epidemiologic studies. I. Indices of agreement and their relation to prevalence. *American Journal of Epidemiology* 83(3):586–592.

Buck AA and Spuyt DJ. (1964). Seroreactivity in the venereal disease research laboratory slide test and the fluorescent treponemal antibody test: a study of patterns in selected disease and control groups in Ethiopia. *American Journal of Hygiene* 80:91–102.

Carey RM, Reid RA, Ayers CR, et al. (1976). The Charlottesville blood-pressure survey. Value of repeated blood-pressure measurements. *JAMA* 236:847–851.

Carter SK. (1987). The Phase I study. In: *Fundamentals of Cancer Chemotherapy*, edited by KK Hellmann and SK Cater, pp. 285–300. New York: McGraw Hill.

Chang MN, Therneau TM, Wiend HS, and Cha SS. (1987). Designs for sequential Phase II clinical trials. *Biometrics* 43:865–874.

Cheng H and Macaluso M. (1997). Comparison of the accuracy of two tests with a confirmatory procedure limited to positive results. *Epidemiology* 11:275–280.

Chevret S. (1993). The continual reassessment method in cancer Phase I clinical trials: a review and results of a Monte Carlo study. *Statistics in Medicine* 12:1093–1108.

Chevret S, editor (2006). *Statistical Methods for Dose-Finding Experiments*. Hoboken, NJ: Wiley.

Chin T, Marine W, Hall E, Gravelle C, and Speers J. (1961). The influence of Salk vaccination on the epidemic pattern and the spread of the virus in the community. *American Journal of Hygiene* 73:67–94.

Chou TC. (1976). Derivation and properties of Michaelis–Menten type and Hill type equations for reference ligands. *Journal of Theoretical Biology* 59:233–276.

Cohen J. (1960). A coefficient of agreement for nominal scale. *Educational and Psychological Measurements* 20:37–46.

Connett JE, Rhame F, Thomas J, and Le CT. (1990). Estimation of infectious potential of blood containing human immunodeficiency virus. *Biometrical Journal* 32:781–789.

Conover WJ. (1974). Some reasons for not using the Yates continuity correction in 2×2 contingency tables. *Journal of the American Statistical Association* 69:374–378.

Cox DR. (1972). Regression models and life tables. *Journal of the Royal Statistical Society, Series B* 34:187–220.

Cox DR and Oakes D. (1984). *Analysis of Survival Data.* New York: Chapman & Hall.

Cox DR and Snell EJ. (1989). *The Analysis of Binary Data*, 2nd editon. London: Chapman and Hall.

D'Angelo LJ, Hierholzer JC, Holman RC, and Smith JD. (1981). Epidemic keratoconjunctivitis caused by adenovirus Type 8: epidemiologic and laboratory aspects of a large outbreak. *American Journal of Epidemiology* 113:44–49.

Daniel WW. (1987). *Biostatistics: A Foundation for Analysis in the Health Sciences.* Hoboken, NJ: Wiley.

Efron B. (1967). The two-sample problem with censored data. *Proceedings of the 5th Berkeley Symposium* 4:831–853.

Efron B. (1978). Regression and ANOVA with zero-one data: measures of residual variation. *Journal of the American Statistical Association* 73:113–121.

Eyster ME, Goedert JJ, Sarngadharan MG, et al. (1985). Development and early natural history of HTLV-III antibodies in persons with hemophilia. *JAMA* 253(15):2219–2223.

Faraggi D, Izikson P, and Reiser B. (2003). Confidence intervals for the 50 per cent response dose. *Statistics in Medicine* 22(12):1977–1988.

Faraggi D, Reiser B, and Schisterman EF. (2003). ROC curve analysis for biomarkers based on pooled assessments. *Statistics in Medicine* 22(15):2515–2527.

Faries D. (1991). The modified continual reassessment method for Phase I cancer clinical trials. *American Statistical Association 1991 Proceedings of the Biopharmaceutical Section*, pp. 269–273. New York: Marcel Dekker.

Fine JP and Gray RJ. (1999). A proportional hazards model for the sub-distribution of a competing risk. *Journal of the American Statistical Association* 94:496–509.

Finney DJ. (1964). *Statistical Method in Biological Assay.* London: C. Griffin.

Fleming TR. (1982). One-sample multiple testing procedure for Phase II clinical trials. *Biometrics* 38:143–151.

Freeman DH. (1980). *Applied Categorical Data Analysis.* New York: Marcel Dekker.

Freireich EJ, Gehan E, Frei E III, et al. (1963). The effect of 6-mercaptopurine on the duration of steroid-induced remissions in acute leukemia: a model for evaluation of other potentially useful therapy. *Blood* 21:699–716.

Frome EL. (1983). The analysis of rates using Poisson regression models. *Biometrics* 39:665–674.

Frome EL and Checkoway H. (1985). Use of Poisson regression models in estimating rates and ratios. *American Journal of Epidemiology* 121:309–323.

Gart JJ. (1969). An exact test for comparing matched proportions in crossover designs. *Biometrika* 56:75–80.

Gart JJ and Buck AA. (1966). Comparison of a screening test and a reference test in epidemiologic studies. II. A probabilistic model for the comparison of diagnostic tests. *American Journal of Epidemiology* 83(3):593–602.

Gaswirth JL. (1987). The statistical precision of medical screening procedures: application to polygraph and AIDS antibodies test data. *Statistical Science* 2:213–238.

Gatsonis C and Greenhouse JB. (1992). Bayesian methods for Phase I clinical trials. *Statistics in Medicine* 11:1377–1389.

Gaynor JJ, Feuer EJ, Tan CC, et al. (1993). On the use of cause-specific failure and conditional failure probabilities: examples from clinical oncology data. *Journal of the American Statistical Association* 88:400–409.

Gehan EA. (1961). The determination of the number of patients required in a preliminary and a follow-up trial of a new chemotherapeutic agent. *Journal of Chronic Diseases* 13:346–353.

Gehan EA. (1965a). A generalized Wilcoxon test for comparing arbitrarily singly-censored samples. *Biometrika* 52:203–223.

Gehan EA. (1965b). A generalized two-sample Wilcoxon test for doubly censored data. *Biometrika* 52:650–653.

Geller N. (1984). Design of Phase I and Phase II clinical trials in cancer: a statistician's view. *Cancer Investigation* 2:483–491.

Goldman AI. (1987). Issues in designing sequential stopping rules for monitoring side effects in clinical trials. *Controlled Clinical Trials* 8:327–337.

Goldman AI and Hannan PJ. (2001). Optimal continuous sequential boundaries for monitoring toxicity in clinical trials: a restricted search algorithm. *Statistics in Medicine* 20:1575–1589.

Goodman SN, Zahurak ML, and Piantadosi S. (1995). Some practical improvements in the continual reassessment method for Phase I studies. *Statistics in Medicine* 14:1149–1161.

Graham S, Marshall J, Haughey B, et al. (1988). Dietary epidemiology of cancer of the colon in western New York. *American Journal of Epidemiology* 128:490–503.

Gray RJ. (1988). A class of *K*-sample tests for comparing the cumulative incidence for a competing risk. *The Annals of Statistics* 16:1141–1154.

Green S, Benedetti JB, and Crowley J. (2003). *Clinical Trials in Oncology*. Boca Raton, FL: Chapman & Hall/CRC Press.

Greenberg RA and Jekel JF. (1969). Some problems in the determination of the false positive and false negative rates of tuberculin tests. *American Reviews in Respiratory Diseases* 100:645–650.

Greenwood M. (1926). The natural duration of cancer. *Reports on Public Health and Medical Subjects, Her Majesty's Stationary Office* 33:1–26.

Hanley JA. (1996). The use of the "binormal" model for parametric ROC analysis of quantitative diagnostic tests. *Statistics in Medicine* 15(14):1575–1585.

Hanley JA and McNeil BJ. (1982). The meaning and use of the area under a receiver operating characteristic (ROC) curve. *Radiology* 143:29–36.

Hanley JA and McNeil BJ. (1982). The meaning and use of the area under a receiver operating characteristic (ROC) curve. *Radiology* 143:29–36.

Hanley JA and McNeil BJ. (1983). Method for comparing the area under the ROC curves derived from the same cases. *Radiology* 148:839.

Helsing KJ and Szklo M. (1981). Mortality after bereavement. *American Journal of Epidemiology* 114:41–52.

Herbst AL, Ulfelder H, and Poskanzer DC. (1971). Adenocarcinoma of the vagina. *New England Journal of Medicine* 284:878–881.

Heyd JM and Carlin BP. (1999). Adaptive design improvements in the continual reassessment method for Phase I studies. *Statistics in Medicine* 18:1307–1321.

Hlatky MA, Pryor DB, Harrell FE, et al. (1984). Factors affecting sensitivity and specificity of exercise electrocardiography—multivariate analysis. *American Journal of Medicine* 77:64–71.

Holford TR. (1982). Covariance analysis for case–control studies with small blocks. *Biometrics* 38:673–683.

Hollows FC and Graham PA. (1966). Intraocular pressure, glaucoma, and glaucoma suspects in a defined population. *British Journal of Ophthalmology* 50:570–586.

Hosmer DW Jr and Lemeshow S. (1989). *Applied Logistic Regression.* Hoboken, NJ: Wiley.

http://thyroid.about.com/cs/basics_starthere/a/thyroid101.htm.

http://www.cancer.gov/templates.

http://www.joslin.org/main.shtml.

http://www.nlm.nih.gov/medlineplus/cervicalcancer.html.

http://www.nlm.nih.gov/medlineplus/ovariancancer.html.

http://www.who.int/hiv/en/.

Hubert JJ. (1984). *Bioassay.* Dubuque, A: Kendall and Hunt Publishing.

Hui XX and Walter XX. (1980). Estimating the error rates of diagnostic tests. *Biometrics* 36:167–171.

Hully SB, Cummings SR, Browner WS, et al. (2001). *Designing Clinical Research.* Philadelphia: Lippincott Williams and Wilkins.

Jackson R, Scragg R, and Beaglehole R. (1992). Does recent alcohol consumption reduce the risk of acute myocardial infarction and coronary death in regular drinkers? *American Journal of Epidemiology* 136:819–824.

Kaplan EL and Meier P. (1958). Nonparametric estimation from incomplete observations. *Journal of the American Statistical Association* 53:457–481.

Kelsey JL, Livolsi VA, Holford TR, et al. (1982). A case–control study of cancer of the endometrium. *American Journal of Epidemiology* 116:333–342.

Kleinbaum DG, Kupper LL, and Muller KE. (1988). *Applied Regression Analysis and Other Multivariate Methods.* Boston: PWS-Kent Publishing.

Kleinman JC and Kopstein A. (1981). Who is being screened for cervical cancer? *American Journal of Public Health* 71:73–76.

Korff FA, Taback MAM, and Beard JH. (1952). A coordinated investigation of a food poisoning outbreak. *Public Health Reports* 67:909–913.

Korn EL and Dorey FJ. (1992). Applications of crude incidence curves. *Statistics in Medicine* 11:813–829.

Korn EL, Midthune D, Chen TT, et al. (1994). A comparison of two Phase I trial designs. *Statistics in Medicine* 13:1799–1806.

Kuhl CK, Schmutzler RK, Leutner CC, et al. (2000). Breast MR imaging screening in 192 women proved or suspected to be carriers of a breast cancer susceptibility gene: preliminary results. *Radiology* 215:267–279.

Larson MG. (1984). Covariate analysis of competing risks models with log-linear models. *Biometrics* 40:459–469.

Le CT. (1981). A new estimator for infection rates using pools of variable size. *American Journal of Epidemiology* 114:132–135.

Le CT. (1997a). *Applied Survival Analysis*. Hoboken, NJ: Wiley.

Le CT. (1997b). Evaluation of confounding effects in ROC studies. *Biometrics* 53: 998–1007.

Le CT. (1997c). Evaluation of confounding effects in ROC studies. *Biometrics* 53:998–1007.

Le CT. (2003). *Introductory Biostatistics*, pp. 5–7, 115–117. Hoboken, NJ: Wiley.

Le CT. (2006). A solution for the most basic optimization problem associated with an ROC curve. *Statistical Methods in Medical Research* 15(6):571–584.

Le CT and Grambsch PM. (2005). Design and analysis of in vitro experiments for combination therapy. *Journal of Biopharmaceutical Statistics* 15:1–9.

Le CT and Lindgren BR. (1988). Computational implementation of the conditional logistic regression model in the analysis of epidemiologic matched studies. *Computers and Biomedical Research* 21:48–52.

Le CT, Daly KA, Margolis RH, Lindgren BR, and Giebink GS. (1992). A clinical profile of otitis media. *Archives of Otolaryngology* 118:1225–1228.

Lee YJ. (1979). Two-stage plans for patients accrual in Phase II cancer clinical trials. *Cancer Treatment Reports* 63:1721–1726.

Li DK, Daling JR, Stergachis AS, Chu J, and Weiss NS. (1990). Prior condom use and the risk of tubal pregnancy. *American Journal of Public Health* 80:964–966.

Liang KY and Self SG. (1985). Tests for homogeneity of odds ratio when data are sparse. *Biometrika* 72:353–358.

Lui KI. (2007). Interval estimation of risk ratio in the simple compliance randomized trial. *Contemporary Clinical Trials* 28:120–129.

Lin Y and Shih WJ. (2001). Statistical properties of the traditional algorithm-based designs for Phase I cancer clinical trials. *Biostatistics* 2:203–215.

Liu PY, LeBlanc M, and Desai M. (1999). False positive rates of randomized Phase II designs. *Control Clinical Trials* 20:343–352.

Lusted LB. (1971a). Decision-making studies in patient management. *New England Journal of Medicine* 284(8):416–424.

Lusted LB. (1971b). Signal detectability and medical decision-making. *Science* 171(977): 1217–1219.

Mack TM, Pike MC, Henderson BE, et al. (1976). Estrogens and endometrial cancer in a retirement community. *New England Journal of Medicine* 294:1262–1267.

Mantel N and Haenszel W. (1959). Statistical aspects of the analysis of data from retrospective studies of disease. *Journal of the National Cancer Institute* 22:719–748.

Matinez FD, Wright AL, Holber CJ, et al. (1992). Maternal age as a risk factor for wheezing lower respiratory illness in the first year of life. *American Journal of Epidemiology* 136:1258–1268.

May D. (1974). Error rates in cervical cytological screening tests. *British Journal of Cancer* 29:106–113.

McCullagh P. (1980). Regression models for ordinal data. *Journal of the Royal Statistical Society Series B* 42:109–142.

McFadden D. (1974). Conditional logit analysis of qualitative choice behavior, In: *Frontiers in Econometrics*, edited by A Zarembka, pp. 105–142. New York: Academic Press.

Mehta CR and Cain KC. (1984). Charts for the early stopping of pilots studies. *Journal of Clinical Oncology* 2:676–692.

Mick R. (1996). Phase I clinical trial design. In: *Principles of Antineoplastic Drug Development and Pharmacology*, edited by RL Schilsky, pp. 29–36. New York: Marcel Dekker.

Miller RG, Efron B, et al. (1980). *Biostatistics Casebook*. Hoboken, NJ: Wiley.

Moller S. (1995). An extension of the continual reassessment methods using a preliminary up-and-down design in a dose finding study in cancer patients, in order to investigate a greater range of doses. *Statistics in Medicine* 14:911–922.

Moskowitz M, Milbrath J, Gartside P, et al. (1976). Lack of efficacy of thermography as a screening tool for minimal and stage I breast cancer. *New England Journal of Medicine* 295 (5):249–252.

Murray DM, Perry CL, O'Connell C, and Schmid L. (1987). Seventh-grade cigarette, alcohol, and marijuana use: distribution in a north central U.S. metropolitan population. *International Journal of the Addictions* 22:357–376.

Negri E, Vecchia CL, Bruzzi P, et al. (1988). Risk factors for breast cancer: pooled results from three Italian case–control studies. *American Journal of Epidemiology* 128:1207–1215.

Nischan P, Ebeling K, and Schindler C. (1988). Smoking and invasive cervical cancer risk: results from a case–control study. *American Journal of Epidemiology* 128:74–77.

O'Brien PC and Fleming TR. (1979). A multiple testing procedure for clinical trials. *Biometrics* 35:549–555.

O'Quigley J. (1992). Estimating the probability of toxicity at the recommended dose following a Phase I clinical trial in cancer. *Biometrics* 46:33–48.

O'Quigley J and Chevret S. (1991). Methods for dose finding studies in cancer clinical trials: a review and results of a Monte Carlo study. *Statistics in Medicine* 10:1647–1664.

O'Quigley J and Reiner E. (1998). A stopping rule for the continual reassessment method. *Biometrika* 85:741–748.

O'Quigley J and Shen LZ. (1996). Continual reassessment method: a likelihood approach. *Biometrics* 52:673–684.

O'Quigley J, Pepe M, and Fisher L. (1990). Continual reassessment method: a practical design for Phase I clinical trials in cancer. *Biometrics* 46:33–48.

Padian NS. (1990). Sexual histories of heterosexual couples with one HIV-infected partner. *American Journal of Public Health* 80:990–991.

Pepe MS. (1991). Inference for events with dependent risks in multiple endpoint studies. *Journal of the American Statistical Association* 86:770–778.

Pepe MS. (2004). *The Statistical Evaluation of Medical Tests for Classification and Prediction*. London: Oxford University Press.

Peto R. (1972). Contribution to discussion of paper by D. R. Cox. *Journal of the Royal Statistical Society, Series B* 34:205–207.

Prentice RL, Kalbfleisch JD, Peterson AV, et al. (1978). The analysis of failure times in the presence of competing risks. *Biometrics* 34:541–554.

Rogan WJ and Gladen B. (1978). Estimating prevalence from the results of a screening test. *American Journal of Epidemiology* 107(1):71–76.

Rosenberg L, Slone D, Shapiro S, et al. (1981). Case–control studies on the acute effects of coffee upon the risk of myocardial infarction: problems in the selection of a hospital control series. *American Journal of Epidemiology* 113:646–652.

Rosner B. (1982). Statistical methods in ophthalmology: an adjustment for the intra-class correlation between eyes. *Biometrics* 38:105–114.

Rousch GC, Kelly JA, Meigs JW, and Flannery JT. (1982). Scrotal carcinoma in Connecticut metal workers: sequel to a study of sinonasal cancer. *American Journal of Epidemiology* 116:76–85.

Sargent DJ and Goldberg RM. (2001). A flexible design for multiple armed screening trials. *Statistics in Medicine* 20:1051–1060.

Schisterman EF, Faraggi D, and Reiser B. (2004). Adjusting the generalized ROC curve for covariates. *Statistics in Medicine* 23(21):3319–3331.

Schisterman EF, Perkins NJ, Liu A, and Bondell H. (2005). Optimal cut-point and its corresponding Youden index to discriminate individuals using pooled blood samples. *Epidemiology* 16(1):73–81.

Schultz JR, Nichol FR, Elfring GL, and Weed SD. (1973). Multi-stage procedures for drug screening. *Biometrics* 29:293–300.

Schwarts BS, Doty RL, Monroe C, et al. (1989). Olfactory function in chemical workers exposed to acrylate and methacrylate vapors. *American Journal of Public Health* 79:613–618.

Shapiro S, Rosenberg L, Slone D, and Kaufman DW. (1979). Oral contraceptive use in relation to myocardial infarction. *Lancet* 1:743–746.

Simon R. (1989). Optimal two-stage designs for Phase II clinical trials. *Controlled Clinical Trials* 10:1–10.

Simon R and Korn E. (1990). Selecting drug combination on total equivalent dose. *Journal of the National Cancer Institute* 82:1469–1476.

Simon R, Wittes RE, and Ellenberg SS. (1985). Randomized Phase II clinical trials. *Cancer Treatment Reports* 69:1375–1381.

Simon R, Freidlin B, Rubinstein L, et al. (1997). Accelerated titration designs for Phase I clinical trials in oncology. *Journal of the National Cancer Institute* 89:1138–1147.

Stamler J, Stamler R, Riedlinger WF, Algera G, and Roberts RH. (1976). Hypertension screening of 1 million Americans. Community Hypertension Evaluation Clinic (CHEC) program, 1973 through 1975. *JAMA* 235(21):2299–2306.

Steck GP. (1971). Rectangle probabilities for uniform order statistics and the probability that the empirical distribution function lies between two distribution functions. *Mathematical Statistics* 42:1–11.

Steck GB and Zimmer WJ. (1972). In: *Proceedings 1972 NATO Conference on Reliability Evaluation and Reliability Testing VIII-B*, The Hague, Netherlands.

Steel GG and Peckham MJ. (1979). Exploitable mechanism in combined radio therapy—chemotherapy: The concept of additivity. *International Jounal of Oncology* 5:85–91.

Storer BE. (1989). Design and analysis of Phase I clinical trials. *Biometrics* 45:925–937.

Storer BE. (2001). An evaluation of Phase I clinical trial designs in the continuous dose-response setting. *Statistics in Medicine* 20:2399–2408.

Strader CH, Weiss WS, and Daling JR. (1988). Vasectomy and the incidence of testicular cancer. *American Journal of Epidemiology* 128:56–63.

Strogatz D. (1990). Use of medical care for chest pain differences between blacks and whites. *American Journal of Public Health* 80:290–293.

Stuart A. (1955). A test for homogeneity of marginal distribution in two-way classification. *Biometrika* 42:412–416.

Swets JA. (1973). The relative operating characteristic in psychology: a technique for isolating effects of response bias finds wide use in the study of perception and cognition. *Science* 182(4116):990–1000.

Swets JA. (1979). ROC analysis applied to the evaluation of medical imaging techniques. *Investigative Radiology* 14(2):109–121.

Tabar L, Chen HH, Duffy SW, et al. (2000). The Swedish two-county trial twenty years later. Updated mortality results and new insights from long-term follow-up. *Radiologic Clinics of North America* 38:625–651.

Tallarida RJ. (2000). *Drug Synergism and Dose–Effect Data Analysis*. New York: Chapman and Hall/CRC.

Tallarida RJ and Murray RB. (1981). *Manual of Pharmacologic Calculations with Computer Programs*. New York: Springer-Verlag.

Tallarida RJ and Raffa RB. (1996). Testing for synergism over a range fixed-ratio drug combinations: Replacing the isobologram. *Life Sciences* 58:23–28.

Tarone RE and Ware J. (1977). On distribution-free tests for equality of survival distributions. *Biometrika* 64:156–160.

Thall PF and Cheng SC. (2001). Optimal two-stage designs for clinical trials based on safety and efficacy. *Statistics in Medicine* 20:1023–1032.

Thall PF, Lee JJ, Tseng CH, and Estey EH. (1999). Accrual strategies for phase I trials with delayed patient outcome. *Statistics in Medicine* 18:1155–1169.

Thall PF, Simon R, and Ellenberg SS. (1988a). Two-stage selection and testing designs for comparative clinical trials. *Biometrika* 75:303–310.

Thall PF, Simon R, Ellenberg SS, and Shrager R. (1988). Optimal two-stage designs for clinical trials with binary response. *Statistics in Medicine* 71:571–579.

Thall PF, Simon RM, and Estey EH. (1996). New statistical strategy for monitoring safety and efficacy in single-arm clinical trials. *Journal of Clinical Oncology* 14:296–303.

Tosteson AN and Begg CB. (1988). A general regression methodology for ROC curve estimation. *Medical Decision Making* 8(3):204–215.

True WR, Golberg J, and Eisen SA. (1988). Stress symptomology among Vietnam veterans. *American Journal of Epidemiology* 128:85–92.

Tuyns AJ, Pequignot G, and Jensen OM. (1977). Esophageal cancer in Ille-et-Vilaine in relation to alcohol and tobacco consumption: multiplicative risks. *Bulletin of Cancer* 64:45–60.

Ury HK. (1972). On distribution-free confidence bounds Pr($Y < X$). *Technometrics* 14:577–581.

Von Hoff DD. (1984). Design and conduct of Phase I trials. In: *Cancer Clinical Trials*. edited by ME Buyse. Oxford, UK: Oxford University Press.

Walter SD, Hildreth SW, and Beaty BJ. (1980). Estimation of infection rates in population of organisms using pools of variable size. *American Journal of Epidemiology* 112:124–128.

Warner E, Plewes DB, Shumak RS, et al. (2001). Comparison of breast magnetic resonance imaging, mammography, and ultrasound for surveillance of women at high risk for hereditary breast cancer. *Journal of Clinical Oncology* 19(15):3524–3531.

Weiss SH, Goedert JJ, Sarngadharan MG, et al. (1985). Screening test for HTLV-III (AIDS agent) antibodies. Specificity, sensitivity, and applications. *JAMA* 253(2):221–225.

Wermuth N. (1976). Explanatory analyses of multidimensional contingency tables. In: *Proceedings of the 9th International Biometrics Conference,* Volume 1; pp. 279–295. Washington, D.C.: The International Biometric Society.

Whittemore AS, Wu ML, Paffenbarger RS Jr, et al. (1988). Personal and environmental characteristics related to epithelial ovarian cancer. *American Journal of Epidemiology* 128:1228–1240.

Whittemore AS, Harris R, Itnyre J, and The Collaborative Ovarian Cancer Group. (1992). Characteristics relating to ovarian cancer risk: collaborative analysis of 12 U. S. case–control studies. *American Journal of Epidemiology* 136:1184–1203.

Wise ME. (1954). A quickly convergent expansion for cumulative hypergeometric probabilities, direct and inverse. *Biometrika* 41:317–329.

Yen S, Hsieh C, and MacMahon B. (1982). Consumption of alcohol and tobacco and other risk factors for pancreatitis. *American Journal of Epidemiology* 116:407–414.

Youden WJ. (1950). Index for rating diagnostic tests. *Cancer* 3(1):32–35.

Zhou XH, Obuchowski NA, and McClish DK. (2002). *Statistical Methods in Diagnostic Medicine*. Hoboken NJ: Wiley.

Zohar S and Chevret S. (2001). The continual reassessment method: comparison of Bayesian stopping rules for dose-ranging Studies. *Statistics in Medicine* 20:2827–2843.

INDEX